VOLUME FIFTY ONE

Advances in
CHEMICAL ENGINEERING
Sorption Enhancement of Chemical Processes

ADVANCES IN CHEMICAL ENGINEERING

VOLUME FIFTY ONE

ADVANCES IN CHEMICAL ENGINEERING

Sorption Enhancement of Chemical Processes

Edited by

ANGELIKI A. LEMONIDOU

Department of Chemical Engineering,
School of Engineering,
Aristotle University of Thessaloniki,
Thessaloniki, Greece

ELSEVIER

ACADEMIC PRESS
An imprint of Elsevier

Academic Press is an imprint of Elsevier
50 Hampshire Street, 5th Floor, Cambridge, MA 02139, United States
525 B Street, Suite 1800, San Diego, CA 92101-4495, United States
The Boulevard, Langford Lane, Kidlington, Oxford OX5 1GB, United Kingdom
125 London Wall, London, EC2Y 5AS, United Kingdom

First edition 2017

ISBN: 978-0-12-813121-3
ISSN: 0065-2377

For information on all Academic Press publications
visit our website at https://www.elsevier.com/books-and-journals

Publisher: Zoe Kruze
Acquisition Editor: Sam Mahfoudh
Editorial Project Manager: Shellie Bryant
Production Project Manager: Vignesh Tamil
Cover Designer: Greg Harris

Typeset by SPi Global, India

CONTENTS

CONTRIBUTORS

Juan C. Abanades
Spanish Research Council, INCAR-CSIC, C/Francisco Pintado Fe, Oviedo, Asturias, Spain

Jurriaan Boon
Energy Research Centre of the Netherlands (ECN), Petten, The Netherlands

Paul Cobden
Energy Research Centre of the Netherlands (ECN), Petten, The Netherlands

Kai Coenen
Eindhoven University of Technology, Eindhoven, The Netherlands

Dânia S.M. Constantino
Laboratory of Separation and Reaction Engineering—Laboratory of Catalysis and Materials (LSRE-LCM), Faculdade de Engenharia, Universidade do Porto, Porto, Portugal

Rui P.V. Faria
Laboratory of Separation and Reaction Engineering—Laboratory of Catalysis and Materials (LSRE-LCM), Faculdade de Engenharia, Universidade do Porto, Porto, Portugal

José R. Fernández
Spanish Research Council, INCAR-CSIC, C/Francisco Pintado Fe, Oviedo, Asturias, Spain

Fausto Gallucci
Eindhoven University of Technology, Eindhoven, The Netherlands

Maria C. Iliuta
Université Laval, Ville de Québec, QC, Canada

Alírio E. Rodrigues
Laboratory of Separation and Reaction Engineering—Laboratory of Catalysis and Materials (LSRE-LCM), Faculdade de Engenharia, Universidade do Porto, Porto, Portugal

Eric van Dijk
Energy Research Centre of the Netherlands (ECN), Petten, The Netherlands

Martin van Sint Annaland
Eindhoven University of Technology, Eindhoven, The Netherlands

PREFACE

The importance of process intensification (PI) in several sectors, such as petrochemicals, bulk chemicals, specialty chemicals, pharmaceuticals, food ingredients, and consumer foods, has been recently recognized by the European Federation of Chemical Engineering (EFCE). According to the European Roadmap in PI drafted by EFCE, PI provides radically innovative principles ("paradigm shift") in the process and equipment design which can benefit—often by more than a factor of 2—process and chain efficiency, capital and operating expenses, quality, wastes, process safety, and so on. Multifunctional reactors constitute one of the most powerful elements, generating creative innovations within the PI concept. The combination of the chemical reaction with product separation (reactive separation) in one step in a single reactor presents great promise, fully conforming to the principles of PI. When applied for thermodynamically limited reactions, the shift of equilibrium to the products' side according to the Le Chatelier's principle allows not only product separation but also higher conversion levels and leads to less complex separation trains and higher efficiency. The so-called sorption-enhanced processes rely on this approach, using as means of product separation reactive adsorption by a solid material. Issues like solid sorbent selection, reactor design, and process configuration scheme are the main challenges typically faced in any process based on the sorption enhancement concept. This volume refers to the application of the sorption enhancement approach in a handful of energy-related processes which in the last decade have attracted great interest, both for the scientific challenges and the potential for technological advancements.

In Chapter 1, van Sint Annaland and coworkers discuss recent developments in sorption-enhanced water–gas shift (SEWGS) as a promising technology for precombustion CO_2 capture with high energy efficiency. The superior performance of hydrotalcite-based adsorbents, in terms of high thermal and mechanical stability and sufficiently high cyclic working capacity and fast adsorption kinetics, is prominent. The work is complemented with a reactor model for the simulation and optimization of SEWGS cycles, using a heterogeneous linear driving force approximation for the intraparticle mass transfer resistance. A comparative analysis with other technologies shows that SEWGS outperforms conventional technologies for H_2/CO_2 separation. In Chapter 2, Iliuta discusses the challenges in hydrogen production via the sorption-enhanced steam reforming (SESR) process. The key element in SESR processes is the selection of appropriate CO_2 sorbent materials for efficient in situ CO_2 removal from the reaction medium. High sorption capacity and kinetics, together with a long-term stability under severe reaction conditions, are the main expected characteristics. The pros and cons of promising potential sorbents, such as Ca-based, hydrotalcite-based, and alkaline ceramic solids, are presented. Particular emphasis is given on the different strategies that aim to improve the sorbents' stability and performance in the cyclic SESR process. Going one step further toward increasing the viability of regenerating the CO_2 sorbent, Fernández and Abanades in Chapter 3 discuss the novel Ca–Cu looping process. The combination of two loops arises as a promising option for sorption-enhanced methods with potentially low energy penalties and costs. This chapter compiles the main recent advances in the Ca–Cu process, focusing on reactor and process design and the experimental work needed to validate the performance of reactors and materials. The significant progress made in simulating this system in power plants and other energy-intensive industrial systems is also presented. In Chapter 4, Rodrigues and coworkers address an innovative approach for the synthesis of oxygenated compounds, acetals and esters, using multifunctional reactors; the simulated moving bed reactor and the simulated moving bed membrane reactor. This technology, which combines reaction with chromatographic separation, reveals a strong ability to increase the equilibrium conversion with high productivities that become even higher when the concept is combined with a selective membrane. The methodology for the development and implementation of these technologies based on experimental data and suitable mathematical models is thoroughly and clearly presented in their contribution.

I would like to thank all authors for the efforts devoted in writing the chapters and all reviewers for their in-depth and constructive comments and suggestions. As the aim of *Advances in Chemical Engineering* is to provide the reader with the views of authorities in the field emphasizing to their contribution and footprint to the field and their perspectives for the future evolution, this volume does not claim to be exhaustive in assessing the state of the art in sorption-enhanced chemical processes.

Angeliki A. Lemonidou
Department of Chemical Engineering
Aristotle University of Thessaloniki

CHAPTER ONE

Sorption-Enhanced Water–Gas Shift

Jurriaan Boon*, Kai Coenen†, Eric van Dijk*, Paul Cobden*, Fausto Gallucci†, Martin van Sint Annaland†,[1]

*Energy Research Centre of the Netherlands (ECN), Petten, The Netherlands
†Eindhoven University of Technology, Eindhoven, The Netherlands
[1]Corresponding author: e-mail address: m.v.sintannaland@tue.nl

Contents

Abstract

Sorption-enhanced water–gas shift (SEWGS) is a promising technology for precombustion CO_2 capture with high energy efficiency. Hydrotalcite-based adsorbents were studied as possible solid sorbents for pressure swing adsorption under SEWGS conditions. They show high thermal and mechanical stability with sufficiently high cyclic working capacity and fast adsorption kinetics. The regenerations step (desorption of CO_2 by feeding steam to the adsorbent) is slower and limits the cyclic working capacity of the adsorbent. It was found that a higher operating temperature is beneficial because of enhanced desorption kinetics. Steam induces the desorption of a second adsorption site available for CO_2 which cannot be desorbed with N_2. Different adsorption sites are

Advances in Chemical Engineering, Volume 51
ISSN 0065-2377
https://doi.org/10.1016/bs.ache.2017.07.004

present on the hydrotalcite material. On the basis of a dedicated set of high-pressure breakthrough experiments an adsorption isotherm has been developed which describes the interaction of CO_2 and H_2O with the hydrotalcite-based adsorbent over the relevant range of partial pressures. Based on the isotherm and a linear driving force approximation for intraparticle mass transfer, a reactor model has been constructed for the simulation and optimization of SEWGS cycles. A parameter study of SEWGS cycles shows that the high-pressure rinse steam improves the CO_2 product purity, while the low-pressure steam purge mainly serves to improve the CO_2 capture ratio. Finally, a comparative analysis of the required work of precombustion separation of CO_2 and H_2 shows that SEWGS outperforms conventional technologies for hydrogen–carbon dioxide separation, partly because of its inherently high CO_2 capture ratio.

NOMENCLATURE

CCR molar carbon capture ratio
CO_xCR molar carbon capture ratio based on CO and CO_2 (i.e., excluding CH_4)
CP molar carbon purity
$C_{p,p}$ particle heat capacity ($J\,kg^{-1}\,K^{-1}$)
CWC cyclic working capacity
d_c column diameter (m)
d_p particle diameter (m)
F_m mass feed flow rate ($kg\,s^{-1}$)
HP molar hydrogen purity
HRF hydrogen recovery factor
HTC hydrotalcite
KMG30 potassium-promoted hydrotalcite with 30 wt.% MgO compared to Al_2O_3 (sorbent)
KMG70 potassium-promoted hydrotalcite with 70 wt.% MgO compared to Al_2O_3 (sorbent)
KSORB potassium-promoted Alumina (sorbent material)
L column length (m)
LHV low eating value ($kJ\,kg^{-1}$)
PBR packed bed reactor
PCO_2 partial pressure of CO_2 (bar)
PH_2O partial pressure of H_2O (bar)
q sorbent loading ($mol\,kg^{-1}$)
S/C_p molar ratio of purge steam to feed carbon oxides
S/C_r molar ratio of rinse steam to feed carbon oxides
T temperature (K)
TGA thermogravimetric analysis
ΔH_a heat of adsorption ($kJ\,mol^{-1}$)
ϵ_b particle bed porosity
ϵ_p intraparticle porosity
η efficiency, as defined in text
ρ_p sorbent particle density ($kg\,m^{-3}$)
ω mass fraction

1. INTRODUCTION

The worldwide demand for hydrogen has steadily increased in recent years. Among the different possible routes to produce hydrogen, conventional steam methane reforming (SMR) is one of the most often used production technologies. Coal gasification is another economically feasible route, because coal is more abundant and cheaper than natural gas (Jang et al., 2012). An important disadvantage of these hydrogen production methods is the associated large emissions of CO_2 into the atmosphere. To reduce the greenhouse gas effect, mainly related to the high atmospheric concentrations of CO_2, carbon capture and storage (CCS) has been proposed as a possible strategy to reduce anthropogenic CO_2 emissions. However, high investments costs related to the CO_2 capture step are hampering the use of CCS technologies at large scale in the industry (Birol, 2013).

Two main reactions are involved in the H_2 production from natural gas. The SMR reaction is a highly endothermic equilibrium reaction which is usually carried out at high temperatures between 800 and 1000°C (see Eq. 1), where CH_4 is converted into CO and H_2. In a second step, CO is converted into CO_2 by the slightly exothermic water–gas shift (WGS) reaction which is usually carried out in two consecutive reactors to achieve high conversion of CO (see Eq. 2).

$$CH_4 + H_2O \rightleftharpoons CO + 3H_2 \quad \Delta H_0^{298\,K} = 206\ \text{kJ/mol} \tag{1}$$

$$CO + H_2O \rightleftharpoons CO_2 + H_2 \quad \Delta H_0^{298\,K} = -41\ \text{kJ/mol} \tag{2}$$

The sorption-enhanced water–gas shift (SEWGS) process is a technology for precombustion CO_2 capture and hydrogen production (van Selow et al., 2009a). SEWGS is a kind of pressure swing adsorption (PSA) process, where the produced CO_2 is in situ adsorbed on solid materials at temperatures between 350 and 550°C during the WGS reaction. Syngas is converted and separated into streams of H_2 and CO_2, which makes SEWGS exceptionally suitable for precombustion CO_2 capture (Boon et al., 2014). SEWGS can yield higher CO_2 capture ratios at lower energy efficiency penalties and at lower costs in comparison with more mature technologies based on solvents (Van Selow et al., 2011). Usually, the second WGS reactor and the PSA gas absorber can be combined in a SEWGS unit reducing the number of process steps and required equipment (Fig. 1).

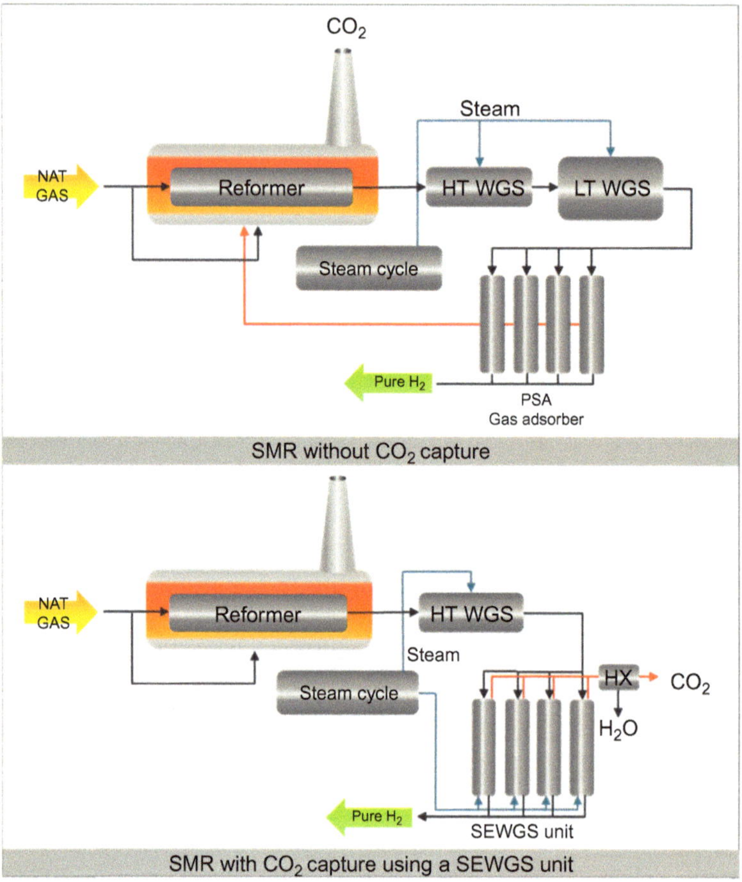

Fig. 1 Steam methane reforming process (*top*: conventional) and with CO_2 capture (*bottom*: with SEWGS unit).

Various materials, such as carbon-based adsorbents, sodium zirconates, lithium zirconates, basic alumina, and CaO have been investigated for use in a SEWGS process (Meis et al., 2010). However, most adsorbents cannot be used because of their poor mechanical stability, sorption kinetics limitations or high costs. Hydrotalcite-based adsorbents have shown high thermal stability (Ebner et al., 2006; Walspurger et al., 2010), fast sorption kinetics, and high selectivity toward CO_2.

Among potassium-promoted mixed metal oxides (based on hydrotalcites), other sorbent materials have been investigated in the past for a possible application in CCS. Zeolites with a uniform pore size distribution have been investigated for CO_2 separation (Choi et al., 2009; Yong et al., 2002).

Alkali cations within the metal framework of zeolites enable them to adsorb acidic gas molecules such as CO_2 in large quantities (Choi et al., 2009; Yong et al., 2002; Zhang et al., 2008). Metal-organic frameworks (MOFs) and carbon molecular sieves (CMSs) are also possible adsorbents with a high affinity toward CO_2 (Hedin et al., 2010). However, CO_2 adsorption capacities are decreased at elevated temperatures and moreover, zeolites are usually not thermally stable under WGS conditions. Additionally, at the presence of H_2O, the CO_2 adsorption is not thermodynamically favorable (Wang et al., 2014).

Activated carbons also showed high CO_2 adsorption capacities under milder operating temperatures and are therefore suitable candidate for CO_2 capture at mild operating temperatures which are drastically reduced at higher temperatures (Pevida et al., 2008).

Calcium oxides have very high adsorption capacities for CO_2 at high operating temperatures. However, deactivation and therefore a decrease in cyclic working capacity for CO_2 of these materials with the number of cycles are a disadvantage of these type of sorbents for the use in a cyclic process (Abanades and Alvarez, 2003; Choi et al., 2009).

Lithium zirconates and lithium orthosilicates have shown high CO_2 adsorption capacities and high operating temperatures. However, slow adsorption kinetics, which are much slower compared to zeolites and activated carbon, are hampering the use of these adsorbents (Choi et al., 2009).

Hydrotalcites are layered double hydroxides (LDHs), which belong to the group of anionic clays. The original structure at room temperature of these materials is close to that of brucite ($Mg(OH)_2$), in which the Mg cation is octahedrally surrounded by hydroxyl groups. These octahedra of Mg^{2+} are joined along their edges, forming infinite sheets which are stacked on top of each other, hold together by hydrogen bonding (Cavani et al., 1991). Partial substitution of the bivalent Mg^{2+} ion by trivalent Al^{3+} generates a net positive charge of the layer. Different anions like CO_3^{2-}, Cl^-, NO_3^{2-}, and SO_4^{2-} usually compensate for this charge imbalance (Debecker et al., 2009; Halabi et al., 2012; Reichle, 1985). Other elements such as Cu, Ni, Mn, and Zn can be substituted for the bivalent Mg^{2+} anion and elements such as Al, Fe, Cr, and V can be substituted for the trivalent anion (M^{3+}). Usually water is present with the compensating anions in the interlayer resulting in the layered structure of hydrotalcites. It was reported that temperature, water vapor pressure, and the type of the anions can influence the number of water molecules present in the interlayer (Vaccari, 1998). The total structure has a rhombohedral orientation and the unit cell composition is $Mg_6Al_2(HO)_{16}CO_3^{2-} \cdot 4H_2O$, where the atomic ratio between Mg/Al can vary for synthetic samples (Di Cosimo et al., 1998a). It has been reported that higher Al content in hydrotalcites leads to the formation of $Al(OH)_3$ because of the increasing number of neighboring Al-containing octahedra. Low Mg/Al ratios lead to a high density of Mg-containing octahedra in the brucite-type sheets and to segregation of $Mg(OH)_2$ (Vaccari, 1998). The basicity of hydrotalcites and their derivates depends on the Mg content. A higher Mg content will result in a higher number of basic sites (Debecker et al., 2009). The adsorption capacity for different acidic gases like CO_2 or H_2S depends on the basic site density. It was reported that the highest concentration of strong basic sites is found for pure MgO, whereas the number of low- and medium-strength sites increases with increasing Al content (Di Cosimo et al., 1998a).

It is well known that the clay layered structure is no longer present at elevated temperatures, which is important for SEWGS, because CO_2 is usually adsorbed at relatively high temperatures. Fresh hydrotalcites are highly hydrated materials and during a heat treatment CO_2 and H_2O is released. Mg–Al hydrotalcites are converted into well-dispersed Mg(Al)O mixed metal oxides. The mixed metal oxide usually has a higher surface area due to the formation of significant porosity (Liu et al., 2007). A first weight loss is detected between 70 and 280°C which is mainly attributed to the loss of interlayer water which is a reversible process (Debecker et al., 2009). CO_2 and H_2O is usually released between 250 and 450°C which is reversible if the temperature is not higher than 550–600°C (Reichle, 1986a). Basic and acidic sites are formed during the heating process and a moderate increase in surface area has been reported, attributed to a cratering mechanism caused by the release of CO_2 and H_2O (Reichle et al., 1986). At high temperatures close to 1000°C the formation of a spinel phase has been reported (Kanezaki, 1998). In general, the decomposition of hydrotalcites and the release of CO_2 and H_2O strongly depends on the preparation method and the used temperature during calcination. It was reported that rapid heating would lead to an amorphous material with mixtures of octahedral and tetrahedral coordinated aluminum sites, where a slow heating rate result in a similar structure to the parent LDH before heating (Rg et al., 1995).

Impregnation of the mixed metal oxide with alkaline metals like K_2CO_3 has shown to improve the adsorption capacity of basic species on the adsorbent (Hanif et al., 2014; Hutson and Attwood, 2008; Lee et al., 2007; Meis et al., 2010; Miguel et al., 2014; Oliveira et al., 2008; Walspurger et al., 2008). K_2CO_3-promoted commercial hydrotalcites have been investigated frequently. Where usually a higher MgO content improves the amount of CO_2 that can be adsorbed, some mechanical stability issues were reported at high partial pressures of CO_2 and H_2O for these type of sorbents due to the formation of bulk potassium carbonate (Van Selow et al., 2011; Walspurger et al., 2010).

This chapter discusses the development of the SEWGS process with a specific focus on the progress made over the last decade. In Section 2, an overview and the state-of-the-art of the SEWGS process is presented, discussing the projects that have been executed and are currently being executed to advance SEWGS. Section 3 focuses on the experimental work that has been performed in the laboratories of Eindhoven University of Technology and ECN to elucidate the complex interactions between CO_2, H_2O, and the SEWGS sorbent material. In Section 4, the models are discussed that have been developed based on these experimental studies. The models are used to translate experimentally observed sorbent behavior into SEWGS column behavior, aiming to predict the efficiency of a SEWGS process for hydrogen–carbon dioxide separations. Finally, the conclusions are summarized in Section 5.

2. SORPTION-ENHANCED WATER–GAS SHIFT (SEWGS)

2.1 The SEWGS Process

In SEWGS, the WGS reaction (3) is combined with in situ adsorption of carbon dioxide (4), typically at about 400°C and 10–40 bar.

$$CO + H_2O \rightleftharpoons CO_2 + H_2 \tag{3}$$

$$CO_2 + \bullet \rightleftharpoons CO_2 - \bullet \tag{4}$$

At the same time, depending on process conditions, steam can also be adsorbed.

$$H_2O + \bullet \rightleftharpoons H_2O - \bullet \tag{5}$$

Potassium-promoted hydrotalcite (K-HTC) catalyzes the WGS reaction and is capable of reversibly adsorbing carbon dioxide (Boon et al., 2014; Hufton et al., 1999; Lee et al., 2008; van Selow et al., 2009b; Yong et al., 2002). Relatively pure carbon dioxide is being released during regeneration of the K-HTC. Because of the periodic loading and regeneration of the sorbent, a state-of-the-art SEWGS process comprises multiple columns which are operated in pressure cycles, resembling the cycles of a PSA

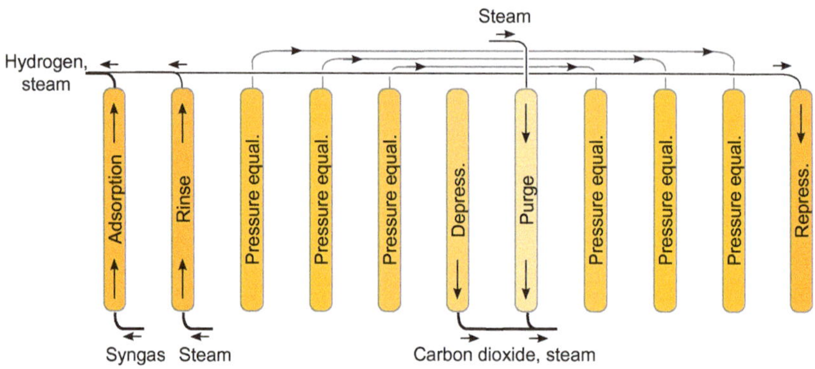

Fig. 2 Eleven-step SEWGS cycle with cocurrent steam rinse. *Reprinted from Boon J, Cobden PD, van Dijk HAJ, van Sint Annaland M: High-temperature pressure swing adsorption cycle design for sorption-enhanced water–gas shift,* Chem Eng Sci *122:219–231, 2015, Copyright (2015) with permission from Elsevier.*

process, and allowing the production of continuous product streams as shown in Fig. 2.

After the adsorption step, a carbon dioxide or steam rinse is performed to improve the carbon dioxide product purity, by preventing the loss of efficiency and pollution of the carbon dioxide product due to the slip of hydrogen or other syngas species into the carbon dioxide product. A number of pressure equalizations (typically 1–3) are performed after the rinse. During these pressure equalizations, a high-pressure column that needs depressurization is connected with a column at lower pressure that needs to be pressurized, thereby conserving the gas pressure. During the pressure equalization, the rinse gas in the high-pressure column expands and pushes the remaining syngas out of the column into the lower pressure column. Finally, in the depressurization or blowdown step, the pressure in the column is released, allowing for regeneration of the loaded particles. The depressurization step is followed by a steam purge to allow a high recovery of carbon dioxide and to enhance sorbent regeneration.

Overall, the SEWGS process directly converts syngas into separate streams of hydrogen at feed pressure and carbon dioxide at regeneration pressure, both at 400°C. The SEWGS process is therefore suited for precombustion carbon dioxide capture, and hence mitigation of greenhouse gas emissions. In fact, Manzolini et al. (2013b) have shown that for a coal-fueled IGCC power plant, the cost of carbon dioxide avoided can be reduced with SEWGS by 15% or more, compared to state-of-the-art precombustion carbon dioxide capture with Selexol, to 31–33 € ton^{-1}

(using K-HTC *sorbent alpha*: −15%) or even 23–25 € ton^{-1} (using K-HTC next-generation *sorbent beta*, −35%). Excluding carbon dioxide compression, the costs of carbon dioxide capture with SEWGS were shown to come mainly from investment costs (55% of the increased cost of electricity) and additional fuel costs for production of high-pressure rinse steam and low-pressure purge steam (23% of the increased cost of electricity)—the economy of the SEWGS process is dominated by investment cost and steam consumption.

2.2 SEWGS State-of-the-Art

The concept of SEWGS is not at all new. Already early in the 20th century, processes were developed for the WGS reaction enhanced by carbon dioxide sorption (Gluud et al., 1931; Gülker, 1927; Verfahren zur Herstellung von Wasserstoff, 1915). Yet, a major breakthrough in efficiency came with the development of K-HTC sorbent by air products and chemicals, a sorbent that can be efficiently regenerated by pressure swing (Hufton et al., 1999).

Important steps were subsequently made in the development of the SEWGS process for precombustion carbon dioxide capture in the European Union's Framework Programs for Research and Technological Development and the Dutch National CATO programs. In the sixth framework program, the integrated project Cachet (2006–09, involving 28 partners from the oil & gas industry, electricity companies, EPC companies, R&D institutes, and universities) SEWGS was one of four precombustion capture technologies that were developed, mainly by Air Products and the Energy Research Centre of the Netherlands (ECN). ECN constructed a single column unit and a multicolumn unit for SEWGS. The results of single column experiments in a column 2 m tall with 38 mm internal diameter lead to a decrease in the required steam consumption for the operation of a SEWGS cycle. Data from the single column were used for model development and subsequently cycle optimization. Several SEWGS cycles were tested in the multicolumn unit consisting of six columns of 6 m tall and 44 mm internal diameter (Fig. 3). An important aspect determining the process operation and efficiency turned out to be the cyclic stability of the K-HTC sorbent. The K-HTC particles used consisted of relatively magnesium-rich KMG70 and showed accumulation of carbon dioxide in the bulk phase of the particle, leading eventually to carbon dioxide slip during the adsorption step (van Selow et al., 2009b) and mechanical degradation (Jansen et al., 2013).

Fig. 3 Multicolumn SEWGS test unit "SEWGS-7" at ECN.

Nonetheless, it was already concluded that SEWGS could improve the energy efficiency by 3.4%-points and the cost of carbon dioxide emissions avoided by 16%, compared to state-of-the-art technology (Beavis, 2011).

In the period 2008–11, the European Union's seventh framework program funded the follow-up project Caesar, in which SEWGS was further developed by a consortium consisting of Air Products UK, BP Exploration Operating, ECN, SINTEF, and Politecnico di Milano (Caesar, 2011). Extensive sorbent development work leads to a K-HTC-based sorbent consisting of KMG30, i.e., which had a lower magnesium content, designated Alkasorb. This new sorbent did not suffer from the accumulation of carbon dioxide in the bulk of the material, see also the discussion on material composition below. Importantly, it was also concluded that the SEWGS sorbents had sufficient activity for the WGS reaction, so that the used of an additional WGS catalyst is not necessary in SEWGS (van Dijk et al., 2011; van Selow et al., 2013). A next-generation sorbent was also developed, designated Alkasorb+, with an adsorption capacity for carbon dioxide about 90% higher than Alkasorb. Alkasorb+ lead to a reduction of near 40% in cost of carbon dioxide emission avoided for IGCC compared to Selexol (Jansen et al., 2013). Another focus was on the effect of different feed gas species on the performance of the sorbent. It was found that the sorbent is robust against hydrogen sulfide, which is in fact cocaptured together with carbon dioxide (van Dijk et al., 2011). Carbonyl sulfide is hydrolyzed to hydrogen sulfide and subsequently cocaptured (Jansen et al., 2013). At the end of Caesar (December 2011), SEWGS had reached a technology readiness level of 5–6 (Jansen et al., 2013).

The ADEM lab Innovation program CMS (catalysis, membranes, and separations) cluster works on technologies for CO_2 capture from fossil fuels, but also looks ahead toward a bio-based economy with a number of projects on efficient conversion of biomass. The link with our industrial partners that are active in the oil, gas, and energy sector is key for the development of efficient sustainable technologies to reduce CO_2 emissions. Within this program, the currently running HOTSORB project deals with sorbent and reactor development on the basis of a better understanding of the kinetics, and mass and heat transport phenomena in sorption-enhanced processes at elevated temperatures.

SEWGS also has significant potential in the decarbonization of gas streams in integrated steelworks (Gazzani et al., 2015). The Stepwise project, funded by the European Union's Horizon 2020 program, started in 2015 and will run for 4 years (STEPWISE, 2015) involving Kisuma Chemicals, Johnson Matthey, ECN, Swerea Mefos, SSAB, Amec Foster Wheeler, Tata Steel Consulting, and the Universities Politecnico de Milano and Universitatea Babes-Bolyai. The main objective of the project is to scale-up the SEWGS technology for the capture of carbon dioxide from blast furnace gas (BFG) with three overall demonstration goals in comparison to state-of-the-art technologies:

- Higher carbon capture rate—i.e., lower carbon intensity, 85% reduction,
- Higher energy efficiency—i.e., lower energy consumption for capture, 60% reduction, and
- Better economy—i.e., lower cost of carbon dioxide emissions avoided, 25% reduction.

The Stepwise project will achieve this by the construction and the operation of a SEWGS pilot test installation at a blast furnace site. The pilot plant will be built at and operated by Swerea MEFOS and the pilot plant will receive BFG from the adjacent steel plant of SSAB in Luleå, Sweden. This will further reduce the risks associated with scaling up of the technology.

Additionally, adsorbent production scale-up and endurance testing will be addressed. For the first time, three key components in the required technology chain will be coupled together: (i) an industrial BFG source, (ii) subsequent BFG preprocessing using advanced WGS technology, and (iii) carbon monoxide and sulfur component cleanup with simultaneous carbon dioxide removal in the SEWGS. The demonstration at a 14 t d^{-1} carbon dioxide removal rate scale is used to prove the technology using real BFG. The performance is used to update the techno-economic evaluation and to assess the environmental impact relative to other state-of-the-art carbon dioxide removal technologies. Moreover, it serves as basis for the design of a full-scale unit and associated costing. In parallel, alternative applications of the hydrogen-rich SEWGS product gas within the steel plant configuration is investigated.

The FReSMe project (From Residual Steel Gases to Methanol, 2016–20), also funded by the European Union's Horizon 2020 program, will produce methanol that will be demonstrated in ship transportation. Building on to the StepWise project, the methanol will be produced from carbon dioxide, recovered through SEWGS from an industrial BFG, and hydrogen which is recovered both from the BFG itself via SEWGS, as well as produced by electrolysis. The project involves 11 different partners, including ECN, Swerea MEFOS, and Politecnico di Milano. The pilot plant will run for a total of 3 months divided over three different runs with a nominal production rate of up to 50 kg/h from an input of 800 m^3/h BFG. This size is commensurate with operation at TRL6, where all the essential steps in the process must be joined together in an industrial environment. FReSMe will address the new integration options that SEWGS has within the Iron and Steel industry and contains supplementary and supporting research of underlying phenomena.

2.3 SEWGS Applications

SEWGS is a typical example of process intensification, combining the WGS reaction, and in situ carbon dioxide capture at high temperature, also enhancing the carbon monoxide conversion. Its inherent advantages have been summarized by Jansen et al. (2013):

- High-temperature hydrogen product (typically around 400°C),
- High-pressure hydrogen product (typically 30 bar),
- Medium purity hydrogen product (typically 90–95 mol%),
- High-temperature carbon dioxide product (typically around 400°C),

- Carbon dioxide product at ambient pressure, and
- High carbon monoxide conversion.

Given these characteristics, it is not surprising that SEWGS has a long-standing history in developing precombustion carbon dioxide capture. It has also been suggested in applications related to distributed refinery hydrogen, refinery process hydrogen, refinery fuel gas, ammonia and urea production, production of methanol, processing of steel plant gas streams, and coal to chemicals or fuels.

3. EXPERIMENTAL DEVELOPMENTS

3.1 Sample Characterization

HTC is a double hydroxide, with the general formula $Mg_{(1-x)}Al_x(OH)_2(CO_3)_{(x/2)}\cdot nH_2O$. It has a layered structure and contains hydroxides, carbonates, and water groups. Between 300 and 500°C, the material undergoes several dehydroxylation and decarbonation steps (Cavani et al., 1991; Reichle, 1986b; Walspurger et al., 2010). The layered structure collapses above approximately 350°C. Therefore, under SEWGS operating conditions the K-HTC sorbent is in transition to a mixed-metal oxide, while still containing carbonate and hydroxide groups, and consequently difficult to fully define. In this section, a detailed overview of recent advances on the sorption of CO_2 and H_2O and their mutual interactions is given, focusing on the adsorption and desorption kinetics, cyclic working capacity and isotherms, and in particular the role of steam and operating pressure using detailed thermogravimetric (TGA) and packed-bed breakthrough experiments.

3.1.1 Adsorption and Desorption Kinetics of CO_2 and H_2O

3.1.1.1 Materials and Methods

A calcined K-HTC-based adsorbent with a Mg/Al molar ratio of 0.54 and a potassium loading of approximately 20 wt.%, produced by Sasol Germany, was used in the experiments and will be further denoted as KMG30 (Puralox). The material was precalcined at 250 and 450°C for 24 h (Maroño et al., 2013). The adsorbent pellets (4.7 mm × 4.7 mm) were milled in a ball mill (Fritsch Pulverisette 6) to produce powder which was used in TGA measurements. After milling, the powder was dried at 250°C for 2 h to remove moisture.

Adsorption experiments were carried out in an in-house developed TGA setup for atmospheric pressure (see Fig. 4). A microbalance

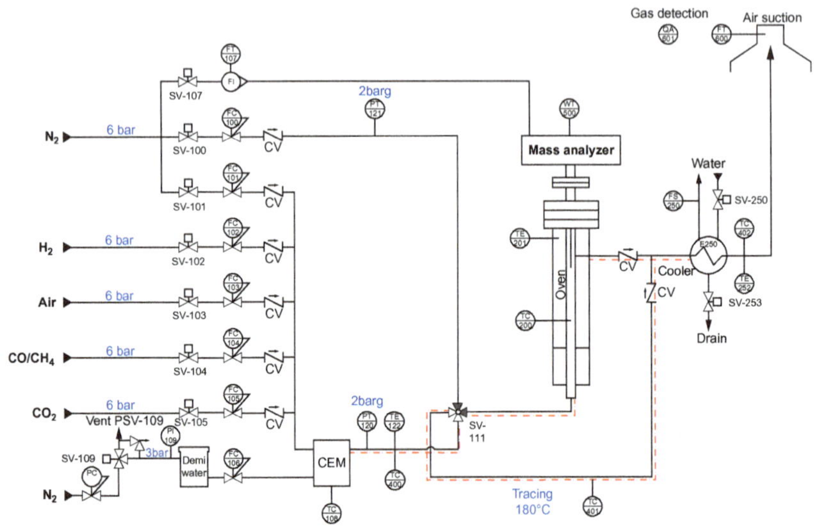

Fig. 4 P&ID of TGA setup used to study the adsorption/desorption kinetics of CO_2 and H_2O at atmospheric pressure at a temperature range between 300 and 500°C. *Reprinted from Coenen K, Gallucci F, Cobden P, van Dijk E, Hensen E, van Sint Annaland M: Chemisorption working capacity and kinetics of CO_2 and H_2O of hydrotalcite-based adsorbents for sorption-enhanced water–gas-shift applications,* Chem Eng J *293:9–23, 2016a, https://doi.org/10.1016/j.cej.2016.02.050, with permission of Elsevier.*

(CI-Precision MK2-5M) with a sensitivity of 0.1 μg and a maximum capacity of 5 g and ±500 mg operating range attached to a quartz reactor allowed accurate weight change measurements. The temperature of the balance head was kept constant and the head was continuously purged with N_2 to prevent reactive gas mixtures entering the balance. The gas feeding system consists of multiple Bronkhorst mass flow controllers (MFCs) and a controlled evaporator mixer (CEM) to introduce desired quantities of steam into the gas mixture. All gas lines are traced and insulated to avoid steam condensation. The temperature in the reactor was controlled using a thermocouple positioned close to the sample to ensure stable conditions. A total flow rate of 480 NmL/min was used during all experiments at atmospheric pressure. The absence of mass transfer limitations (both internal and external) at these conditions has been confirmed with separate experiments. Experiments with different gas flow rates have ruled out influence of external mass transfer limitations, whereas by comparing the adsorption rates for both CO_2 and H_2O measured for a powder with a mean diameter of about 40 μm and for pellets with a size of a few millimeter, absence of internal mass transfer limitations was demonstrated.

3.1.1.2 Results and Discussion

The literature 400°C (range of WGS) is a typical value where a lot of adsorption capacities have been measured, which was therefore chosen as standard reference temperature (Boon et al., 2014; Halabi et al., 2012; Hutson and Attwood, 2008; Oliveira et al., 2008; van Selow et al., 2009a; Van Selow et al., 2011; Walspurger et al., 2008).

The sample was heated to 600°C under N_2 to desorb water and CO_2, according to literature a good pretreatment temperature for K-HTC-based adsorbents (Maroño et al., 2014). A higher pretreatment temperature was not considered to avoid decomposition of potassium carbonate species, which has been reported for some K-HTCs and Al_2O_3 (Walspurger et al., 2008). Hydrotalcites are usually calcined between 500 and 700°C, which is the range where hydroxyl'carbonate completely decomposes, releasing CO_2 and H_2O without forming crystalline spinel (Maroño et al., 2014). Therefore, all samples were heated to 600°C under N_2 using a heating rate of 10°C/min followed by isothermal treatment at the same temperature for 120 min.

Adsorption and desorption capacities are defined as:

$$q_{CO_2}(\text{half cycle}) = \frac{\Delta m(\text{half cycle})}{M_{CO_2} \times m_{\text{sample}}(\text{after pretreatment})} \tag{6}$$

in which the weight change occurring during either the adsorption or the desorption part of the cycle, is divided by the molar mass of CO_2 or H_2O and the weight of the sample after the pretreatment.

The cyclic working capacity is defined as the average of the adsorption and desorption capacities:

$$q_{CO_2}(\text{cylcic}) = \frac{\Delta m(\text{ads}) + \Delta m(\text{des})}{2{*}M_{CO_2} \times m_{\text{sample}}(\text{after pretreatment})} \tag{7}$$

Fig. 5A shows the measured weight change during the first experiment listed in Table 1. After pretreatment, the sorbent lost about 10% of its initial mass resulting in a large contribution to the reported cyclic sorption capacity, depending on whether the difference between the initial mass and the mass after pretreatment is taken into account. Note that the plotted mass change in Fig. 5A shows the uncorrected weight loss based on the initial sample mass which was corrected with a blank experiment. It can be seen that the first cycles show a higher CO_2 adsorption capacity until a pseudo steady state between adsorption and desorption is reached. The shape of

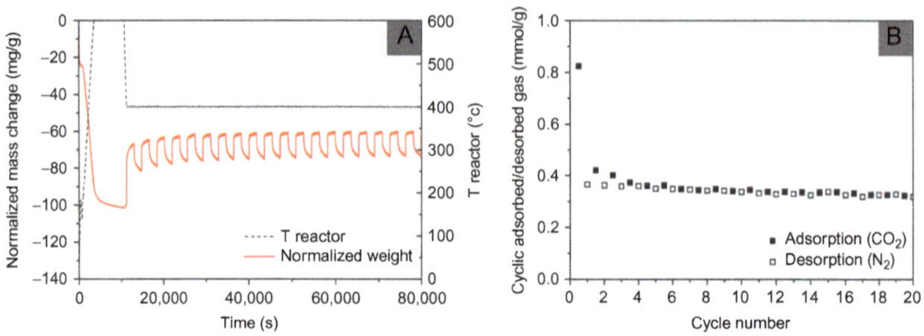

Fig. 5 TGA results for 30 min cycle experiments at 400°C and atmospheric pressure: (A) normalized weight change and reactor temperature and (B) cyclic CO_2 sorption capacity. *Reprinted from Coenen K, Gallucci F, Cobden P, van Dijk E, Hensen E, van Sint Annaland M: Chemisorption working capacity and kinetics of CO_2 and H_2O of hydrotalcite-based adsorbents for sorption-enhanced water–gas-shift applications,* Chem Eng J *293:9–23, 2016a, https://doi.org/10.1016/j.cej.2016.02.050, with permission of Elsevier.*

Table 1 Set of Experiments to Investigate the Adsorption of CO_2 on KMG30 at Atmospheric Pressure

No.	Name of Experiment	CO_2 Fraction	Temperature (°C)	Adsorption Time (min)	Desorption Time (min)	Cycles
1	30 min cycles	1	400	30	30	20
2	60 min cycles	1	400	60	60	10
3	120 min cycles	1	400	120	120	5
4	300 min cycles	1	400	300	300	5
5	600 min cycles	1	400	600	600	5
6	Varying adsorption time cycles	1	400	30, 60, 120	300	5
7	Varying desorption time cycles	1	400	300	30, 60, 120	5
8	300°C cycles	1	300	60	60	10
9	500°C cycles	1	500	60	60	10

Reprinted from Coenen K, Gallucci F, Cobden P, van Dijk E, Hensen E, van Sint Annaland M: Chemisorption working capacity and kinetics of CO_2 and H_2O of hydrotalcite-based adsorbents for sorption-enhanced water–gas-shift applications, *Chem Eng J* 293:9–23, 2016a, https://doi.org/10.1016/j.cej.2016.02.050, with permission of Elsevier.

the adsorption kinetics is changing during the first few cycles compared to cycles at the end of the experiment. Both observations are related to the initial state of the material after the pretreatment, since the CO_2 loading of the material is very low. A half-cycle time of 30 min at 400°C is not sufficient to desorb the same amount of CO_2 which was adsorbed in the previous adsorption step. This is also valid for experiments with longer half-cycle times, where even 600 min of desorption is still insufficient to reach the same CO_2 loading prior to CO_2 adsorption. The cyclic working capacity of $\approx$0.33 mmol/g determined according to Eq. (7) was determined based on the cyclic measured weight change of the last two cycles shown in Fig. 5B. It seems that the cyclic working capacity is still decreasing very slowly after 20 cycles.

The cyclic working capacity for CO_2 was determined for different half-cycle times and different operating temperatures as listed in Table 1 and are plotted in Fig. 6. We differentiate between the adsorption capacity during the first cycle (red bars) and the cyclic working capacity (black bars), where always the average of the last two cycles was taken into account. It is evident, that a longer half-cycle time can increase both the adsorption capacity and cyclic working capacity for CO_2, where the difference between the two is getting smaller for longer half-cycle times. It can also be observed in Fig. 6B that an increase in operating temperature leads to an increase in the cyclic working capacity, where the adsorption capacity during the first adsorption step remains constant.

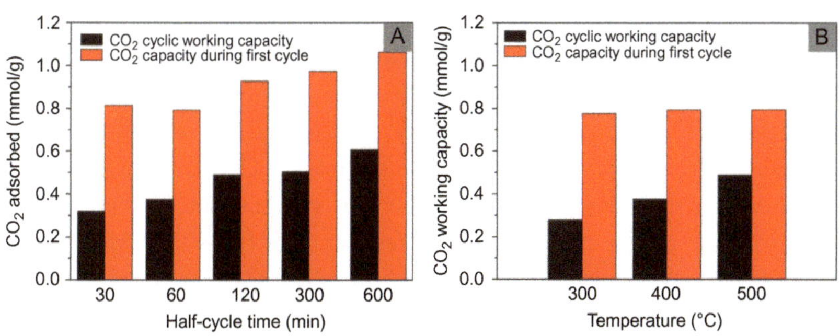

Fig. 6 (A) Cyclic CO_2 sorption capacity (averaged ADS and DES) as a function of half-cycle time and (B) cyclic CO_2 sorption capacity as a function of temperature for 60 min half-cycle time at atmospheric pressure. *Reprinted from Coenen K, Gallucci F, Cobden P, van Dijk E, Hensen E, van Sint Annaland M: Chemisorption working capacity and kinetics of CO_2 and H_2O of hydrotalcite-based adsorbents for sorption-enhanced water–gas-shift applications,* Chem Eng J *293:9–23, 2016a, https://doi.org/10.1016/j.cej.2016.02.050, with permission of Elsevier.*

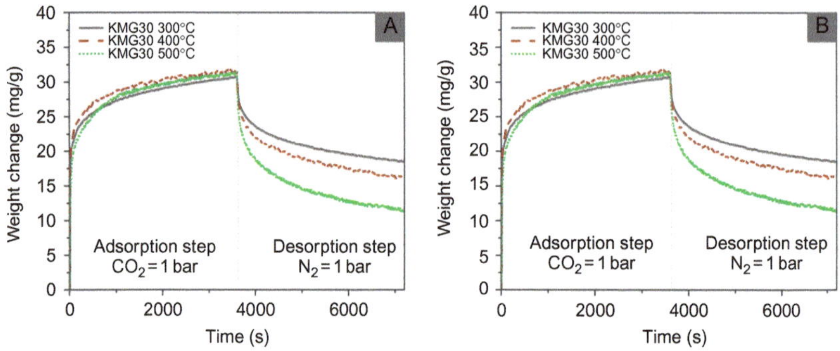

Fig. 7 (A) Weight change obtained at different operating temperatures with $PCO_2=1$ bar during adsorption and $PN_2=1$ bar during desorption at atmospheric pressure and (B) obtained weight change for adsorption and desorption during cycles 9 and 10 of Experiment 1. *Reprinted from Coenen K, Gallucci F, Cobden P, van Dijk E, Hensen E, van Sint Annaland M: Chemisorption working capacity and kinetics of CO_2 and H_2O of hydrotalcite-based adsorbents for sorption-enhanced water–gas-shift applications,* Chem Eng J *293:9–23, 2016a, https://doi.org/10.1016/j.cej.2016.02.050, with permission of Elsevier.*

The reason for this behavior can be explained by the adsorption and desorption rates of CO_2 under the chosen operating conditions. Fig. 7A shows that the adsorption capacity during the first cycle is constant (adsorption step), where during the desorption step with N_2 at higher temperature the desorption rate is increased. This leads to a lower CO_2 loading on the sorbent at the beginning of the next adsorption step and to the obtained increase in cyclic working capacity at a higher operating temperature. When a pseudo steady state between adsorption and desorption steps is established, the desorption kinetics are slower than the adsorption kinetics (see Fig. 7B). This is the reason why the desorption step is the limiting step mainly influencing the cyclic working capacity. The initial CO_2 loading is lower at the start of the next adsorption cycle so more CO_2 can be adsorbed.

In general, the adsorption rate of CO_2 is very fast and a sharp drop in adsorption rate can be obtained after a few seconds (Fig. 7B). After this fast adsorption, the sorbent continues to take up more CO_2 during the adsorption step resulting in the measured increase in adsorption capacity during the first cycle (Fig. 6A). The ability of the sorbent to continuously increase its CO_2 loading, even after long exposure times, has also been noted in the literature. It is reported that the uptake of CO_2 continues even after 700 min of adsorption for a K-HTC-based adsorbent at 400°C (Du et al., 2010).

In order to prove, that indeed the adsorption capacity is mainly related to the desorption kinetics, we have plotted the weight change obtained during

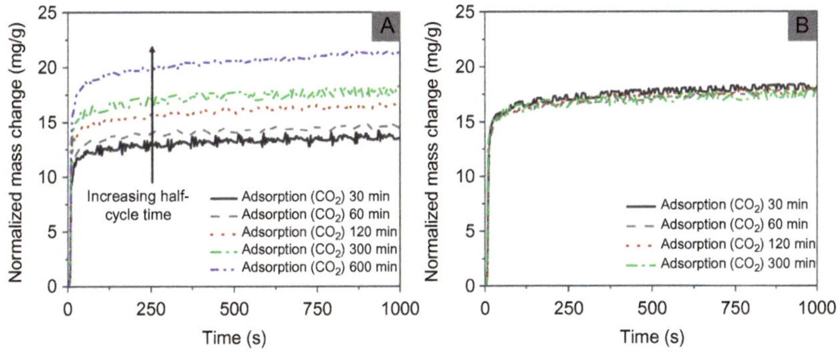

Fig. 8 (A) Mass change obtained during the first 1000 s of the adsorption step with different half-cycle times used (Experiments 1–5) with $PCO_2=1$ bar and (B) mass change obtained during the first 1000 s of the adsorption step with $PCO_2=1$ bar with constant desorption time of 300 min (Experiment 7). *Reprinted from Coenen K, Gallucci F, Cobden P, van Dijk E, Hensen E, van Sint Annaland M: Chemisorption working capacity and kinetics of CO_2 and H_2O of hydrotalcite-based adsorbents for sorption-enhanced water–gas-shift applications,* Chem Eng J *293:9–23, 2016a, https://doi.org/10.1016/j.cej.2016.02.050, with permission of Elsevier.*

the adsorption for different half-cycle times which were measured in Experiments 1–5 in Fig. 8A. In Fig. 8B, we have plotted the normalized weight change of Experiment 7, where we used a constant desorption time of 300 min and different adsorption times. Fig. 8A clearly shows that indeed the decrease in adsorption rate strongly depends on the chosen half-cycle time, which is caused by the difference in initial CO_2 loading at the beginning of the adsorption step. Indeed, if we keep the desorption time constant at 300 min and compare the adsorption rate for experiments with different adsorption times, the weight change obtained is identical and the same as in Experiment 4 (see Fig. 8B). To study the performance of a sorbent for CO_2 adsorption by comparing the cyclic working capacities of different sorbents, it is therefore essential that the regeneration conditions are carefully reported and kept the same.

The same procedure as reported for experiments 1–9 listed in Table 1 was carried out using a H_2O partial pressure of 0.34 bar (balanced with N_2) during the adsorption steps and a pure N_2 stream for the desorption step.

The results of the water adsorption and desorption at 400°C including the pretreatment is plotted in Fig. 9A. The measured H_2O adsorption capacity for the first adsorption is much higher than for the rest of the adsorption cycles, which is similar to what was observed for CO_2 adsorption. Again, this can be attributed to the initial sorbate loading of the material after the

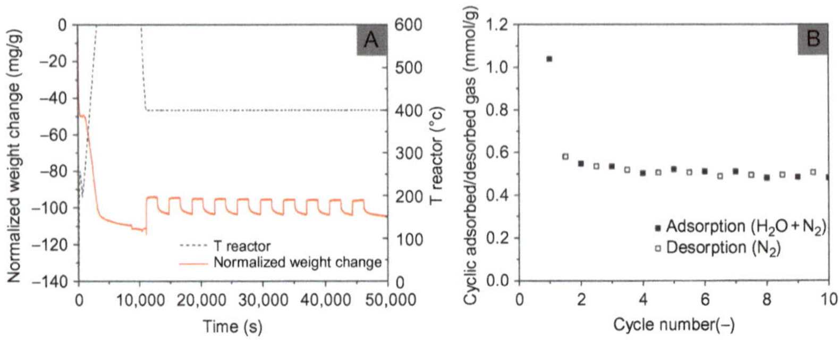

Fig. 9 TGA results for 30 min cycle experiments at 400°C and atmospheric pressure: (A) normalized weight change and reactor temperature and (B) cyclic CO_2 sorption capacity. *Reprinted from Coenen K, Gallucci F, Cobden P, van Dijk E, Hensen E, van Sint Annaland M: Chemisorption working capacity and kinetics of CO_2 and H_2O of hydrotalcite-based adsorbents for sorption-enhanced water–gas-shift applications,* Chem Eng J *293:9–23, 2016a, https://doi.org/10.1016/j.cej.2016.02.050, with permission of Elsevier.*

pretreatment. The maximum adsorption capacity is reached after 30 min, which implies that all available adsorption sites for H_2O are covered (equilibrium). This was not observed for CO_2 adsorption. A pseudo steady state for H_2O adsorption and desorption is already established after the first cycle. If we compare the cyclic working capacity for H_2O of 0.5 mmol/g to the cyclic working capacity for CO_2 (0.32 mmol/g), it is clear that the sorbent has an even higher cyclic working capacity for H_2O at the used experimental conditions. A lower H_2O adsorption capacity of 0.23 mmol/g has been reported in the literature using breakthrough experiments for the same material (Boon et al., 2014). The difference in the measured adsorption capacity for H_2O may be due to differences in pretreatment, differences in the local partial pressure of gases and difficulties in the interpretation of the breakthrough experiments in case of gas bypass.

The cyclic working capacity for H_2O can be increased by using longer half-cycle times (Fig. 10A), but the increase is less pronounced as for CO_2. During the first adsorption cycle, the capacity even seems to decrease slightly when the half-cycle time is increased. It could be observed that during the first cycle after the first rapid weight increase due to adsorption of H_2O the weight was decreasing slightly. This is due to further desorption of some CO_2 left on the sorbent which can be removed with H_2O. The slight increase in cyclic working capacity is again caused by the slower desorption kinetics. A lower H_2O loading at the end of the desorption step results in more H_2O adsorption during the subsequent adsorption step.

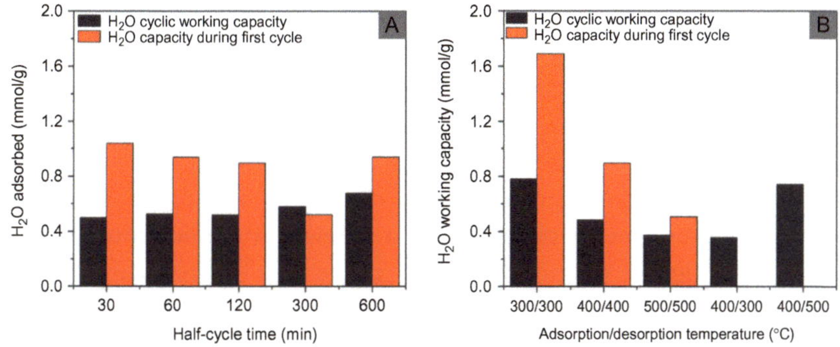

Fig. 10 Cyclic H_2O sorption capacity (averaged ADS and DES) as a function of half-cycle time with $PH_2O=0.34$ bar and (B) cyclic H_2O sorption capacity as a function of temperature for 60 min half-cycle time at atmospheric pressure with $PH_2O=0.34$ bar. *Reprinted from Coenen K, Gallucci F, Cobden P, van Dijk E, Hensen E, van Sint Annaland M: Chemisorption working capacity and kinetics of CO_2 and H_2O of hydrotalcite-based adsorbents for sorption-enhanced water–gas-shift applications,* Chem Eng J *293:9–23, 2016a, https://doi.org/10.1016/j.cej.2016.02.050, with permission of Elsevier.*

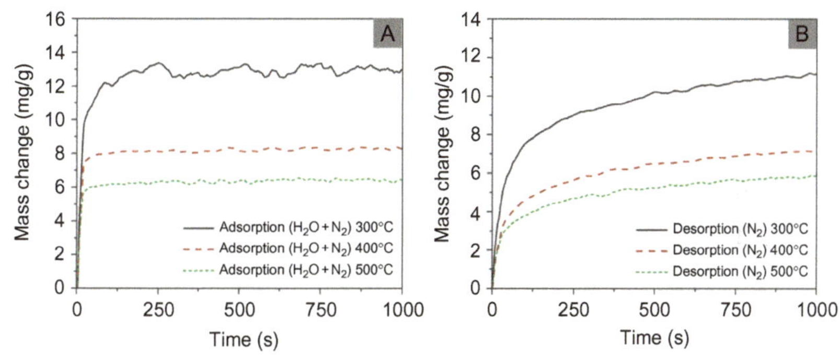

Fig. 11 (A) Mass change obtained during the first 1000 s of the adsorption step at different operating temperature (Experiments 2, 8, and 9) with $PH_2O=0.34$ bar and (B) mass change obtained during the first 1000 s of the desorption step with $PN_2=1$ bar at different operating temperature. *Reprinted from Coenen K, Gallucci F, Cobden P, van Dijk E, Hensen E, van Sint Annaland M: Chemisorption working capacity and kinetics of CO_2 and H_2O of hydrotalcite-based adsorbents for sorption-enhanced water–gas-shift applications,* Chem Eng J *293:9–23, 2016a, https://doi.org/10.1016/j.cej.2016.02.050, with permission of Elsevier.*

An increase in the operating temperature leads to a strong decrease in both cyclic working capacity and adsorption capacity during the first cycle, contrary to what was observed for CO_2. It was found that for H_2O the adsorption capacity is strongly reduced at higher operating temperatures. Fig. 11A shows that at lower temperatures indeed more H_2O can be adsorbed, and that the adsorption rate for H_2O is approaching the steady state very fast after the initial fast adsorption rate. No slow continuous uptake of H_2O was observed, as we found for CO_2. When comparing the desorption rate for the first 1000 s (Fig. 11B), it appears that H_2O desorption is faster at lower temperatures. One should however consider, that the weight change obtained during the experiments is always a function of the initial sorbate loading at the beginning of an adsorption/desorption step. Since more H_2O is present on the material, it seems that the desorption rate is enhanced at lower temperatures. To prove that indeed (like already described for CO_2) at higher operating temperatures the desorption rate is increased, two additional experiments were carried out changing the operating temperature only during the desorption step (last two experiments in Fig. 10B). Indeed, the desorption kinetics can be significantly enhanced by increasing the operating temperature during the desorption step leading to a higher cyclic working capacity, as shown in Fig. 10B.

3.1.2 The Role of Steam

3.1.2.1 Materials and Methods

TGA experiments were performed using an in-house designed setup for operation up to 10 bar (Fig. 12). A microbalance (Sartorius M25D) with a sensitivity of 1 μg and 200 mg of operating range is connected to a reactor designed for TGA experiments. The maximum operating temperature of this reactor is 1100°C. A N_2 stream is used to purge the balance and the reactor heating elements. MFCs and a CEM system is installed to produce the desired quantities of steam (either with N_2 or CO_2 as carrier gas) and other gas mixtures. All lines are traced and can be heated up to 450°C to avoid steam condensation even at high pressures. A porous ceramic basket is used with approximately 100 mg of sample mass for each experiment. At every pressure, the gas flow rate was adjusted such that mass transfer limitations due to the reduced volumetric flow rate in the reactor were avoided.

The weight change obtained from the TGA experiments to study the cyclic sorption capacity cannot be directly attributed to specific adsorbing/desorbing species for gas mixtures. TGA cycles containing different consecutive adsorption and regeneration steps with different gas compositions are used to link the weight change to a certain gaseous species adsorbed or desorbed. To systematically study the influence of H_2O on the adsorption of CO_2, a basic set of experiments was designed and performed in the TGA. The CO_2 adsorption was measured with a CO_2 partial pressure of 0.66 bar keeping the H_2O concentration at 34% as before (van Dijk et al., 2011) the duration of 30 min for each step is used as it was

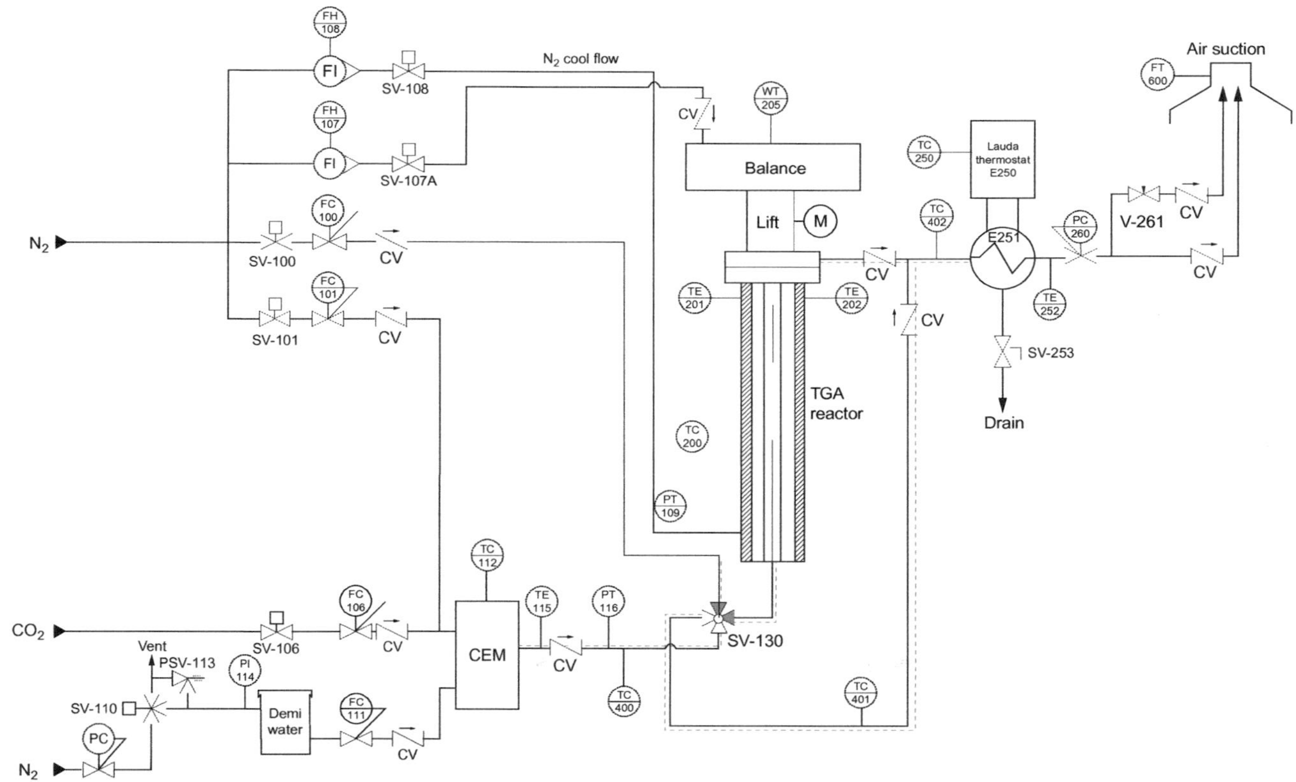

Fig. 12 P&ID of the TGA setup used to investigate the effect of H_2O on CO_2 adsorption on a potassium-promoted HTC. *Reprinted from Coenen K, Gallucci F, Pio G, Cobden P, Dijk Ev, Annaland MvS: On the influence of steam on the CO_2 chemisorption capacity of a hydrotalcite-based adsorbent for SEWGS applications,* Chem Eng J *314:554–569, 2016b. https://doi.org/10.1016/j.cej.2016.12.013, with permission of Elsevier.*

Table 2 Base Set of Experiments to Study the Influence of Steam on the Sorption Capacity of KMG30 for CO_2 at 400°C With $PCO_2 = 0.66$ bar and $PH_2O = 0.34$ bar

Experiment Number	Experimental Cycle Description	Steps in Cycle	Total Cycle Time (min)
1	$H_2O/N_2 \rightarrow N_2$	2	60
2	$CO_2 \rightarrow N_2$	2	60
3	$CO_2 \rightarrow N_2 \rightarrow N_2/H_2O \rightarrow N_2$	4	120
4	$CO_2/H_2O \rightarrow CO_2 \rightarrow N_2 \rightarrow N_2/H_2O \rightarrow N_2$	5	150
5	$CO_2 \rightarrow CO_2/H_2O \rightarrow N_2 \rightarrow N_2/H_2O \rightarrow N_2$	5	150
6	$CO_2/H_2O \rightarrow N_2 \rightarrow N_2/H_2O \rightarrow N_2$	4	120
7	$CO_2/H_2O \rightarrow N_2/H_2O$	2	60
8	$CO_2/H_2O \rightarrow N_2/H_2O \rightarrow N_2$	3	90
9	$N_2/H_2O \rightarrow CO_2 \rightarrow N_2$	3	90
10	$CO_2/H_2O \rightarrow CO_2 \rightarrow N_2/H_2O$	3	90
11	$CO_2/H_2O \rightarrow N_2/H_2O \rightarrow CO_2$	3	90
12	$CO_2 \rightarrow H_2O$	2	60

Reprinted from Coenen K, Gallucci F, Pio G, Cobden P, Dijk Ev, Annaland MvS: On the influence of steam on the CO_2 chemisorption capacity of a hydrotalcite-based adsorbent for SEWGS applications, *Chem Eng J* 314:554–569, 2016b. https://doi.org/10.1016/j.cej.2016.12.013, with permission of Elsevier.

established earlier that a half-cycle time of 30 min is sufficient to study the prevailing phenomena in the reactive system (Coenen et al., 2016a). Table 2 shows the basic set of experiments performed in the TGA. One experiment consists of 2–5 different consecutive steps. In the table a step is indicated as follows: Step 1 → Step 2, where the different gases in the reactor feed are mentioned for each step. Every experiment was performed five times in a cyclic way, starting again with the first step, before the next experiment was conducted. The average of the last three measurements for every experiment is reported as the cyclic working capacity. Experiments 1–8 were carried out at different operating temperature of 300 and 500°C (see Table 2).

In addition, packed bed reactor (PBR) experiments were carried out using a small PBR with a diameter of 27 and 350 mm height (AISI 316L). A gas distributor plate (sintered porous plate) with a pore size of 40 μm was installed at a height of 50 mm from the bottom. The reactor was filled with 53.6 g of hydrotalcite particles with a diameter of

1.8–3.15 mm. The effective length of the packed bed was determined at 176 mm. The reactor was installed in an electrical furnace. A multipoint thermocouple (10 measuring points at a distance of 20 mm) was installed to measure the axial temperature profile in the bed to observe temperature fronts due to sorption effects. A gas feeding system with MFCs and a CEM system was used to supply the desired gas mixtures including steam. All gas lines to the reactor were trace heated to avoid steam condensation. The reactor could be bypassed to check the gas compositions before exposing the material to the gas mixture. Two independent gas analyzing systems were used to monitor the gas compositions in the outlet during the experiments. A SICK GMS 800 gas analyzer for CO_2, CO, CH_4, H_2, and O_2 was used to monitor the gas streams continuously. To measure the steam content in the gas streams an Agilent Technologies Cary 630 FTIR with CaF_2 windows was used together with a RED-SHIFT gas sampling system. The FTIR was calibrated prior to the experiments using the classical Lambert–Beer law for the gases CO_2, CH_4, and H_2O. CH_4 was used as a tracer gas during the experiments to determine the total gas flow rate to convert the measured gas quantities to molar flow rates. It was proven that CH_4 does not interact with the material or steam under the considered operating conditions. A total gas flow rate of 0.5 NL/min was used during the experiments. The used molar fractions for CO_2, H_2O, and CH_4 were 0.025, 0.10, and 0.10, respectively, balanced with N_2. The gas composition was changed by bypassing the reactor for 5 min and measuring the concentration (to check the feed gas composition). After this stabilization time, the feed was sent to the reactor from bottom to top while monitoring the outlet composition of the reactor. Experiments were conducted at atmospheric pressure and 400°C. The empty volume of the reactor and the tubing was determined previously with blank measurements to correct the breakthrough times for both the FTIR and SICK gas analyzers. The reported results are the results obtained by the FTIR gas analyzer. A process flow diagram of the setup is given in Fig. 13.

3.1.2.2 Results and Discussion

Before the adsorption experiments, the sample was pretreated in N_2 at 600°C for 2 h in order to desorb CO_2 and H_2O which may have been adsorbed on the sorbent during storage (Coenen et al., 2016a). The cycling working capacity was determined for each adsorption/desorption step using the obtained mass change average over the last three consecutive experiments (see Table 2). Note that we use the mass-based cyclic working capacity,

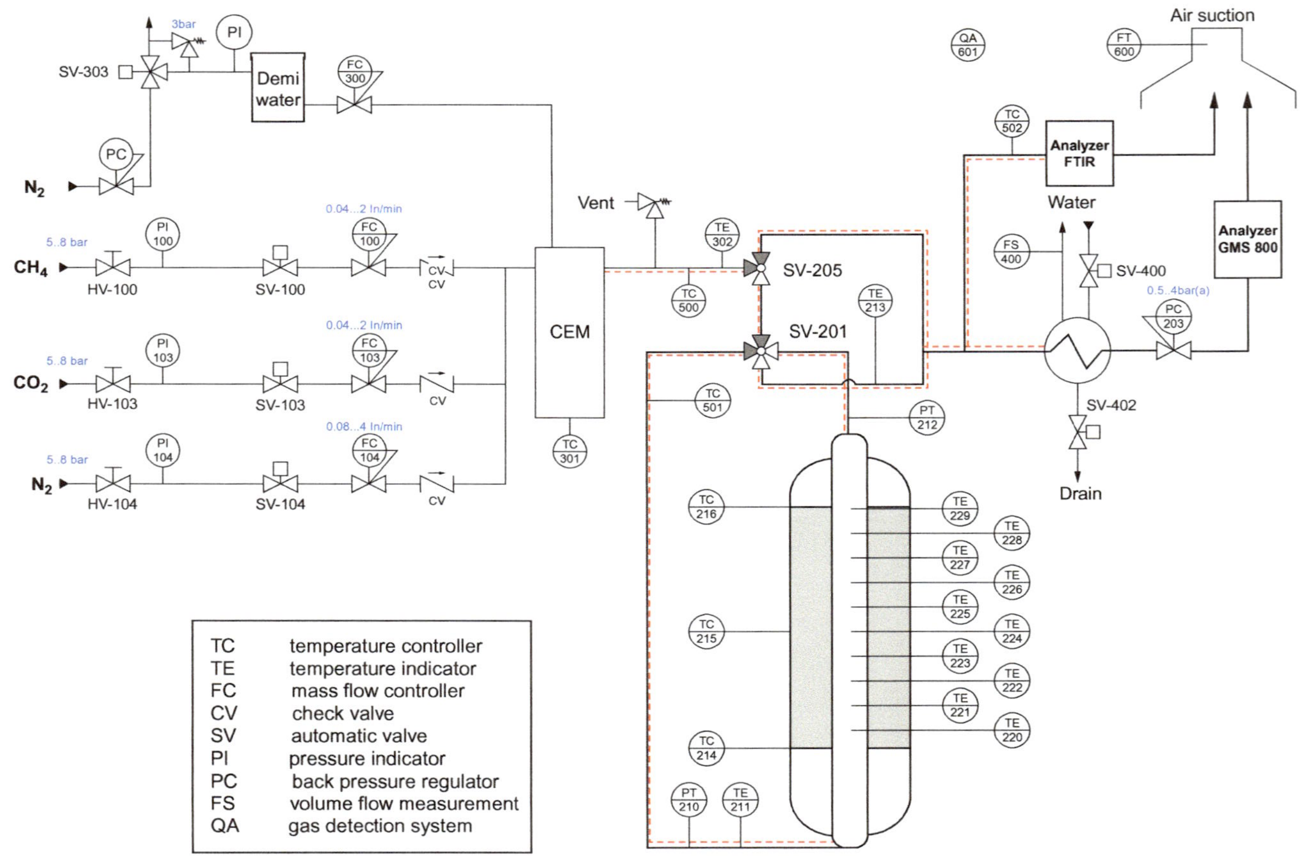

Fig. 13 P&ID of packed the bed reactor for the breakthrough experiments. *Reprinted from Coenen K, Gallucci F, Pio G, Cobden P, Dijk Ev, Annaland MvS: On the influence of steam on the* CO_2 *chemisorption capacity of a hydrotalcite-based adsorbent for SEWGS applications,* Chem Eng J *314:554–569, 2016b. https://doi.org/10.1016/j.cej.2016.12.013, with permission of Elsevier.*

as the TGA does not provide direct information on the sorbate being absorbed/desorbed. The development of an adsorption mechanism for the adsorption and desorption of CO_2 and H_2O will be explained step by step in the next paragraphs. Note that the experimental conditions in the PBR are somewhat different (in terms of partial pressures of the sorbate). These results are used to qualitatively describe the changes obtained from TGA measurements. TGA and PBR experiments will be described by the use of two different types of figures:

1. TGA (weight change)

To explain different adsorption sites which are responsible for the measured mass change in the TGA experiments arrows are used that indicate the normalized mass change measured in the TGA. The mass at the beginning and end of each step is indicated by a dotted line. Red arrows are used to represent CO_2 and blue arrows represent the exchange of H_2O. Adsorption is indicated by a solid arrow and desorption by a shaded arrow.

2. PBR (outlet concentration of CO_2 and H_2O)

The signal for CO_2 and the corresponding areas are plotted in red (integration of analyzer signal with respect to the baseline collected during the bypass of the reactor as described in Section 3.1.2.1), whereas the signal and areas for H_2O are plotted in blue. Again, we distinguish between adsorption (solid area above the signal) and desorption (shaded area below the analyzer signal) (Fig. 14; Coenen et al., 2016b).

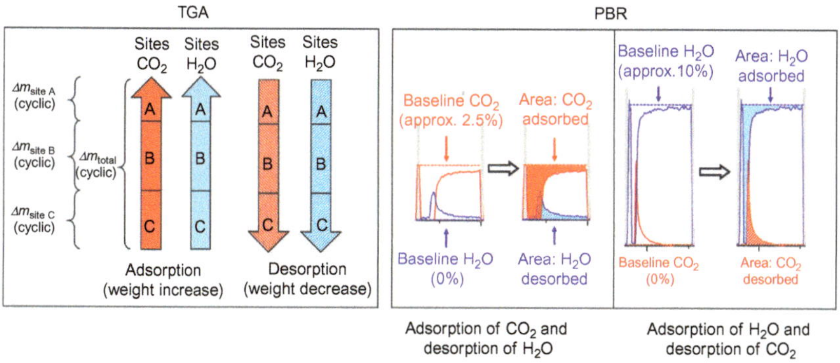

Fig. 14 Explanation of visual help to identify the different sites involved in CO_2 and H_2O adsorption on KMG30 based on TGA and PBR experiments. *Reprinted from Coenen K, Gallucci F, Pio G, Cobden P, Dijk Ev, Annaland MvS: On the influence of steam on the CO_2 chemisorption capacity of a hydrotalcite-based adsorbent for SEWGS applications,* Chem Eng J *314:554–569, 2016b. https://doi.org/10.1016/j.cej.2016.12.013, with permission of Elsevier.*

3.1.2.2.1 Adsorption of H_2O on KMG30 The adsorption of H_2O was studied in Experiment 1 using five adsorption/desorption cycles as listed in Table 2. It can be observed in Fig. 15A that the first cycle shows a higher change in mass compared to the following adsorption steps. The cyclic working capacity of the sorbent for H_2O was determined based on the last three cycles to 8 mg/g. The results obtained by analyzing the breakthrough time from PBR experiments confirm the results from the TGA. Also in the PBR the first adsorption cycle (1) shows a higher breakthrough time compared to the second adsorption step (3), where the breakthrough of H_2O during the desorption step remains the same during the two cycles. This adsorption site for H_2O will be called site A in order to describe the cyclic working capacity.

3.1.2.2.2 Adsorption of CO_2 on KMG30 In the same way as for H_2O, Fig. 16 shows the results obtained for CO_2 in the TGA and PBR experiments. Like for H_2O, the cyclic working capacity of this site (called site B) was determined based on the cyclic working capacity of the last three cycles, resulting in 12 mg/g. Also in this case the first cycle shows a higher adsorption capacity which was confirmed with the PBR experiments (see Fig. 16B). It is evident from both TGA and PBR experiments that the desorption kinetics and capacity are the same after the first adsorption step, as described in the previous section. Exposing the sorbent to either CO_2 or H_2O in a cyclic manner could be described with two sites introduced until now. It will become clear in the next section, that if both CO_2 and H_2O are used in one cycle the cyclic working capacity can no longer be described with only these two adsorption sites.

3.1.2.2.3 Simultaneous Adsorption of CO_2 and H_2O on KMG30 We will start to describe the experimental results obtained with the TGA and PBR experiments, shown in Fig. 17, with the sorbent being exposed to CO_2 (implying that site B is covered with CO_2 according to the results described earlier). During the first step with N_2, the amount of CO_2 adsorbed onto site B desorbs in the same manner as described during Experiment 2. In the second step, the sorbent was exposed to H_2O using the same conditions as in Experiment 1. If only the two introduced adsortpion sites were present, one would expect the adsorption of H_2O on site A resulting in a mass increase of about 8 mg/g. It becomes clear from Fig. 17A that after a first rapid weight increase the mass is actually decreasing. It becomes evident from the PBR experiments (Step 2 in Fig. 17B) that the sorbent releases CO_2 when exposed to H_2O. This implies that apparently during this step

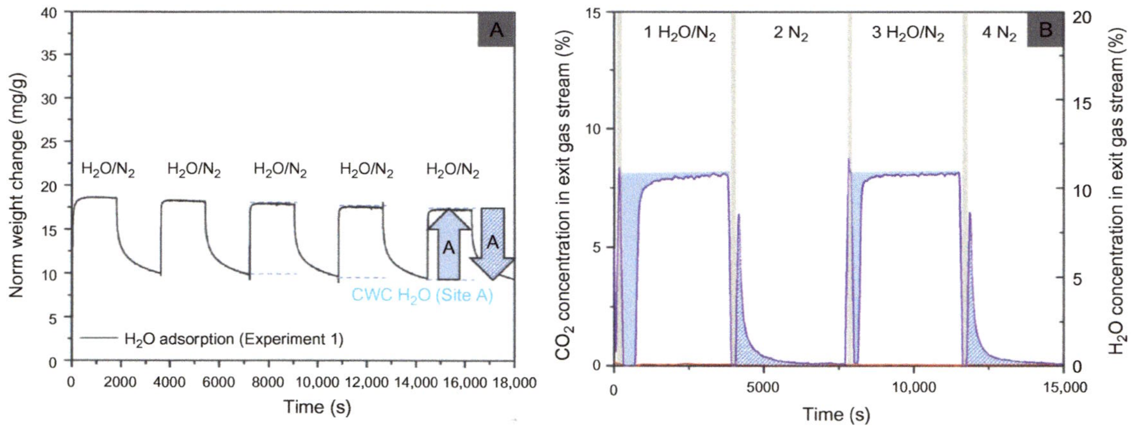

Fig. 15 (A) TGA Experiment 1: Adsorption/desorption of H_2O at 400°C and $PH_2O = 0.34$ bar and (B) PBR Experiment 1 at 400°C and $PH_2O = 0.1$ bar. *Reprinted from Coenen K, Gallucci F, Pio G, Cobden P, Dijk Ev, Annaland MvS: On the influence of steam on the CO_2 chemisorption capacity of a hydrotalcite-based adsorbent for SEWGS applications*, Chem Eng J *314:554–569, 2016b. https://doi.org/10.1016/j.cej.2016.12.013, with permission of Elsevier.*

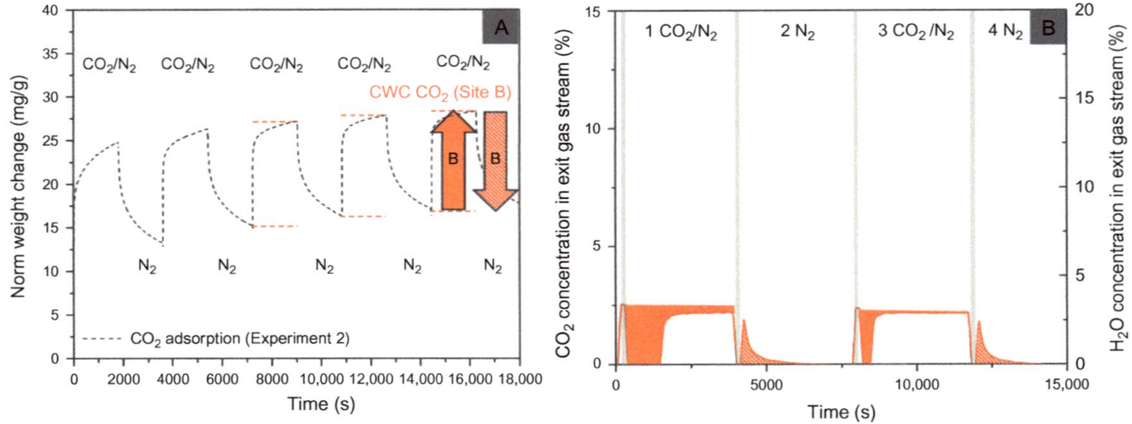

Fig. 16 (A) TGA Experiment 2: Adsorption/desorption of CO_2 at 400°C and $PCO_2=0.66$ bar and (B) PBR Experiment 2 at 400°C and $PCO_2=0.025$ bar. *Reprinted from Coenen K, Gallucci F, Pio G, Cobden P, Dijk Ev, Annaland MvS: On the influence of steam on the CO_2 chemisorption capacity of a hydrotalcite-based adsorbent for SEWGS applications,* Chem Eng J *314:554–569, 2016b. https://doi.org/10.1016/j.cej.2016.12.013, with permission of Elsevier.*

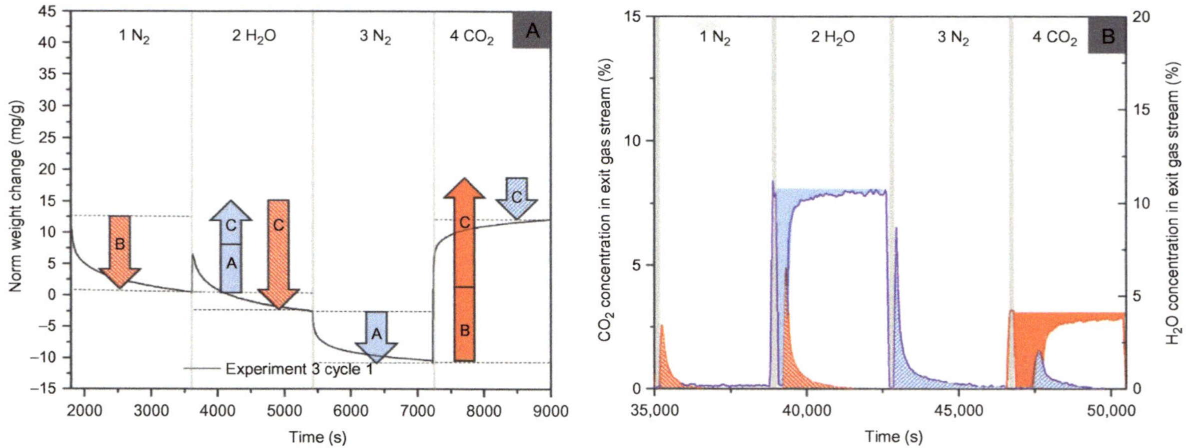

Fig. 17 TGA Experiment 3: Adsorption of CO_2 with dry (N_2) and wet regeneration (H_2O) at 400°C and $PH_2O = 0.34$ bar $PCO_2 = 0.66$ bar and (B) PBR Experiment 3 at 400°C and $PH_2O = 0.1$ bar $PCO_2 = 0.025$ bar. *Reprinted from Coenen K, Gallucci F, Pio G, Cobden P, Dijk Ev, Annaland MvS: On the influence of steam on the CO_2 chemisorption capacity of a hydrotalcite-based adsorbent for SEWGS applications*, Chem Eng J *314:554–569, 2016b. https://doi.org/10.1016/j.cej.2016.12.013, with permission of Elsevier.*

some CO_2 can be desorbed that was still in the adsorbed state after the desorption step with N_2 (Step 1). During the following desorption step (Step 3) only H_2O is released by the sorbent and the same mass loss is measured as in Experiment 1, confirming that H_2O from site A is desorbed. Once the material is exposed again to CO_2 during the Step 4 of Experiment 3 a higher mass change is measured than during the previous experiment. It can be seen in Fig. 17B that during this step, while CO_2 is being adsorbed, H_2O is released. Comparing the breakthrough-time of this experiment to previous experiments, it was noticed that the amount of CO_2 being adsorbed is higher and comparable to the first adsorption of CO_2 during Experiment 2. The same observation can be made for the mass change measured by the TGA. During Step 4 of Experiment 3 (Fig. 17A) a mass change of about 23 mg/g is measured, which is very similar to the measured mass change of the first step during Experiment 2 (Fig. 16A).

Based on these observations, one can conclude that steam adsorption during Step 2 induces desorption of CO_2. We propose that this adsorbed CO_2 must be assigned to a second site for CO_2 and H_2O (referred to as site C) were adsorption of one sorbate species induced the desorption of the adsorbed species which cannot be desorbed with N_2. Based on these observations, we can describe site C assuming a general metal oxide, forming a metal hydroxide, or carbonate considering the following reactions:

$$C[\bullet]O + H_2O \rightarrow C\left[(OH)_2\right] \tag{8}$$

$$C\left[(OH)_2\right] + CO_2 \rightleftharpoons C[CO_3] + H_2O \tag{9}$$

$$C[\bullet]O + CO_2 \rightarrow C[CO_3] \tag{10}$$

Based on experiments starting with either CO_2 or H_2O, we assume an equimolar exchange between the sorbate species at the chosen experimental conditions which is described in more detail elsewhere (Coenen et al., 2016b). The total capacity of this site was determined at 11 mg/g considering reaction (9). This implies the exchange of ~0.42 mmol/g considering the difference in molar mass between H_2O and CO_2 being 26 g/mol. Based on this we can calculate the expected weight change for reactions 8 and 10 as 7.5 and 18.5 mg/g, respectively. Note that reactions 1 and 3 can only be present when the sorbent is exposed to either CO_2 or H_2O after the pretreatment. After this only reaction 9 will occur at isothermal conditions.

3.1.2.2.4 Equilibrium Adsorption of CO_2 and H_2O on KMG30 Experiments 4 and 5 were carried out in order to investigate the

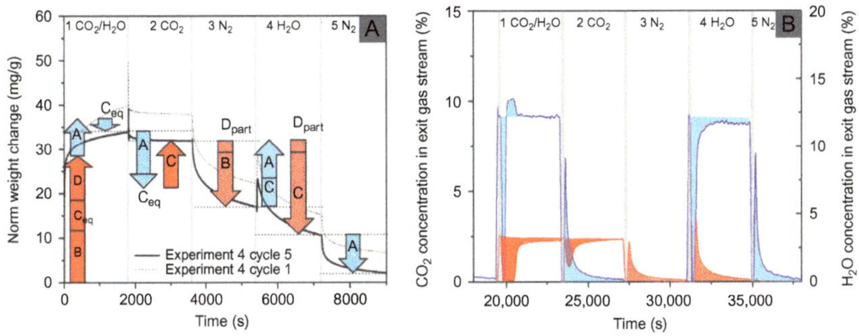

Fig. 18 (A) TGA Experiment 4: Adsorption of CO_2/H_2O followed by CO_2 and desorption with dry (N_2) and wet regeneration (H_2O) at 400°C and $PH_2O=0.34$ bar and $PCO_2=0.66$ bar and (B) PBR Experiment 4 at 400°C and $PH_2O=0.1$ bar and $PCO_2=0.025$ bar. *Reprinted from Coenen K, Gallucci F, Pio G, Cobden P, Dijk Ev, Annaland MvS: On the influence of steam on the CO_2 chemisorption capacity of a hydrotalcite-based adsorbent for SEWGS applications,* Chem Eng J *314:554–569, 2016b. https://doi.org/10.1016/j.cej.2016.12.013, with permission of Elsevier.*

coadsorption of CO_2 and H_2O. The order of a dry CO_2 feed and a mixture of CO_2 and H_2O was therefore reversed, where the remaining steps of the experiment were kept the same. During the first step of Experiment 4 (Fig. 18A), CO_2 adsorbs on sites B and C. During Step 2 a small weight decrease of 2 mg/g is observed. It can be obtained from the PBR experiments (Fig. 18B) that additional CO_2 is absorbed, even if the CO_2 partial pressure was kept constant during Steps 1 and 2, indicating that the equlibrium of site C is shifted toward the formation of carbonates. Considering that site B remains unchanged and site A desorbs (−8 mg/g) upon exposure to the dry CO_2, it is possible to calculate the weight change due to the shift in equilibrium. Since the measured weight loss is 2 mg/g, an increase of 6 mg/g can be attributed to site C. The mass change of 6 mg/g corresponds to an exchange of 0.23 mmol/g hydroxide to carbonate of site C. Therefore, in Step 1 (due to the equilibrium reaction 9) 0.186 mmol/g of site C was already occupied by CO_2.

Finally, the expected increase in mass during Step 1 should thus be 25 mg/g: 8 mg/g of H_2O for site A, 12 mg/g of CO_2 for site B, and of 4.8 mg/g for site C as explained above. However, the measured mass change during this step equals 31 mg/g, which can be described with an additional site (D) responsible for a 6 mg/g mass change in Step 1. During Step 3 only CO_2 is desorbing. The measured mass change is somewhat higher than in Experiment 3, which is explained by partial release of site D. In Step 4

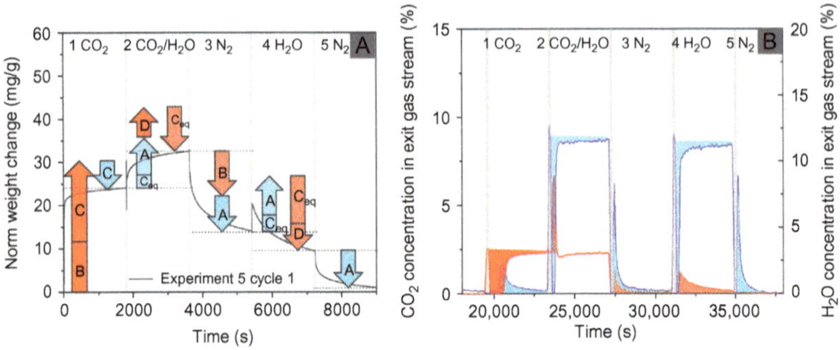

Fig. 19 (A) TGA Experiment 5: Adsorption of CO_2 followed by CO_2/H_2O and desorption with dry (N_2) and wet regeneration (H_2O) at 400°C and $PH_2O = 0.34$ bar $PCO_2 = 0.66$ bar and (B) PBR Experiment 5 at 400°C and $PH_2O = 0.1$ bar and $PCO_2 = 0.025$ bar. *Reprinted from Coenen K, Gallucci F, Pio G, Cobden P, Dijk Ev, Annaland MvS: On the influence of steam on the CO_2 chemisorption capacity of a hydrotalcite-based adsorbent for SEWGS applications,* Chem Eng J *314:554–569, 2016b. https://doi.org/10.1016/j.cej.2016.12.013, with permission of Elsevier.*

H_2O adsorbs on sites A and C (carbonate is converted into a hydroxide). This also implies that CO_2 is desorbed from site C and partly from site D. In the last step, the expected mass change of 8 mg/g was observed due to H_2O desorption on site A.

The results for Experiment 5 are plotted in Fig. 19. Step 1 of Experiment 5 is the same as Step 4 of Experiment 3 and can be described in the same way: CO_2 adsorbs on sites A and C which is transformed from a hydroxide into a carbonate, resulting in H_2O desorption. In the second step, CO_2 is desorbing due to the shift in equilibrium of site C (Eq. 9) which leads to a mass decrease of approx. 6 mg/g. Additionally, site D is occupied again with CO_2 and site A with H_2O resulting in the measured increase in sample mass. Exposure to N_2 only leads to desorption of CO_2 from site A and H_2O from site B. H_2O adsorption on sites A and C in Step 4 also causes site D to release adsorbed CO_2. The total amount of CO_2 released in this step however is smaller compared to that in Experiment 4 because site C was occupied by both CO_2 and H_2O according to the equilibrium reaction (9).

Fig. 20 shows the results of Experiment 6, which can be described in the same way as before for Experiments 4 and 5. During a wet fed of CO_2 site A is occupied by H_2O and B and D by CO_2, where site C is occupied by both CO_2 and H_2O according to the equilibrium of the site C. The subsequent steps are in total agreement with the previous experiments and can be explained with the four site model developed until here.

During Experiment 6 the partial pressure of steam was kept constant, where the partial pressure of CO_2 was varied between 0.66 and 0 bar.

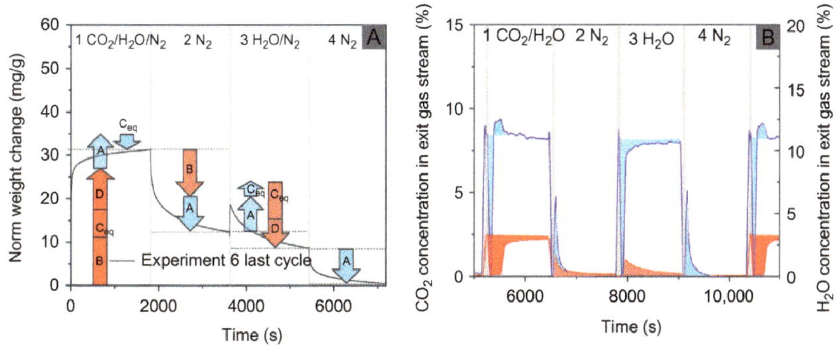

Fig. 20 (A) TGA Experiment 6: Adsorption of CO_2/H_2O and desorption with dry (N_2) and wet regeneration (H_2O) at 400°C and $PH_2O = 0.34$ bar and $PCO_2 = 0.66$ bar and (B) PBR Experiment 6 at 400°C and $PH_2O = 0.1$ bar and $PCO_2 = 0.025$ bar. *Reprinted from Coenen K, Gallucci F, Pio G, Cobden P, Dijk Ev, Annaland MvS: On the influence of steam on the CO_2 chemisorption capacity of a hydrotalcite-based adsorbent for SEWGS applications,* Chem Eng J *314:554–569, 2016b. https://doi.org/10.1016/j.cej.2016.12.013, with permission of Elsevier.*

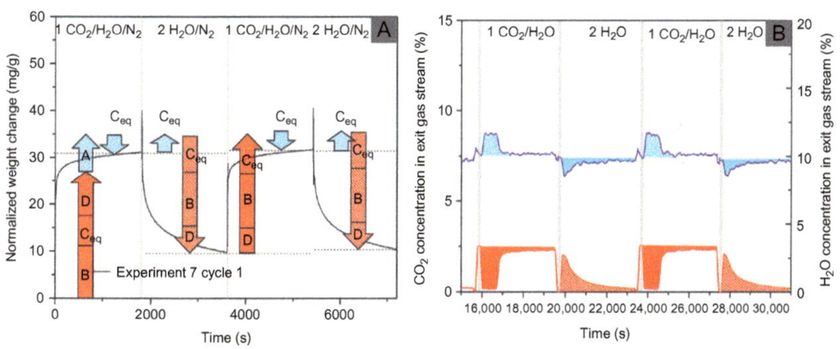

Fig. 21 (A) TGA Experiment 7: Adsorption of CO_2/H_2O and wet regeneration (H_2O) at 400°C and $PH_2O = 0.34$ bar and $PCO_2 = 0.66$ bar and (B) PBR Experiment 7 at 400°C and $PH_2O = 0.1$ bar and $PCO_2 = 0.025$ bar. *Reprinted from Coenen K, Gallucci F, Pio G, Cobden P, Dijk Ev, Annaland MvS: On the influence of steam on the CO_2 chemisorption capacity of a hydrotalcite-based adsorbent for SEWGS applications,* Chem Eng J *314:554–569, 2016b. https://doi.org/10.1016/j.cej.2016.12.013, with permission of Elsevier.*

The results for both TGA and PBR experiments are shown in Fig. 21. It can be confirmed that indeed due to the equilibrium reaction of site C, H_2O is desorbed when CO_2 is adsorbed in Step 1 and vice versa in Step 2.

Experiments 9–12 (shown in Fig. 22) are used to prove that the model can well describe the weight change obtained in the TGA experiments.

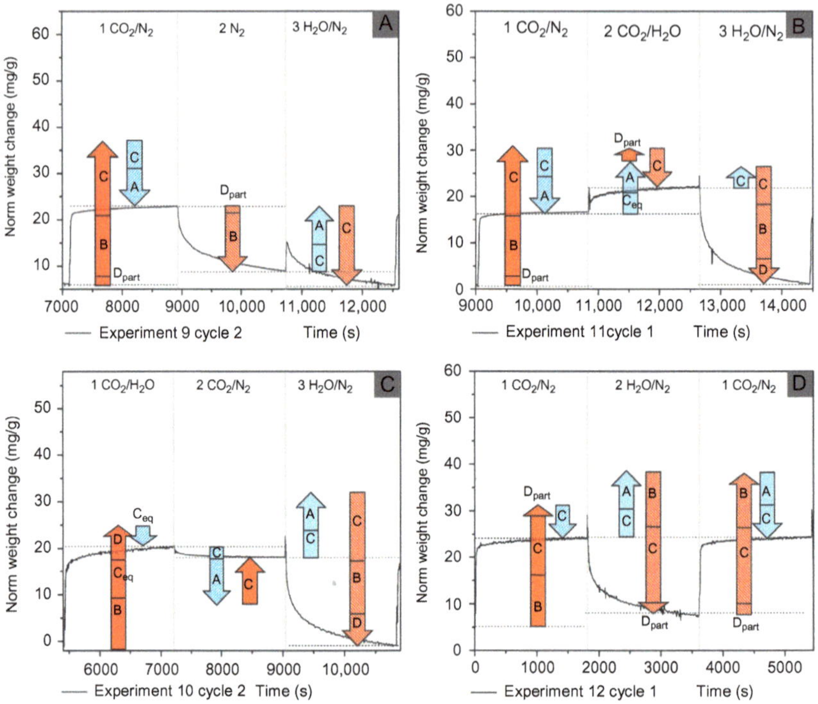

Fig. 22 (A–D) TGA results of Experiments 9–12 conducted at 400°C, $PH_2O = 0.34$ bar and $PCO_2 = 0.66$ bar to support the developed mechanism for the CO_2/H_2O interaction on KMG30. *Reprinted from Coenen K, Gallucci F, Pio G, Cobden P, Dijk Ev, Annaland MvS: On the influence of steam on the CO_2 chemisorption capacity of a hydrotalcite-based adsorbent for SEWGS applications,* Chem Eng J *314:554–569, 2016b. https://doi.org/10.1016/j.cej.2016.12.013, with permission of Elsevier.*

Based on more than 30 different experiments conducted with the TGA, the cyclic working capacities of the different adsorption sites were determined. The capacities were fitted to achieve a minimum deviation between the model and the experimental values. A summary of the proposed model and the involved adsorption sites is given in Fig. 23 together with the determined values for the cyclic working capacity. It shows that the developed model can describe all experiments quite accurately. It is evident that the cyclic working capacity of KMG30 for CO_2 is strongly dependent on the cycle design of the experiment. Steam enhances not only the adsorption capacity but also especially the ability to desorb more CO_2. If the sorbent is regenerated with a stream of N_2 (considering only site

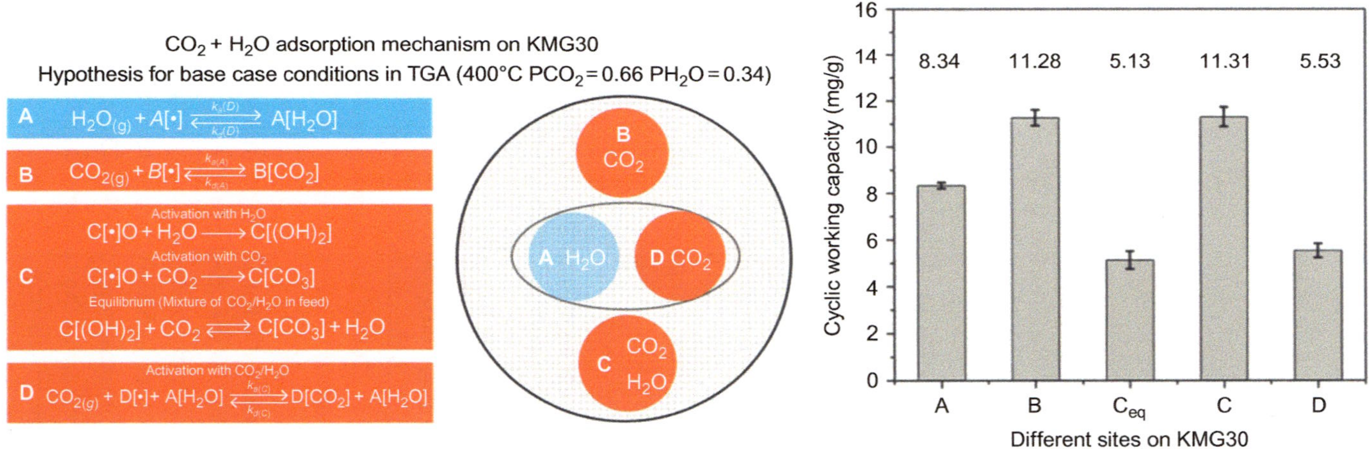

Fig. 23 Four-site model to describe the cyclic working capacity for both CO_2 and H_2O on KMG30 and capacities determined for experiments at 400°C and $PCO_2 = 0.66$ and $PH_2O = 0.34$. *Reprinted from Coenen K, Gallucci F, Pio G, Cobden P, Dijk Ev, Annaland MvS: On the influence of steam on the* CO_2 *chemisorption capacity of a hydrotalcite-based adsorbent for SEWGS applications*, Chem Eng J *314:554–569, 2016b. https://doi.org/10.1016/j.cej.2016.12.013, with permission of Elsevier.*

B available) only 40% of the capacity for CO_2 at the measured conditions is actually used.

3.2 KMG30: High-Pressure Testing, Isotherm, and Kinetics

Ideally, mixture adsorption is measured in a dedicated setup that allows for highly accurate measurements, as described above. However, for adsorption of carbon dioxide and steam on K-HTC at 400°C and industrially relevant pressures over 10 bar, the standard approach faces particular complications. First of all, it is important to ensure a proper reference state at the start of each experiment. As discussed above, under SEWGS operating conditions, K-HTC is in the middle of a transition to mixed-metal oxide, while still containing carbonate and hydroxide groups, and consequently difficult to fully define. Therefore, in order to insure that the material is always in the same state, a standard regeneration method was used in between the measurements, as described below. Second, measurements with a high partial pressure of steam is challenging. At 25 bar partial pressure, steam has a dew point of approximately 224°C, requiring flow control and all parts of the setup to be at least at this temperature in order to prevent condensation and have a stable flow for accurate measurements. Third, sampling and quantification of various concentrations of steam are also not straightforward. One solution to the issues mentioned is to resort to breakthrough experiments to measure the multicomponent isotherms. Hence, breakthrough experiments have been performed at bench scale with mass spectrometer (MS) gas sampling—yielding new insights that are not readily obtained at laboratory scale.

3.2.1 Breakthrough and Regeneration Experiments

Experiments were performed in the multicolumn SEWGS test rig, described in detail by Van Selow et al. (2009c). It comprises a total of six columns of 6 m tall and 38 mm internal diameter. Each column is filled with 4.7 mm × 4.7 mm pellets of a K-HTC, produced by Sasol Germany. The hydrotalcite has a Mg/Al ratio of 0.6 and was promoted with approximately 20 wt.% potassium carbonate, as specified by the vendor. At the time of the breakthrough tests, the pellets had been in the column for more than 2 years and proven to be stable in various SEWGS cycles for thousands of cycles in total. In order to measure breakthrough of carbon dioxide and steam with an accuracy as high as possible, one of the columns has been adapted for breakthrough measurements by installation of a sample system. The system, shown schematically in Fig. 24, provides two sample lines that connect

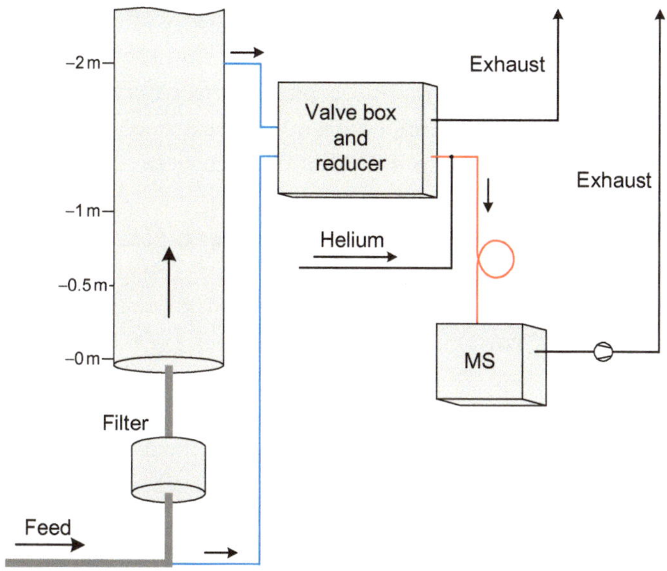

Fig. 24 Bottom part of SEWGS column and installed sample system. *Reprinted from Boon J, Cobden PD, van Dijk HAJ, Hoogland C, van Selow ER, van Sint Annaland M: Isotherm model for high-temperature, high-pressure adsorption of CO_2 and H_2O on K-promoted hydrotalcite,* Chem Eng J *248:406–414, 2014, Copyright (2014) with permission from Elsevier.*

the setup to a MS. The lower line allows the withdrawal of samples of the feed gas and the top line allows extraction of sample gas from the SEWGS column, 2 m downstream the feed connection. A sample flow of approximately 0.5 slpm is drawn from the system, reduced to atmospheric pressure and then mixed with 0.5 slpm of helium, which proved to be sufficiently small to prevent loss of MS signal fidelity, yet sufficiently large to prevent problems with condensation. The sample flow is then fed to the MS at 80°C, measuring the responses for argon ($m/z=40$), water ($m/z=18$), carbon dioxide ($m/z=44$), and nitrogen ($m/z=28$).

During the experiments in the temperature range of 400–450°C, the K-HTC sorbent is in transition from LDH to a mixed-metal oxide, with its structure containing carbonate and hydroxide groups. Consequently, the proper definition of the reference state of the material deserves specific attention. To make sure that the sorbent was in the same reference state with respect to carbon dioxide and steam loading at the start of each of the breakthrough experiments, the sorbent was cleaned and dried in a number of steps in between each of the breakthrough experiments. (1) Initially, the sorbent was saturated with steam at 400°C and 27 bar(a), purging remaining carbon dioxide from the material. (2) The purge continued while heating the reactor to 450°C, further removing carbon dioxide that remained on the sorbent. (3) The sorbent was dried with nitrogen from cryogenic storage at 450°C and 4 bar(a) pressure, until no steam was measured in the column outlet gas, followed by cooling to 400°C in dry nitrogen.

3.2.1.1 Single Component Breakthrough Experiments of Carbon Dioxide and Steam

The steam breakthrough capacity was measured at three levels of total pressure: 8, 17.5, and 27 bar(a). At each pressure level, breakthrough measurements with 11, 43, and 78 vol.% of steam in argon–nitrogen gas mixture were performed sequentially. In order to minimize the number of time-consuming regeneration steps, both a second and a third concentration levels were measured relative to the first concentration level. Accordingly, 11 vol.% steam was measured first, followed directly by measurement of the additional capacity at 43 vol.%. Afterward, the sorbent in the column was saturated with steam and then dried with nitrogen, followed by reequilibration with 11 vol.% steam. Then, the final point at 78 vol.% steam was measured, again relative to the uptake at 11 vol.%. In between experiments at different pressure levels and before measuring 11 vol.% steam, the sorbent was dried at 450°C and cooled back to 400°C in dry nitrogen. Upon completion of the steam breakthrough measurements, pure carbon dioxide breakthrough experiments were performed. These were carried out at 7, 12, 17, 22, and 27 bar(a) total pressure. In between the pressure levels, the sorbent was purged with steam at 27 bar(a) and dried in nitrogen at 4 bar(a). At each pressure setting, measurements were performed with 3 vol.% carbon dioxide, directly followed by 33 and 71 vol.% carbon dioxide in argon–nitrogen mixture, measuring the additional uptake after the preceding concentration.

3.2.1.2 Carbon Dioxide–Steam Mixture Breakthrough Experiments

The carbon dioxide–steam combined breakthrough experiments were aimed at measuring the excess adsorption of carbon dioxide and steam at 400°C and 27 bar(a). To this end, the column was operated in cyclic mode, with each single breakthrough experiment being preceded by a sequence of preconditioning steps: (1) saturation of the column with 18.5 vol.% carbon dioxide and 37 vol.% steam in nitrogen at 27 bar(a) total pressure, (2) depressurization to atmospheric pressure, (3) purge by steam until no more carbon dioxide was measured in the column outlet, followed by switching off the purge gas to depressurize the column (which had been slightly pressurized due to the flow of purge steam) to atmospheric pressure again, and (4) depressurization of the column to 27 bar(a) with dry nitrogen. Thus, the column is as much as technically possible in the same state at the start of each breakthrough experiment, containing 1 bar of steam and balance nitrogen at 27 bar(a) total pressure. By design of experiments, 32 conditions had been defined (that were all measured twice in randomized order except for the center point, which was measure nine times throughout the experimental program), with carbon dioxide partial pressures in the range of 0.5–24 bar and steam partial pressures in the range of 4–24 bar, all with argon tracer and nitrogen balance at 27 bar(a) total pressure. The argon tracer serves to describe the hydrodynamic characteristics of the column and sampling system.

3.2.1.3 Regeneration Experiments

For the purpose of evaluating the kinetics of regeneration of the adsorbent, a series of regeneration experiments has been carried out. In these experiments, the column was initially saturated with 17 vol.% of carbon dioxide and 25 vol.% of steam in balance nitrogen at 27 bar(a) and 400°C. Then, a counter-current rinse was performed with 30 slpm steam, followed by depressurization to atmospheric pressure, and finally a purge step with 18 slpm of steam. During these steps, the dry outlet flow rate was measured after the gas cooling and condensate removal. The same experiment was also carried out with 38 slpm purge flow rate.

3.2.2 High-Pressure Testing, Results, and Discussion

3.2.2.1 Equilibrium Data From Breakthrough Experiments

Setting up a material balance for component i over the column, accumulation of component i between $t=0$ and complete breakthrough ($t=t_e$) must equal the difference between the molar inflow and outflow rates, i.e.,

$$\frac{y_i(t_e)p}{RT}V_g + q_i(t_e)m_s - \frac{y_i(0)p}{RT}V_g - q_i(0)m_s = t_e F_{i,\text{in}} - \int_0^{t_e} (y_i F)_{\text{out}} dt \quad (11)$$

The trapezoidal rule has been used for approximating the integral by a summation over discrete measurement data. Prior to breakthrough of carbon dioxide and steam, the breakthrough of argon (assuming no argon is being adsorbed, $q_{\text{Ar}}=0$) is integrated to obtain gas volume V_g, the total interparticle and intraparticle gas volume in the column. After breakthrough, the argon signal is used for quantifying the outlet flow rate prior to and during breakthrough of steam and carbon dioxide. Eq. (11) can then be used to compute the excess amount of carbon dioxide and steam adsorbed, q_{CO_2} and q_{H_2O}, respectively.

3.2.2.2 Breakthrough Capacities for Carbon Dioxide, Steam, and Mixtures Thereof

Adsorption capacities for carbon dioxide and steam have been derived from the breakthrough measurements. The data points in Figs. 25 and 26 show the pure component breakthrough capacities measured for carbon dioxide or steam, respectively. Note that all lines are drawn based on the isotherm as derived in Section 4.1 from the full dataset, so including carbon dioxide–steam mixture data. The data points in Fig. 27 show actual mixture capacities. The pure component capacities in Fig. 25 for carbon dioxide follow a two-step adsorption mechanism, of which the first contribution up to 0.4 mol kg^{-1} is attributed to surface sites and the additional adsorption at partial pressures of 3 bar and over corresponds to nanopore adsorption, as discussed in more detail in Section 4.1. Given the current range of measurements, it is not possible to discern a more detailed interaction at the lower concentration levels, as for example presented by Lee et al.

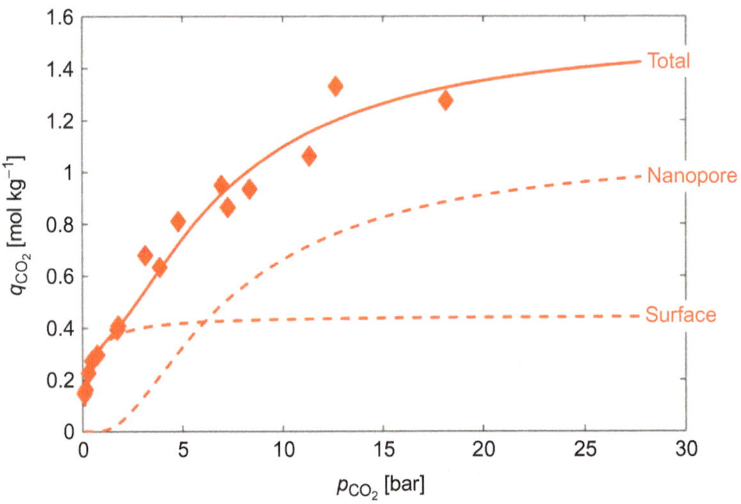

Fig. 25 Pure carbon dioxide breakthrough capacities vs carbon dioxide partial pressure (*diamonds*), predicted capacity (*line*), and contribution of different sites (*dashed lines*). *Reprinted from Boon J, Cobden PD, van Dijk HAJ, Hoogland C, van Selow ER, van Sint Annaland M: Isotherm model for high-temperature, high-pressure adsorption of CO_2 and H_2O on K-promoted hydrotalcite,* Chem Eng J *248:406–414, 2014, Copyright (2014) with permission from Elsevier.*

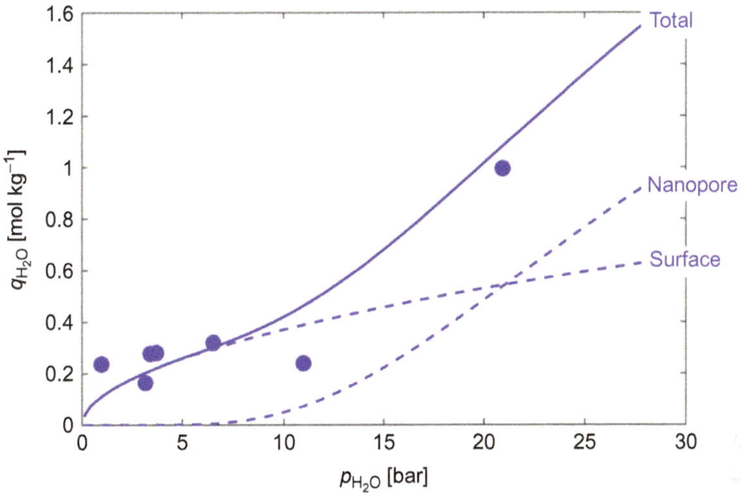

Fig. 26 Pure steam breakthrough capacities vs steam partial pressure (*circles*), predicted capacity (*line*), and contribution of different sites (*dashed lines*). *Reprinted from Boon J, Cobden PD, van Dijk HAJ, Hoogland C, van Selow ER, van Sint Annaland M: Isotherm model for high-temperature, high-pressure adsorption of* CO_2 *and* H_2O *on K-promoted hydrotalcite,* Chem Eng J *248:406–414, 2014, Copyright (2014) with permission from Elsevier.*

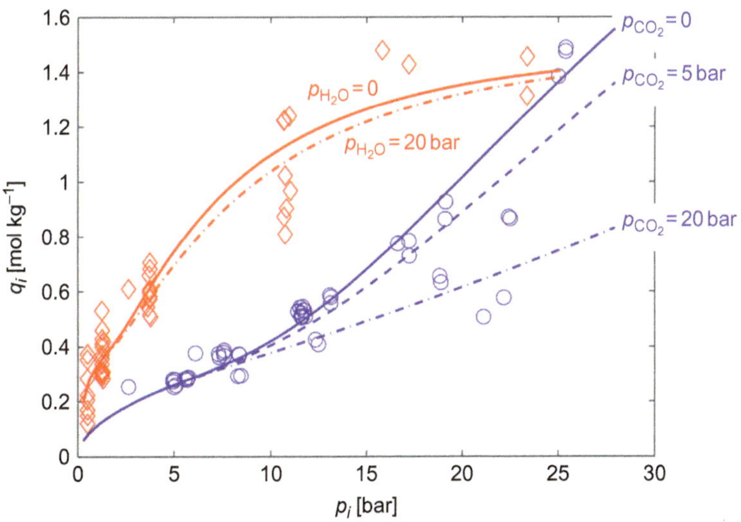

Fig. 27 Carbon dioxide–steam mixture breakthrough capacities for carbon dioxide (*red, open diamonds*) and steam (*blue, open circles*, corrected for the predicted remaining amount of steam adsorbed after regeneration) vs partial pressure, predicted pure component capacities (*line*), and mixture capacities (*dashed lines*). *Reprinted from Boon J, Cobden PD, van Dijk HAJ, Hoogland C, van Selow ER, van Sint Annaland M: Isotherm model for high-temperature, high-pressure adsorption of* CO_2 *and* H_2O *on K-promoted hydrotalcite,* Chem Eng J *248:406–414, 2014, Copyright (2014) with permission from Elsevier.*

(2007). For steam, similar trends can be observed. Up until 12 bar partial pressure, the measured capacities for steam show a leveling off of the surface contribution around 0.3 mol kg^{-1}, which is in line with the Langmuir isotherm proposed by Van Selow et al. (2009c). Noticeably, however, the single data point above 20 bar suggests an additional nanopore adsorption mechanism. Over the full partial pressure range, the trend in steam adsorption data is similar to the trend observed by Hatch et al. (2012) and supports the hypothesis of a binary mechanism.

3.2.2.3 Adsorption Isotherm

The entire experimental dataset was used to fit a multicomponent adsorption isotherm for the system. A good fit was obtained with two double isotherms, consisting of a surface and nanopore contribution (discussed in more detail below). The regressed parameters and statistics are shown in Table 3. The isotherms have also been added to the figures shown above for easy comparison. Parity plots of the predicted vs measured adsorption capacity, both for the single components and mixtures, are shown in Figs. 28 and 29.

Obtaining data with high accuracy from a bench scale setup as the one used in this study presents a challenge and consequently there is significant scatter in the data, especially in the mixture data, and the number of data

Table 3 Regressed Multicomponent Isotherm Parameters, With 90% Confidence Interval

Parameter	**Estimate**	
$q_{CO_2}^s$	0.45 ± 0.13	mol kg^{-1}
K_{CO_2}	28 ± 26	MPa^{-1}
E_{CO_2}	23 ± 1	kJ mol^{-1}
m_{CO_2}	5.2 ± 1.5	
K_{H_2O}	0.37 ± 0.07	mol kg^{-1} MPa$^{-1/n}$
n	1.9 ± 0.5	
E_{H_2O}	12 ± 1	kJ mol^{-1}
m_{H_2O}	3.6 ± 0.8	
V_0	74 ± 17	cm^3 kg^{-1}

Reprinted from Boon J, Cobden PD, van Dijk HAJ, Hoogland C, van Selow ER, van Sint Annaland M: Isotherm model for high-temperature, high-pressure adsorption of CO_2 and H_2O on K-promoted hydrotalcite, *Chem Eng J* 248:406–414, 2014, Copyright (2014) with permission from Elsevier.

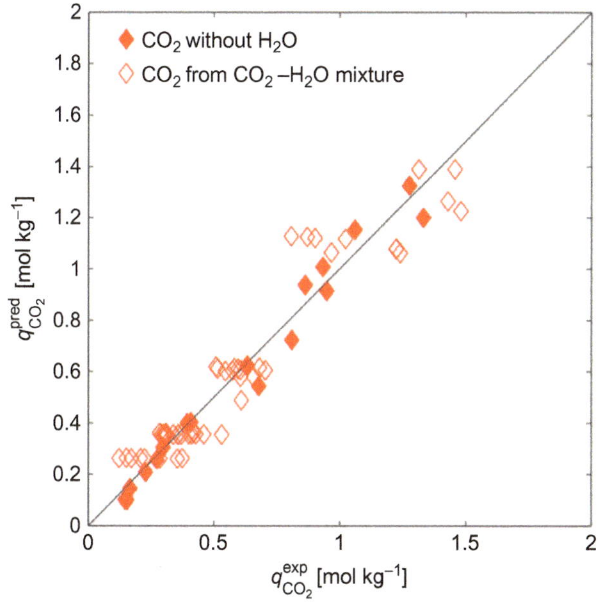

Fig. 28 Parity plot of predicted carbon dioxide adsorption capacities vs measured carbon dioxide adsorption capacities. *Reprinted from Boon J, Cobden PD, van Dijk HAJ, Hoogland C, van Selow ER, van Sint Annaland M: Isotherm model for high-temperature, high-pressure adsorption of CO_2 and H_2O on K-promoted hydrotalcite*, Chem Eng J *248:406–414, 2014, Copyright (2014) with permission from Elsevier.*

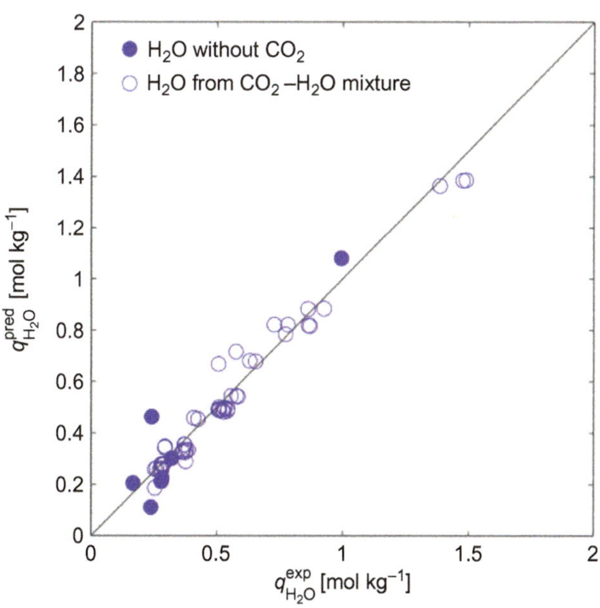

Fig. 29 Parity plot of predicted steam adsorption capacities vs measured steam adsorption capacities. *Reprinted from Boon J, Cobden PD, van Dijk HAJ, Hoogland C, van Selow ER, van Sint Annaland M: Isotherm model for high-temperature, high-pressure adsorption of CO_2 and H_2O on K-promoted hydrotalcite*, Chem Eng J *248:406–414, 2014, Copyright (2014) with permission from Elsevier.*

points for low concentrations is small. In this respect, the use of alternative experimental techniques can generate important results for the modeling of the low-pressure interaction, as has been discussed above. Alternatively, tapered element oscillating microbalance experiments have been successfully employed to measure the thermal composition of hydrotalcites at temperatures up to 700°C (Pérez-Ramírez and Abelló, 2006), to measure equilibria and kinetics of *n*-hexane and *n*-heptane adsorption on catalyst particles (Alpay et al., 2003), and to study the effect of steam upon carbon dioxide capture by several high-temperature sorbents at sorption-enhanced SMR conditions (Ochoa-Fernández et al., 2009). Alternatively, full binary equilibrium characterization could be performed by equilibrating the solid under flow conditions followed by complete desorption and characterization of total column contents, or by a combination of gravimetric and volumetric techniques (Talu, 2011). Still, the current set of experiments provides values that would be very difficult to measure otherwise. The scatter gives rise to relatively large confidence intervals in the parameter estimates and the small number of data points in the low concentration region further adds to the uncertainty for surface adsorption parameters, particularly K_{CO_2} (see Table 3). Nevertheless, all model parameter estimates are statistically significant at 90% confidence interval. The match of the model with less scattered single component data, Figs. 25 and 26, is good. The trends observed in the parity plot for carbon dioxide (Fig. 28) are randomly scattered in time and with steam concentration and reflect the accuracy of measurement for carbon dioxide–steam mixtures using the present techniques. The absence of trends for the pure carbon dioxide data plot with steam variations (Fig. 29) gives further confidence to the assumptions made in deriving the adsorption model.

Within the range of uncertainties, the values determined here are in agreement with values published by other groups. For adsorption of carbon dioxide on K-HTC, Lee et al. report monolayer capacities in the range of 0.25–0.45 mol kg^{-1} (compared to $q_{CO_2}{}^{s}=0.45$ mol kg^{-1} reported here) yet much higher gas–solid interaction parameters at 400°C in the range of 167–369 MPa^{-1} (Lee et al., 2007). The difference might be due to the uncertainty in the current Langmuir parameters but could also be caused by differences in surface characteristics of the sorbent. In a study with limited data at lower partial pressures Van Selow et al. report 44.5 MPa^{-1} (van Selow et al., 2013), which is larger but yet within the confidence interval of the present result. For steam adsorption on clay-like inorganic materials, literature data are scarce. In Section 3.1, the capacity for water adsorption al low partial pressure has been discussed in more detail already. With respect to the high-pressure capacity, the Freundlich exponent $n = 1.9$, that quantifies the nonlinearity of the isotherm, is in the range of 1.7–5.1 reported for the adsorption of steam on various clays (Hatch et al., 2012). For the nanopore adsorption model, few relevant studies are available. The nanopore volume in the current work is estimated at $74\,cm^3\,kg^{-1}$, while Soares et al. reported $73.3\,cm^3\,kg^{-1}$ for MG30 and Maroño et al. (2014) found similar pore volumes in the range of $50–60\,cm^3\,kg^{-1}$ for KMG30, depending on the temperature of calcination. The parameter m (5.2 for CO_2 and 3.6 for H_2O) characterizes nanopore heterogeneity, where values generally vary between 1 and 6 (Do, 1998). The characteristic energies E (Table 3: $E_{CO_2}/RT = 4.1$ and $E_{H_2O}/RT = 2.1$) are typical values for nanopore–sorbate interaction energies using the Dubinin–Astakhov model (Do, 1998; Doong and Yang, 1988).

The trends in the adsorption of CO_2 and H_2O on K-HTC observed here have been reported by other authors as well, although there are important differences that arise mainly because of the difference in partial pressures of the adsorbates. In contrast with the present model, the capacity for CO_2 has been reported to increase for low partial pressures of steam, compared to dry conditions (Ding and Alpay, 2000; Ram Reddy et al., 2008; Reijers et al., 2006; Yong et al., 2001). The presence of steam has been suggested to create adsorption sites through the formation of hydroxides, which in turn activate adsorption sites (Ding and Alpay, 2000), form bicarbonates (Ram Reddy et al., 2008), or carbonates (Walspurger et al., 2011) in the presence of CO_2. At lower partial pressures, steam will probably enhance the adsorption of CO_2 in the present system as well, but the smallest steam partial pressure in these experiments was 4 bar and the set of experimental data thus does not allow to discern between these interactions. Again, additional experimental work, such as Fourier transform infrared spectroscopy, might prove useful in elucidating the low partial pressure mechanisms in more detail (Ram Reddy et al., 2008). At higher steam partial pressures, however, competitive adsorption has been observed (van Selow et al., 2013). For CO_2 adsorption itself, a number of different mechanisms has been suggested to contribute to the adsorption capacity. A combination of physisorption and chemisorption has been suggested for CO_2 (Ebner et al., 2007; Oliveira et al., 2008; Wu et al., 2013). Lee et al. observed a binary mechanism up to 3 bar CO_2 partial pressure, suggesting a mechanism of chemisorption and surface complexation (Lee et al., 2007). Based on experiments with 1 bar CO_2, Wang et al. found physical and chemical contributions to the adsorption capacity and

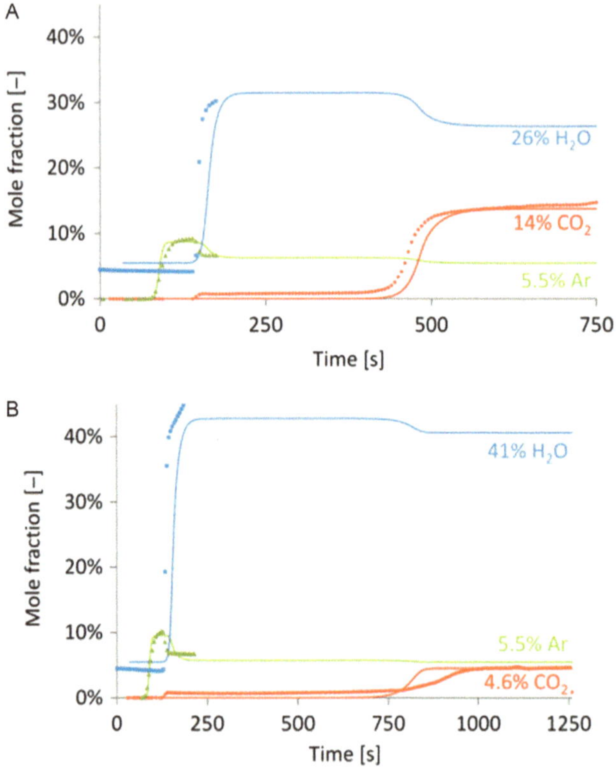

Fig. 30 Predicted (*lines*) and measured (*symbols*) breakthrough of Ar, H_2O, and CO_2 at 400°C, 27 bar(a) and different feed compositions. (A) Feed composition: 14 vol.% CO_2, 26 vol.% H_2O, 5.5 vol.% Ar, balance N_2 and (B) feed composition: 5.5 vol.% CO_2, 41 vol.% H_2O, 4.6 vol.% Ar, balance N_2. *Reprinted from Boon J, Cobden PD, van Dijk HAJ, Hoogland C, van Selow ER, van Sint Annaland M: Isotherm model for high-temperature, high-pressure adsorption of CO_2 and H_2O on K-promoted hydrotalcite,* Chem Eng J *248:406–414, 2014, Copyright (2014) with permission from Elsevier.*

concluded that not only the surface, but also morphology and pore structure are essential to achieve a high adsorption capacity (Wang et al., 2012). The present experiment has produced little data in the low-pressure range that was studied by these authors. However, it should be noted that high-pressure H_2O is able to desorb traces of CO_2 that remain on the sorbent after low-pressure regeneration, as shown in Fig. 30, after breakthrough of H_2O and before CO_2 breakthrough. This is indeed an indication of interaction between high-pressure H_2O and small amounts of CO_2 adsorbed. Therefore, it seems probable that the surface contribution for CO_2, and possibly for H_2O, is the sum of contributions from several different mechanisms where interaction may also play a role, as already discussed extensively in Section 3.1. Additionally, the present work reaffirms the importance of the nanostructure of the sorbent, predominantly at higher partial pressures. In this respect, a clear indicator is the observation that the impact of CO_2 on the capacity for H_2O is much stronger than vice versa. This can hardly be explained by a site-specific model but is well accounted for by the nanopore model. The larger molar volume of CO_2 ($\nu_{CO_2} = 69\,\mathrm{cm^3\,mol^{-1}}$) compared to H_2O ($\nu_{H_2O} = 31\,\mathrm{cm^3\,mol^{-1}}$) ensures that the available nanopore volume is more strongly attenuated by CO_2 than by H_2O.

4. MODELING AND CYCLE DESIGN

4.1 Adsorption Isotherm

Adsorption isotherms are derived on the basis of the underlying physics of sorbent–sorbate interactions. As presented in Section 3.2.2.2, the main body of literature is based on Langmuir and Freundlich isotherms. These are derived from adsorption on specific surface sites. Indeed, a transformation occurs of potassium carbonate at the surface of HTC and that aluminum oxide—potassium carbonate interaction creates the strongly basic active sites for CO_2 adsorption (Walspurger et al., 2008). This, however, is not to say that the surface sites on K-HTC are homogeneous, nor that each site can adsorb only a single molecule. In fact, many authors have proposed multisite adsorption models for lower partial pressures and the sorbates may interact as well (Coenen et al., 2016a; Lee et al., 2007; Ram Reddy et al., 2008; van Selow et al., 2013; Walspurger et al., 2011; Wang et al., 2012; Yong et al., 2001). The current focus, however, is on higher partial pressures and multisite or multicomponent interaction mechanisms at low partial pressures are ignored. In addition to surface sites, nanopores may play an important role at higher pressures. Significant volumes of nanopores have been measured (Maroño et al., 2013; Soares et al., 2005) that may additionally contribute to the adsorption capacity for CO_2. Nanopore adsorption occurs above a certain partial pressure threshold when surface adsorption turns into pore volume filling (Do, 1998). For H_2O, little is known about adsorption on K-HTC as discussed in Section 3.

A model is derived here that is based on contributions of surface and nanopores for the adsorption of CO_2 and H_2O. Whereas the surface sites are specific in the sense that they adsorb either CO_2 or H_2O, competitive adsorption is expected in the nanopores. Surface adsorption of CO_2 is described by a Langmuir isotherm, concurring with most studies in literature. For H_2O, a Freundlich isotherm is used for the surface contribution, reflecting the presumed heterogeneous character of the sites for H_2O. Given the range of measurement and within the range of uncertainty, the choice is arbitrary. The nanopore contribution is described using a Dubinin–Astakhov equation based on volume filling theory. In such a case, assuming no lateral interactions between adsorbed species, the amount of gas adsorbed is determined by the limiting nanopore volume V_0, pore-size distribution parameter m, and the nanopore–sorbate interaction energy E (Do, 1998; Doong and Yang, 1988; Ozawa et al., 1976). Thus, for CO_2,

$$q^*_{CO_2} = \frac{q^s_{CO_2} K_{CO_2} p_{CO_2}}{1 + K_{CO_2} p_{CO_2}} + \frac{A_{CO_2}(V_0 - V_0 A_{H_2O})}{v_{m,CO_2}(1 - A_{CO_2} A_{H_2O})} \tag{12}$$

and for H_2O,

$$q^*_{H_2O} = K_{H_2O} p_{H_2O}^{1/n} + \frac{A_{H_2O}(V_0 - V_0 A_{CO_2})}{v_{m,H_2O}(1 - A_{CO_2} A_{H_2O})} \tag{13}$$

with

$$A_i = \exp\left(-\left[\frac{RT}{E_i} \ln\left(\frac{p_{0,i}}{p_i}\right)\right]^{m_i}\right) \tag{14}$$

$$p_{0,i} = p_{c,i}\left(\frac{T}{T_{c,i}}\right)^2 \tag{15}$$

$$v_{m,i} = \frac{RT_{c,i}}{8p_{c,i}}\left(\frac{T}{T_{c,i}}\right)^{0.6} \tag{16}$$

The required model parameters have been determined from breakthrough experiments as explained in Section 3.2.

4.2 Particle Modeling

The equations for the intraparticle model are shown in Table 5. They are derived from the common linear driving force assumption, for porous spherical particles (Ruthven, 1984). The adsorption isotherm (4.8) has been derived and discussed in Section 3.2. Constitutive equations have been taken from literature (Bird et al., 1960; Poling et al., 2001; Ruthven, 1984; Westerterp et al., 1987). Model parameters are shown in Table 4.

Table 4 SEWGS Process Parameters

	2	m (breakthrough)
L	6	m (regeneration)
d_r	38	mm
d_p	4.4	mm
ρ_p	1074	$kg\,m^{-3}$
ϵ_b	0.4	
ϵ_p	0.3	
T_{wall}	400	°C
p	27	bar (breakthrough)
u_{feed}	0.029	$m\,s^{-1}$

Reprinted from Boon J, Cobden PD, van Dijk HAJ, Hoogland C, van Selow ER, van Sint Annaland M: Isotherm model for high-temperature, high-pressure adsorption of CO_2 and H_2O on K-promoted hydrotalcite, *Chem Eng J* 248:406–414, 2014, Copyright (2014) with permission from Elsevier.

4.3 Column Modeling and Kinetics

A reactor model was developed, based on a previous reactor model for low-pressure sorption-enhanced reforming (Boon et al., 2014; Reijers et al., 2009a,b). The transport equations solved are shown in Table 6, now including accurate description of transport of concentrated species and density changes. The feed flow rate is specified at the column inlet, the pressure at the column outlet. Danckwerts boundary conditions are used for the heat and mass balances. Using the method of lines, the equations are discretized on a uniform grid in the axial direction with a Barton flux delimited second-order upwind scheme for the convective terms, second-order implicit central differencing for the dispersion terms, and semiimplicit linearized source terms. The resulting time-dependent ordinary differential equations are integrated in time with first-order Euler scheme with time step adaptation.

4.3.1 Kinetics of Adsorption and Desorption

The heterogeneous linear driving force relation used in this work is based on the intraparticle concentrations as discussed in Section 4.2 (Eqs. 17 and 18, Table 5). A similar relation has been used by Wu et al. (2013). The equation accounts for contributions of intraparticle gas phase (through $c_{int,i}$) as well as intraparticle surface and nanopore accumulation terms (through q_i). It approximates the rigorous intraparticle diffusion equations to a numerically

Table 5 SEWGS Particle Model Equations

Intraparticle mass balance	$\frac{d\langle c_i \rangle}{dt} = k_{\mathrm{LDF},i}\left(c_{\mathrm{int},i} - \langle c_i \rangle\right)$	(17)
LDF mass transfer coefficient	$k_{\mathrm{LDF},i} = \dfrac{15\mathcal{D}_{p,i}}{r_p^2\left(\epsilon_p + \rho_p \dfrac{\partial q_i}{\partial c_i}\right)}$	(18)
Multicomponent isotherm	$\langle q_i \rangle = f(\langle c_{1\ldots N} \rangle)$	(19)

Reprinted from Boon J, Cobden PD, van Dijk HAJ, Hoogland C, van Selow ER, van Sint Annaland M: Isotherm model for high-temperature, high-pressure adsorption of CO_2 and H_2O on K-promoted hydrotalcite, *Chem Eng J* 248:406–414, 2014, Copyright (2014) with permission from Elsevier.

efficient ordinary differential equation, an approach which is common practice in PSA modeling (Yang, 1987). The heterogeneous linear driving force relation only approximates intraparticle resistances to mass transfer. It can be expanded or elaborated in more detail by accounting for the resistance in the crystals, or multicomponent mass transfer (Wu et al., 2013). However, such an approach does not seem to be justified in the present study, unless the internal structure of the particles was to be characterized with much more detailed information on the particle morphology. Besides, the match between experiment and model is fair, as shown above. During breakthrough, as shown in Fig. 18, the stoichiometric breakthrough time of each of the species is predicted well. This indicates that the sorption capacity for these conditions is accurately predicted by the proposed multicomponent isotherm. The transient of breakthrough is also represented well by the model, except for CO_2 in case (b) with 4.6 vol.% CO_2 in the feed, where the transient is somewhat faster in the model than in the measurement. In this context, however, it should be noted that measurements with CO_2–H_2O mixtures at these low CO_2 concentrations are nontrivial in the current setup. A fact that also shows in the relatively large scatter for the data points for CO_2 partial pressures below 2 bar (Fig. 27). For the desorption experiments shown in Fig. 30, the model predicts a slower desorption of CO_2 than measured experimentally. Apart from a difference in absolute amount of CO_2 adsorbed, this may be at least partly caused by pressure dynamics in the bench scale system that are not fully represented in the model (Table 6).

4.4 Cycle Modeling

The state-of-the-art SEWGS cycle follows a number of steps that govern the performance of the process. The steps are schematically shown in Fig. 2 for

Table 6 SEWGS Reactor Model Equations

Continuity	$\frac{\partial \rho}{\partial t} = -\frac{\partial \rho v}{\partial z} + \frac{1-\epsilon_b}{\epsilon_b} a_p \sum_i M_i N_i$	(20)
Momentum	$0 = -\frac{\partial p}{\partial z} - f\frac{\rho\vert u\vert u}{d_p}$	(21)
Heat balance	$\left(\epsilon_b \rho C_p + (1-\epsilon_b)\rho_p C_{p,p}\right)\frac{\partial T}{\partial t} = -\rho C_p u\frac{\partial T}{\partial z} + \frac{\partial}{\partial z}\left(\lambda \frac{\partial T}{\partial z}\right) + \frac{4U(T_{\text{wall}}-T)}{d_r} - (1-\epsilon_b)\rho_p\left((-\Delta H_r) r_{\text{WGS}} + \sum_i \left(-\Delta H_{a,i}\frac{d\langle q_i\rangle}{dt}\right)\right)$	(22)
Mass balance	$\frac{\partial \rho\omega_i}{\partial t} = -\frac{\partial \rho v \omega_i}{\partial z} + \frac{\partial}{\partial z}\left(D_z \rho \frac{\partial \omega_i}{\partial z}\right) + \frac{1-\epsilon_b}{\epsilon_b} a_p M_i N_i$	(23)
State	$pM = \rho RT$	(24)

Reprinted from Boon J, Cobden PD, van Dijk HAJ, Hoogland C, van Selow ER, van Sint Annaland M: Isotherm model for high-temperature, high-pressure adsorption of CO_2 and H_2O on K-promoted hydrotalcite, *Chem Eng J* 248:406–414, 2014, Copyright (2014) with permission from Elsevier.

the SEWGS cycle proposed in the current work, aiming at a high carbon capture ratio and CO_2 purity for a power plant configuration. First, during the adsorption step, syngas is converted at high pressure through WGS, while CO_2 is being adsorbed. Then, a high-pressure rinse is performed, in which part of the unconverted syngas in the column is replaced by H_2O. The use of rinse originates from PSA cycle design (Yang, 1997) and is known to improve the CO_2 purity. During the rinse, additional H_2 product is produced. After the rinse step, a number of pressure equalization steps is carried out, in which a high-pressure column is connected to a lower pressure column, in order to exchange gas from columns at higher pressure to columns in a lower pressure part of the cycle. This serves to reduce the amount of gas fed to the column for repressurization. Additionally, the rinse gas is expanded causing syngas, that would otherwise become impurities in the CO_2 product, to be transported to another column that can use it in the upcoming adsorption step. During the depressurization step, relatively pure CO_2 product is collected. In order to further desorb CO_2, a low-pressure purge step follows. Once sufficient CO_2 has been recovered, the column is repressurized, first by pressure equalization and finally by repressurization with H_2 product, or alternatively with syngas. For the overall SEWGS cycle, the design criteria are generally formulated in terms of the carbon capture ratio, the amount of CO_2 product recovered divided by the amount of CO and CO_2 fed, and the CO_2 purity. The former quantifies the amount of carbon that can be captured from the feed. The latter is important mainly because any H_2 and CO that end up in the CO_2 product will reduce the efficiency of the process. The performance of the cycle within these design criteria can be expressed in three dependent variables: (1) productivity, the amount of CO_2 produced per unit time per amount of sorbent, (2) rinse steam consumption (S/C_r), the amount of steam used in the rinse relative to the amount of CO and CO_2 fed, and (3) purge steam consumption (S/C_p), the amount of steam used in the purge relative to the amount of CO and CO_2 fed. The optimal SEWGS cycle would consequently minimize cost (CAPEX) and energy penalty (OPEX), by optimization of the total steam consumption and productivity.

The development of a SEWGS cycle requires a prolonged endeavor, based on the three fields designated above. In terms of sorbent, the cause and conditions of detrimental $MgCO_3$ formation (van Selow et al., 2009b; Walspurger et al., 2010, 2011) have been studied, and a stable successor was developed (sorbent alpha, Gazzani et al., 2013). It has been shown that the K-HTC sorbent is sufficiently catalytically active for the WGS reaction, so an additional catalyst can be omitted (van Dijk et al., 2011; van Selow et al., 2013), and that the CO_2–sorbent interaction is not affected by H_2S and other impurities (Jansen et al., 2013; van Dijk et al., 2011). In terms of understanding the sorption equilibrium, the role of potassium sites has been elucidated (Walspurger et al., 2008). As discussed above, a new adsorption isotherm has been presented that encompasses the SEWGS partial pressure range for CO_2 and H_2O on K-HTC, based on preferential surface adsorption and competitive nanopore adsorption (Boon et al., 2014). In terms of cycle design, the improved performance of the H_2O rinse, when compared to the CO_2 rinse has been addressed (Van Selow et al., 2009c), and the importance of several operating parameters has been assessed (Reijers et al., 2011). A logical next step in SEWGS development is an evaluation of the cycle performance with the improved adsorption isotherm.

This section aims to evaluate the performance of the SEWGS cycle with the newly developed isotherm, i.e., including both surface adsorption and competitive nanopore adsorption for CO_2 and H_2O. A state-of-the-art SEWGS cycle is used, with nine columns, cocurrent rinse, and three pressure equalizations. Simulation results will be discussed in terms of the effect of operating conditions (cycle time, rinse steam flow rate, and purge steam flow rate) on the cycle performance (carbon capture ratio and CO_2 purity). Based on these data, improved operating conditions in terms of productivity and steam consumption are identified. Finally, the impact of competitive adsorption of CO_2 and H_2O on the performance of the SEWGS cycle is discussed (Fig. 31).

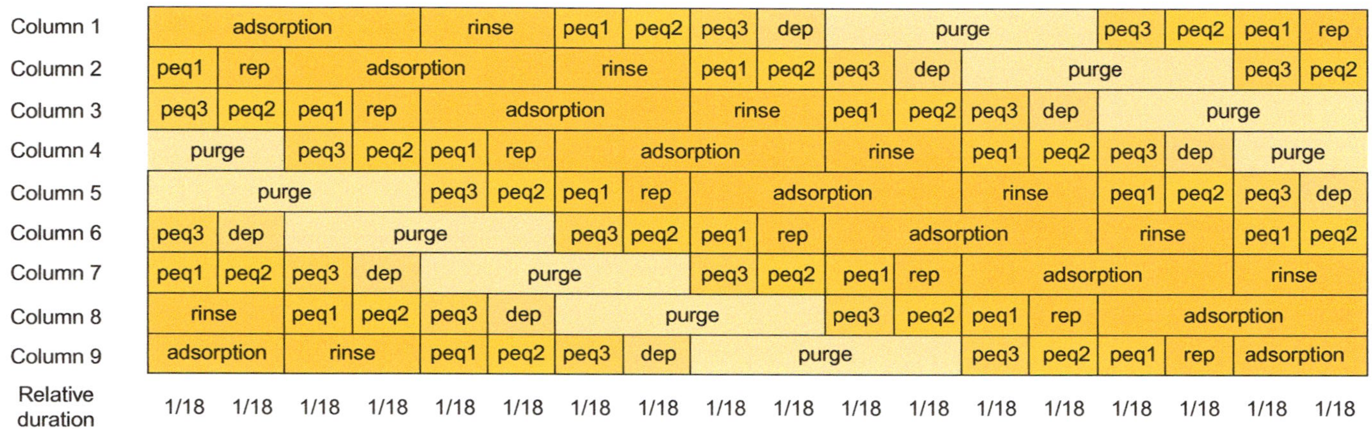

Fig. 31 Nine columns for 11-step SEWGS cycle with cocurrent H_2O rinse and counter-current H_2O purge. *Reprinted from Boon J, Cobden PD, van Dijk HAJ, van Sint Annaland M: High-temperature pressure swing adsorption cycle design for sorption-enhanced water–gas shift,* Chem Eng Sci *122:219–231, 2015, Copyright (2015) with permission from Elsevier.*

4.4.1 Pressure Swing SEWGS Cycle

The single column model was extended to simulate consecutively all steps of a complete SEWGS cycle for CO_2 capture in an IGCC power plant. The SEWGS cycle consisted of 11 steps:

1. adsorption,
2. cocurrent steam rinse,
3. pressure equalization 1 (providing step),
4. pressure equalization 2 (providing step),
5. pressure equalization 3 (providing step),
6. depressurization,
7. purge,
8. pressure equalization 3 (receiving step),
9. pressure equalization 2 (receiving step),
10. pressure equalization 1 (receiving step), and
11. depressurization.

Cycle timing for nine columns in parallel is depicted in Fig. 32, showing two columns concurrently in adsorption mode and timing of the pressure equalization steps. Similar cycles with cocurrent CO_2 rinse and counter-current H_2O rinse were proposed by Van Selow et al. (2009c). Boundary conditions for the large-scale process are shown in Table 7. Isotherm parameters used have been discussed in Section 3.2. Power plant model data presented by Gazzani et al. (2013) were used for the feed conditions.

H_2S and other impurities that can be present in the feed as well as in the rinse and purge steam are not accounted for explicitly. H_2S adsorbs on basic metal oxides such as K-HTC. Toops and Crocker (2008) have elucidated different mechanisms to contribute to the capacity of HTC for H_2S adsorption at room temperature, including dissociative surface adsorption and incorporation of sulfur in the metal oxide matrix. A similar mechanism to the latter was observed for the adsorption of COS, among the formation of other, more complex sulfur species. van Dijk et al. (2011) have shown in lab scale tests at 400°C that K-HTC captures CO_2 and H_2S simultaneously, proposing two types of sites. On one type, either CO_2 or H_2S can adsorb, on the other only H_2S can adsorb, potentially involving the formation of metal sulfides. The net effect for a SEWGS cycle is that H_2S will end up in the CO_2 product. Indeed, the combined capture of H_2S and CO_2 from syngas has been demonstrated in bench scale tests (van Selow et al., 2013). Finally, experiments with various impurities under SEWGS conditions have shown that COS is hydrolyzed and adsorbed as H_2S, CH_4 is neither converted nor adsorbed and will therefore end up in the H_2 product

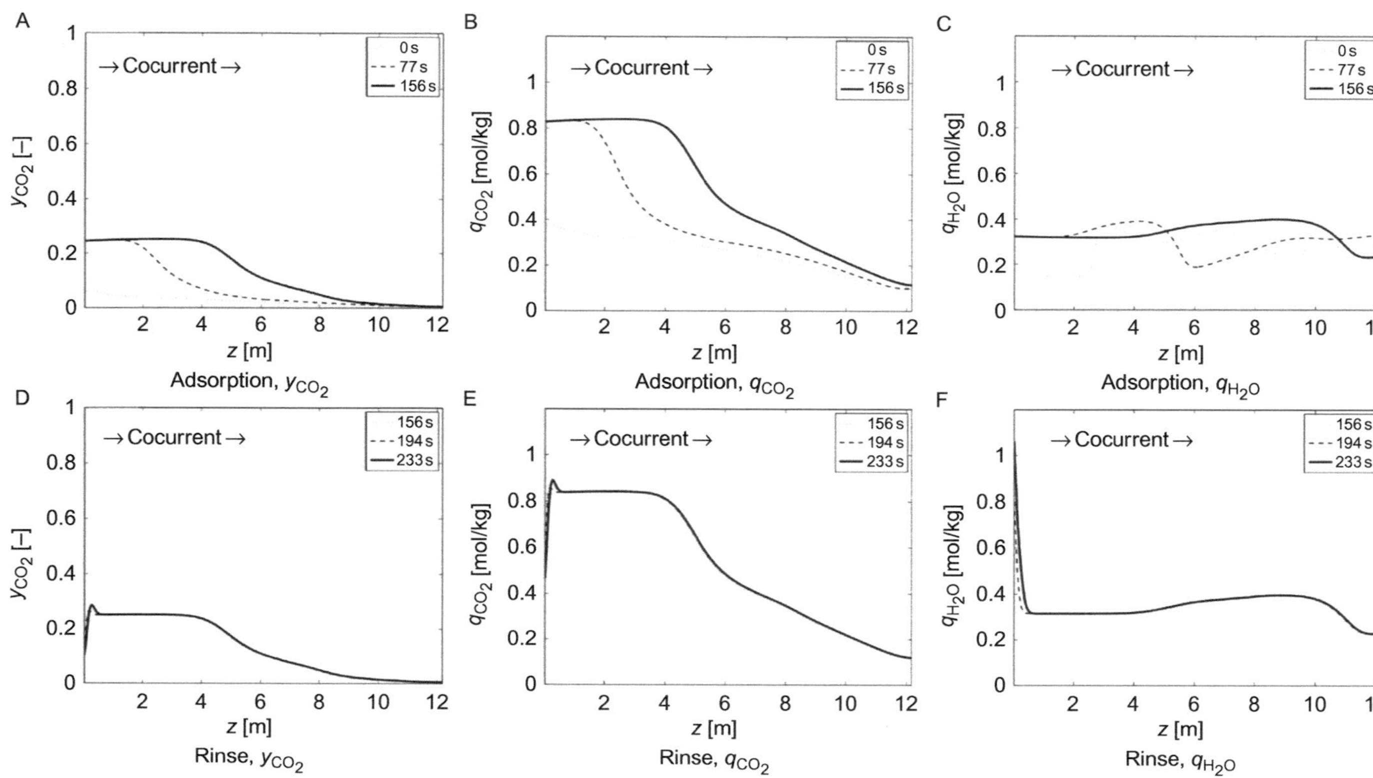

Fig. 32 Concentration profiles during SEWGS cycle, Case I. *Reprinted from Boon J, Cobden PD, van Dijk HAJ, van Sint Annaland M: High-temperature pressure swing adsorption cycle design for sorption-enhanced water–gas shift,* Chem Eng Sci *122:219–231, 2015, Copyright (2015) with permission from Elsevier.*

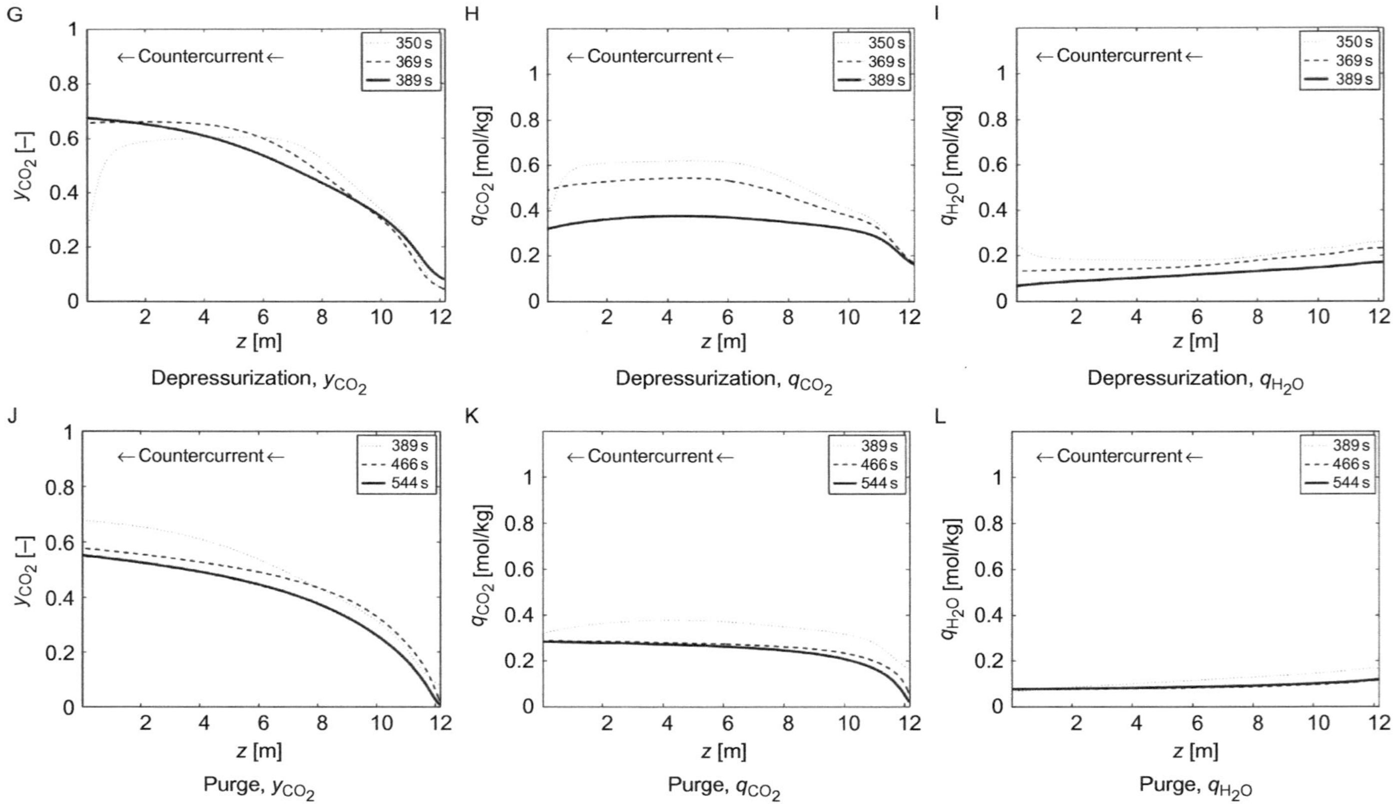
G
H
I
J
K
L
← Countercurrent ←
350 s
369 s
389 s
466 s
544 s
y_{CO_2} [–]
q_{CO_2} [mol/kg]
q_{H_2O} [mol/kg]
z [m]
Depressurization, y_{CO_2}
Depressurization, q_{CO_2}
Depressurization, q_{H_2O}
Purge, y_{CO_2}
Purge, q_{CO_2}
Purge, q_{H_2O}

Fig. 32—cont'd

Table 7 Large-Scale SEWGS Process Parameters

L	12.2	m
d_c	3.66	m
T_{wall}	400	°C
ρ_p	1329	$kg\,m^{-3}$
ϵ_b	0.38	
ϵ_p	0.65	
d_p	4.75	mm
$C_{p,p}$	1048	$J\,kg^{-1}\,K^{-1}$
$\Delta H_{a,CO_2}$	−65	$kJ\,mol^{-1}$
$\Delta H_{a,H_2O}$	−50	$kJ\,mol^{-1}$

Reprinted from Boon J, Cobden PD, van Dijk HAJ, van Sint Annaland M: High-temperature pressure swing adsorption cycle design for sorption-enhanced water–gas shift, *Chem Eng Sci* 122:219–231, 2015, Copyright (2015) with permission from Elsevier.

stream, and NH_3 and HCN are partially captured (Jansen et al., 2013). In conclusion, neither of these impurities will upset the SEWGS cycle, which will produce a H_2 product stream with low contents of CO_2, COS, and H_2S and a CO_2 product stream with H_2S (Fig. 35).

Numerically, a single column is simulated in time while connecting steps are temporarily stored in a spreadsheet file. Timing is controlled by setting the total cycle time and division over the steps as indicated by the relative duration in Table 8. The simulation continues for a number of cycles until cyclic steady state is reached. Depending on the initial state of the column, 20–30 cycles are typically required for the amount of CO_2 in the column and the pressure levels of the equalization steps to converge and for the SEWGS process to reach cyclic steady state. In order to save time, the column is divided relatively coarsely into 200 finite differences, but it was verified that a more refined grid of 400 points yielded the same results.

After the adsorption step, a cocurrent H_2O rinse displaces part of the unconverted syngas near the column inlet and produces additional H_2 product (Fig. 2). Then, three subsequent top-to-top pressure equalization steps are performed, in which the high-pressure columns are connected to the low-pressure columns. After the pressure equalization steps, the CO_2 product is recovered in the depressurization and purge steps. During the purge, steam is fed to the column to lower the CO_2 partial pressure and enhance

Table 8 SEWGS Cycle Boundary Conditions for a Single Train of Nine Columns

Feed	
Relative duration	4/18
F_m	15.05 kg s^{-1}
P_{out}	24 bar
ω_{Ar}	0.008
ω_{H_2O}	0.276
ω_{H_2}	0.036
ω_{CO}	0.071
ω_{CO_2}	0.544
ω_{N_2}	0.065
Rinse	
Relative duration	2/18
F_m	Varied
P_{out}	24 bar
ω_{H_2O}	1
Pressure equalization 1–3	
Relative duration	1/18
P_{out}	Interpolated
Depressurization	
Relative duration	1/18
P_{out}	Interpolated
Purge	
Relative duration	4/18
F_m	Varied
P_{out}	1.1 bar
ω_{H_2O}	1
Pressure equalization 3–1	
Relative duration	1/18
F_m	From providing steps
Repressurization	
Relative duration	1/18
F_m	Interpolated

sorbent regeneration. Then, the pressure in the column is increased in steps. First, in three pressure equalizations, the column is fed from the providing high-pressure columns. The final pressures of the columns in these steps are the result of the simulation and depend on the initial pressure, feed conditions, and mass exchange with the sorbent. Finally, the repressurization step is done by taking part of the H_2 product.

The simulation is run with the standard parameters detailed above. As explained above, the performance indicators of the SEWGS process are formulated in terms of carbon capture ratio and CO_2 purity. After identifying a case near the optimum, additional simulations were performed to test the behavior of the SEWGS process as a function of the adsorption isotherm. First, a simulation was performed without adsorption of steam ($q_{H_2O}=0$). Then, a series of simulations was performed in order to study the effect of the linear driving force intraparticle mass transfer coefficient on the performance of the SEWGS cycle.

4.4.2 Parameter Study

A parameter study was done in order to evaluate the performance of the SEWGS cycle under different operating conditions. Aiming at 95% carbon capture ratio and 99% CO_2 purity, the performance data from EU FP7 Caesar project reported by Gazzani et al. (2013) is used as a starting point: $S/C_r=0.44$, and $S/C_p=1.06$ (Case 0). Around the starting point, simulations are run for cycle times in the range of 200–750 s, S/C_r 0.001–0.5, and S/C_p 0.001–2. When changing the total cycle time, the relative duration of each step and the feed flow rate are not changed. Each of the conditions yields a cyclic steady state with carbon capture ratio and CO_2 purity values. The performance data are then interpolated in order to find solutions that match the criteria of 95% carbon capture ratio and 99% CO_2 purity. Rather than a rigorous optimization in a mathematical sense, this is a heuristic parametric analysis of the SEWGS cycle with a limited set of variables.

4.4.3 SEWGS Cycle Design

4.4.3.1 Pressure Swing SEWGS Cycle

The evolution of the CO_2 mole fraction and CO_2 and H_2O solid loading during the Case 1 SEWGS cycle in cyclic steady state is detailed in Fig. 33, the pressure profile is shown in Fig. 34. During adsorption, as syngas is fed, the CO_2 profile advances through the column while CO_2 adsorbs. Before large amounts of CO_2 can reach the column outlet, in order to prevent a deterioration in carbon capture ratio, the adsorption is stopped and a

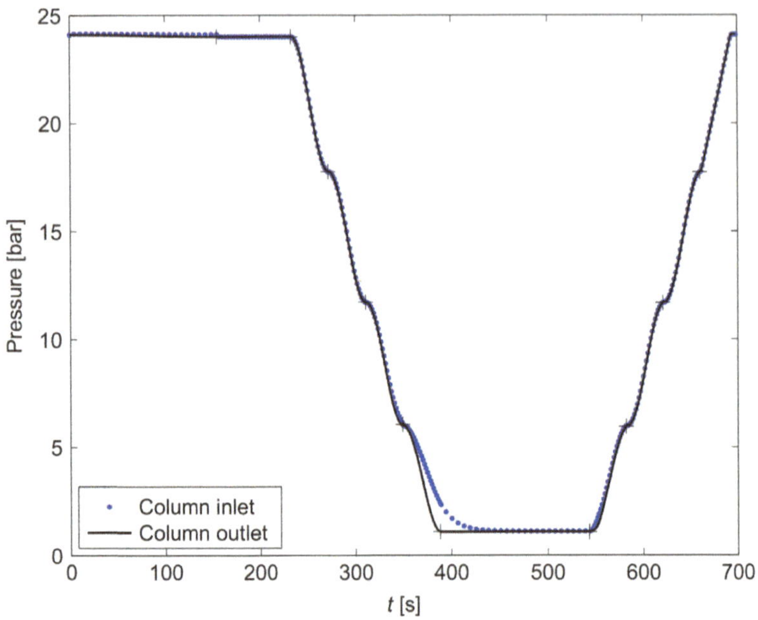

Fig. 33 Column pressure development over the 11-step SEWGS cycle, Case I. *Reprinted from Boon J, Cobden PD, van Dijk HAJ, van Sint Annaland M: High-temperature pressure swing adsorption cycle design for sorption-enhanced water–gas shift,* Chem Eng Sci *122:219–231, 2015, Copyright (2015) with permission from Elsevier.*

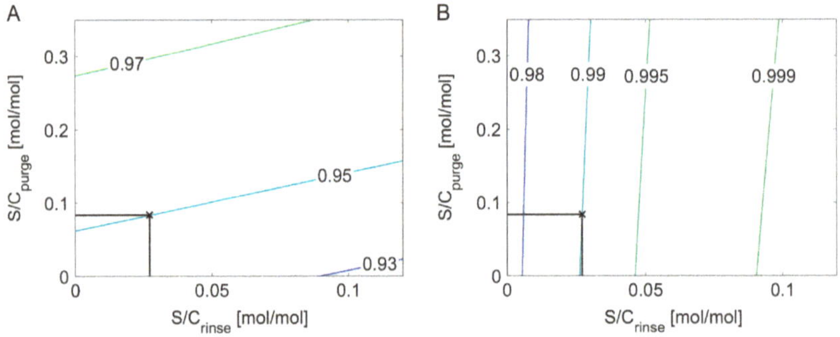

Fig. 34 Interpolated cycle performance vs S/C_r and S/C_p for a 700 s SEWGS cycle. (A) Carbon capture ratio and (B) CO_2 purity. *Reprinted from Boon J, Cobden PD, van Dijk HAJ, van Sint Annaland M: High-temperature pressure swing adsorption cycle design for sorption-enhanced water–gas shift,* Chem Eng Sci *122:219–231, 2015, Copyright (2015) with permission from Elsevier.*

cocurrent H_2O rinse is performed. The H_2O mainly adsorbs, yet some of the CO_2 in the column are already displaced. The well-known roll-up (roll-over) effect, in which a weaker adsorbate is displaced by a stronger adsorbate (Yang, 1997), is witnessed by the increase in CO_2 concentration just before the H_2O front. The roll-up effect is rather subtle, which is due to the relatively modest impact of steam upon the amount of CO_2 adsorbed (see Section 3.2, Fig. 15) when increasing the steam partial pressure from 7.3 bar in the feed to 24 bar in the rinse. A significant amount of steam is adsorbed, severely limiting the progress of the H_2O front in the column. During the subsequent pressure equalizations, however, the column pressure decreases and significant amounts of the adsorbed CO_2 and H_2O are released as can be seen from the adsorbed concentrations at the start of the depressurization. Remarkably, little of the CO_2 released has been transferred to another column during the pressure equalization steps. Rather, it has been readsorbed further downstream in the column. This, of course, is an important precondition in order to achieve a high carbon capture ratio during the next cycle. Finally, the bulk of the CO_2 desorbs and the product CO_2 is collected during the depressurization and purge steps that follow. The majority of the CO_2 is predicted to be recovered during depressurization. During the steam purge still more CO_2 desorbs, and is largely recovered as CO_2 product, although significant amounts of CO_2 remain both in the gas phase and adsorbed. However, the top of the column is well cleaned, with $q_{CO_2} \rightarrow 0$. In the following steps, pressure equalizations (without CO_2, as explained above) and repressurization with H_2 product leave the top of the column free from CO_2 while building up the pressure in the column. CO_2 that was still adsorbed at the end of the purge step remains adsorbed. The gas phase concentration and solid loading at the start of the adsorption step show that the increasing pressure causes readsorption of the CO_2 from the gas phase. The top of the column remains CO_2 lean, and H_2 can be produced again in the adsorption step at a high carbon capture ratio. So, the purge steam effectively functions to free the top part of the column from CO_2, allowing to maintain a high carbon capture ratio. Except for the depressurization and the start of the purge step, there is no significant pressure drop over the column (Fig. 34).

The single column performance, discussed above, forms the basis for the performance of the nine column SEWGS train. In terms of feed and product streams, a single column operates in transient mode. Nine columns in parallel, however, always have two columns in adsorption and in purge (Fig. 32). Continuous streams of H_2 and CO_2 are being produced. Nevertheless, transients remain because of the depressurization and repressurization steps. These transients need to be resolved for downstream processes (specifically the gas turbine section), which can be done by buffering using gas storage tanks or "surge tanks" (Yang, 1997). Note that volumes in externals, piping, etc. (that would cause backmixing and buffering to some extent) are not accounted for in the simulations. The performance of the SEWGS cycle is expressed in terms of the time average CO_2 capture rate, CO_2 purity, and productivity required for achieving 95% carbon capture ratio and 99% CO_2 purity. The cycle performance data are further discussed below.

4.4.3.2 Parameter Study

Starting from the cases outlined above, a series of 62 simulations was done for 200–750 s cycle time, S/C_r 0.001–0.5, and S/C_p 0.001–2. Each of the conditions yields a cyclic steady state with carbon capture ratio and CO_2 purity values. The cycles had a carbon capture ratio in the range of 85%–100%, and a CO_2 purity in the range of 62%–100%, causing the productivity to be in the range of 1.5–1.7 $\text{mol}\,\text{kg}^{-1}\,\text{h}^{-1}$ and a cyclic working capacity of 0.32 $\text{mol}\,\text{kg}^{-1}$. The total steam consumption ($S/C_r + S/C_p$) was varied between 0.4 and 43. Using generalized linear regression (MATLAB 8.1), two quadratic equations are fitted to $-\log(1 - CCR)$ and $-\log(1 - CP)$ from the simulation results. The trends described by these equations are depicted in Fig. 34.

The trends in Fig. 34 clearly show that the carbon capture ratio is mainly determined by the purge: an increase in S/C_p leads to an increase in carbon capture ratio. This is caused by the fact that the purge flow rate determines the extent to which the column is cleaned during the purge step and consequently the amount of CO_2 captured in the subsequent adsorption step. Indeed, the same trend has been reported earlier (Jansen et al., 2013; Reijers et al., 2011). There is a slight opposite trend with rinse, an increase of which causes the carbon capture ratio to decrease slightly. This also corresponds to the trends reported by Reijers et al. (2011). By feeding more rinse gas to the column, the chance increases of CO_2 being carried over during pressure equalization to the top of the receiving column (see Fig. 2). It will then end up in the H_2 product, effectively lowering the carbon capture ratio. Fig. 34 shows that the CO_2 purity is determined by the rinse, in line with Reijers et al. (2011) and Jansen et al. (2013). An increase in rinse steam causes more syngas to be removed from the column during the rinse and the pressure equalization steps and consequently increases the CO_2 purity. Conversely, decreasing or even omitting rinse will decrease the CO_2 purity. An increase in purge, on the other hand, leads to a slight decrease in CO_2 purity, similar to the trend reported by Reijers et al. (2011). In contrast, Jansen et al. (2013) have observed an increase in CO_2 purity with an increase in purge steam. The difference might be caused by the kinetics of desorption. Everything else being equal, and provided desorption kinetics is sufficiently fast, an increase in S/C_p will lead to a CO_2 leaner sorbent at the start of the adsorption step. With a leaner sorbent, and given the amount of syngas fed in the adsorption step, the CO_2 front will progress much less far into the column. Consequently, the column will contain more H_2 product gas at the end of the adsorption step and the start of the rinse, resulting in a lower CO_2 purity. In contrast, if desorption kinetics is slower, the CO_2 loading profile in the column will be more homogeneous (flat) throughout the column at the start of the adsorption step (compared to the current process, see Fig. 32B and K). The adsorption step will then have a less steep front and syngas will be present in a larger part of the column at the start of the rinse, already decreasing the CO_2 purity. In such a case, using more purge gas will improve the extent to which the column is regenerated near the H_2 product side of the column. This then causes a better separation during the adsorption, rinse, and pressure equalization steps. The CO_2 purity is then improved by increasing the amount of purge steam. Fig. 35 shows that purge as well as rinse are required to achieve the targets for carbon capture ratio and CO_2 purity. In conclusion, the amount of rinse steam effectively determines the CO_2 purity and the amount of purge steam effectively determines the carbon capture ratio, they are determined independently by the targets of 99% CO_2 purity and 95% carbon capture ratio.

Given the currently employed isotherm and linear driving force kinetics, the model predicts the SEWGS cycle to perform at specified carbon capture ratio and CO_2 purity of 95% and 99%, respectively, with S/C_r of 0.03 and S/C_p of 0.08 indicated by the × mark in Fig. 35, corresponding to a productivity of $1.6\,\mathrm{mol\,kg^{-1}\,h^{-1}}$. This is a marked reduction in the predicted steam consumption compared to available literature data (Gazzani et al., 2013; Jansen et al., 2013; Reijers et al., 2011; Wright et al., 2011), both for the rinse and the purge. Improvements to the predicted cycle performance are due to the new isotherm and better description of intraparticle mass transfer, that were discussed above. Of course, the cycle proposed in the current work remains to be validated experimentally. The fate of H_2S in the SEWGS cycle, which is presently assumed to be cocaptured with CO_2, also needs to be verified. The present study aimed at finding the optimum operating conditions, as well as important sensitivities, for the proposed 11-step cycle for 9 columns. Other designs are possible and have been presented in literature. The H_2O rinse could be replaced by a CO_2 rinse, and repressurization could be done with syngas instead of H_2 product. Given the fact that the steam consumption for the current design is very low, it might be feasible to reduce the number of pressure equalization steps in the

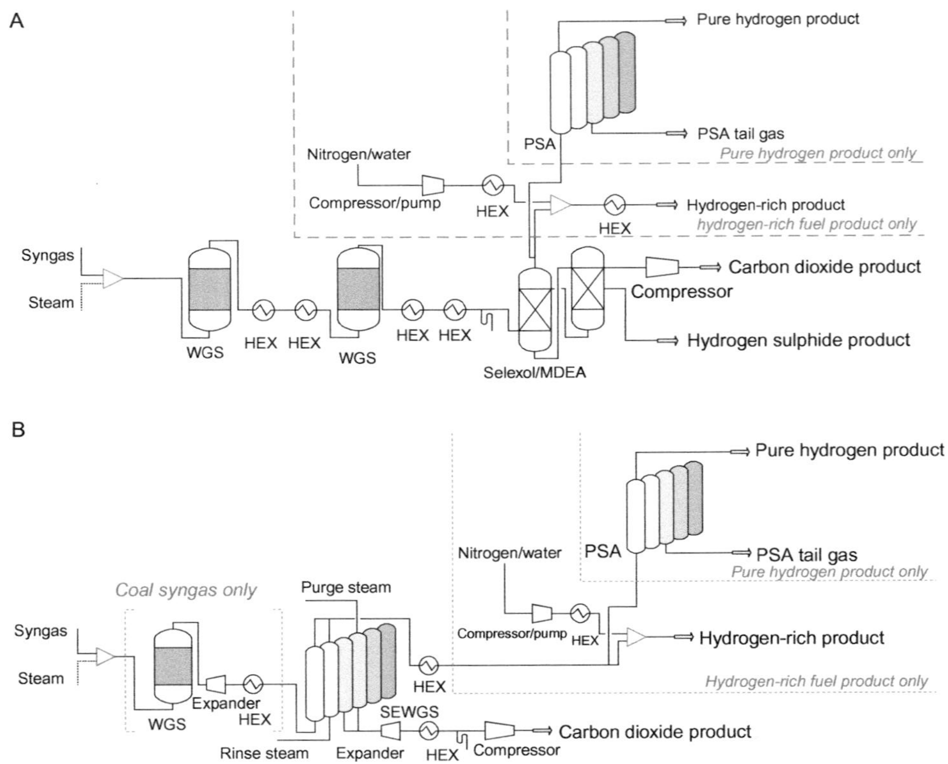

Fig. 35 Process flow diagrams, *dotted lines* indicate optional unit operations and process routes, depending on the case. (A) Conventional separation processes Selexol and MDEA and (B) SEWGS-integrated water–gas shift and separation. *Reprinted from Boon J, Cobden PD, van Dijk HAJ, van Sint Annaland M: High-temperature pressure swing adsorption cycle design for sorption-enhanced water–gas shift,* Chem Eng Sci *122:219–231, 2015, Copyright (2015) with permission from Elsevier.*

cycle, in order to reduce the required number of columns and CAPEX for the system. This is being discussed below.

4.5 SEWGS: Efficiency of Carbon Capture

Starting with syngas from either coal gasification or natural gas reforming, this section quantifies the work of the separation of carbon dioxide for conventional low-temperature technology (Selexol, MDEA) on the one hand and for SEWGS on the other hand. Initially, a total of eight different process schemes are discussed to produce various qualities of hydrogen while capturing carbon dioxide, including schemes with conventional low-temperature separations as well as schemes with high-temperature separation. Then, this section discusses the thermodynamic framework for assessment of the work required for the separation of carbon dioxide and hydrogen. The advantage of SEWGS will be assessed in terms of the efficiencies according to the first and second law of thermodynamics.

4.5.1 Methods

Three basic process options are considered in this work:

- low-temperature separation by Selexol,
- low-temperature separation by methyl diethanolamine solution (MDEA), and
- high-temperature separation by SEWGS.

The three process schemes are used with syngas either from coal gasification or from natural gas reforming. With respect to the feeds streams, shown in Table 9, the main difference is in the carbon oxides content. Coal syngas has a higher carbon content (>20 mol% $CO+CO_2$, note that syngas from a Shell gasifier is used after a first stage of high-temperature WGS reaction) while natural gas-derived syngas has a lower carbon oxides content (<15 mol% $CO+CO_2$). This is a principal difference, because the partial pressure of carbon dioxide in the feed determines the driving force for capture in Selexol, MDEA, and SEWGS, and thereby to a large extent the performance of the capture processes. Therefore, the obtained results for coal syngas also represent the potential to separate carbon dioxide from industrial gases with high carbon oxides content, such as BFG in steelworks and producer gas from biomass gasification, possibly after compression (Carbo et al., 2011; Gazzani et al., 2015; Gielen, 2003). Similarly, the results for natural gas-derived syngas indicate the potential to separate carbon dioxide from industrial gases with low carbon oxides content, such as coke oven gas in steelworks (Gazzani et al., 2015).

Table 9 Syngas Feed Compositions Considered

		Syngas From Coal Gasification, Water–Gas Shift	Syngas From Natural Gas Reforming
Pressure	bar	40	29
Temperature	°C	400	400
Composition	Mole fraction		
H_2		0.36	0.49
CO_2		0.24	0.084
H_2O		0.30	0.34
CO		0.050	0.056
CH_4			0.033
N_2		0.044	0.002
Ar		0.004	
H_2S		0.001	

Reprinted from Boon J, Cobden PD, van Dijk HAJ, van Sint Annaland M: High-temperature pressure swing adsorption cycle design for sorption-enhanced water–gas shift, *Chem Eng Sci* 122:219–231, 2015, Copyright (2015) with permission from Elsevier.

Note that sulfur, specifically hydrogen sulfide, is not accounted for quantitatively, except for the Rectisol process. For SEWGS, hydrogen sulfide will be cocaptured with carbon dioxide and leave the column with the carbon dioxide product in the purge step (Boon et al., 2015; van Dijk et al., 2011).

Table 9 shows the considered process routes. The schemes have been designed to produce two different qualities of hydrogen, and a carbon dioxide stream of sufficient purity for sequestration (Table 10). Pure hydrogen product (>99.9 vol.%) has strict purity constraints and can be produced by either a conventional PSA system (Kohl and Nielsen, 1997). In contrast, there is no strict purity constraint for the hydrogen product in case it is used as fuel, e.g., in a precombustion carbon dioxide capture scheme. In fact, the hydrogen product requires dilution with inert (nitrogen when available, in case of coal syngas, or steam) before its use as fuel in a hydrogen-rich fueled gas turbine for power production (Gazzani et al., 2013; Manzolini et al., 2013a). Gazzani et al. (2014) have well addressed and discussed the fuel specifications for the use of hydrogen as gas turbine fuel in precombustion carbon

Table 10 Process Routes Considered

Case	Syngas	Technology	Product	Byproduct
I	Coal	Sour WGS, Selexol	H_2 fuel	CO_2
II	Coal	Sour WGS, Selexol, PSA	H_2	CO_2
III	Coal	Sour WGS, SEWGS	H_2 fuel	CO_2
IV	Coal	Sour WGS, SEWGS, PSA	H_2	CO_2
VII	NG	HTS, LTS, MDEA	H_2 fuel	CO_2
VIII	NG	HTS, LTS, MDEA, PSA	H_2	CO_2
IX	NG	SEWGS	H_2 fuel	CO_2
X	NG	SEWGS, PSA	H_2	CO_2

Reprinted from Boon J, Cobden PD, van Dijk HAJ, van Sint Annaland M: High-temperature pressure swing adsorption cycle design for sorption-enhanced water–gas shift, *Chem Eng Sci* 122:219–231, 2015, Copyright (2015) with permission from Elsevier.

dioxide capture schemes. The carbon dioxide purity constraint for compression and sequestration is taken from the European Benchmarking Task Force to be 90 vol.%, yet with more stringent purity constraints for many relevant impurities (Franco, 2009). The pressure for sequestration is assumed to be 110 bar(a).

4.5.1.1 Water–Gas Shift (WGS)

The WGS reaction is used to convert carbon monoxide with steam into additional hydrogen product. Three types of industrially mature WGS processes are employed in the schemes above: sour WGS, high-temperature WGS (HTS), and low-temperature WGS (LTS) (Carbo et al., 2009; Newsome, 1980). First important distinction between the processes is in terms of the sulfur content and the corresponding catalyst stability, the second distinction is in the operating temperature regime. In sour WGS, high'sulfur syngas is converted using sulfided catalysts (typically cobalt molybdenum sulfide). A fairly high level of sulfur is required in the syngas in order for the catalyst to remain in the sulfided state and the reactor operates in the range of 244–250°C inlet and up to 480°C outlet temperature. The HTS process at 350°C inlet and up to 395°C outlet temperature uses iron oxide-based catalysts and can tolerate a fair amount of sulfur with only a mild reduction in activity (Boon et al., 2009; van Dijk et al., 2014). The LTS process requires sulfur-free syngas (<0.1 ppmv) and operates from 200°C inlet to 217°C outlet temperature.

The low-temperature carbon dioxide capture cases (I, II, VII, and VIII) use conventional technology (Selexol, MDEA) and represent state-of-the-art. Advanced split flow configuration of the WGS section could significantly lower the steam demand, albeit at the cost of an increased number of reactors and total catalyst volume (Carbo et al., 2009), which is not considered here.

4.5.1.2 Selexol, MDEA

Absorption is widely used industrially for the separation of acid species (such as carbon dioxide and hydrogen sulfide) from gas streams (Boll et al., 2000; Jansen et al., 2015; Kohl and Nielsen, 1997). The gas phase is contacted with a liquid which either dissolves (physical absorption) or chemically binds (chemical absorption) the acid gases. Regeneration is done by pressure swing or temperature swing (reboiling). At high partial pressures, physical solvents outperform chemical solvents, while the reverse is true for low partial pressures (Boll et al., 2000). As state-of-the-art, physical absorption by Selexol is selected for the capture of carbon dioxide and hydrogen sulfide from coal syngas (Cases I and II) and chemical absorption by MDEA is selected for the capture of carbon dioxide from natural gas reformate (Cases VII and VIII).

The Selexol section for separation of carbon dioxide and hydrogen sulfide is modeled after (Spallina et al., 2014). The energy consumption is 189 kJ $kg_{CO_2\text{-rich}}^{-1}$ for pumping and 303 kJ $kg_{CO_2\text{-rich}}^{-1}$ at 200°C for the reboiler. Compression of the carbon dioxide product consumes 252 kJ $kg_{CO_2\text{-rich}}^{-1}$ of electricity. The MDEA section for separation of carbon dioxide is modeled after (Chiesa et al., 2013). The energy consumption is 84 kJ $kg_{CO_2\text{-rich}}^{-1}$ of electricity for pumping and 887 kJ $kg_{CO_2\text{-rich}}^{-1}$ at 200°C for the reboiler. Compression of the carbon dioxide product is modeled according to the EBTF (Franco, 2009).

4.5.1.3 Sorption-Enhanced Water–Gas Shift (SEWGS)

SEWGS is modeled as discussed above. In SEWGS, steam is added in the rinse (high-pressure rinse steam, expressed on a molar basis as S/C_r) and in the purge (low-pressure purge steam, expressed on a molar basis as S/C_p). Above, extensive results have been presented of a SEWGS cycle design study, based on recently developed expressions for the interaction of carbon dioxide and steam with K-HTC at high pressure, yielding a SEWGS cycle that consumes significantly less steam than cycle designs previously reported in literature. For IGCC syngas, a SEWGS system of nine columns was presented which could reach a carbon capture ratio and carbon dioxide purity of 95% and 99%, respectively, with S/C_r of 0.03 and S/C_p of 0.08. For the performance in a carbon dioxide capture process, it is important to stress that the sorbent itself can perform substantial WGS conversions (it requires no additional catalyst), and that it is able to capture hydrogen sulfide together with carbon dioxide. Also, SEWGS easily achieves very high capture ratios.

Based on these results, new SEWGS cycle designs have been studied in the present work using the previously presented sorption-enhanced reactor model. The S/C_p has been fixed at 0.1 in order to achieve a carbon dioxide capture ratio of around 98%. The main parameter variations in the current study are S/C_r and the number of pressure equalizations. The former mainly affects the carbon dioxide purity while it represents the largest work lost. The reduction in pressure equalizations from 3 to 2 and 1 reduces the number of SEWGS columns from 9 to 8 and 7, respectively. Thus, the steam consumption and the number of pressure equalizations can be varied in order to achieve the desired carbon dioxide purity. Simulations for coal syngas have been extended from the previous section, while adding simulations for natural gas reformate.

4.5.1.4 Pressure Swing Adsorption (PSA)

State-of-the-art technology for the production of pure hydrogen is PSA (Kohl and Nielsen, 1997). It produces a high-pressure stream of high-purity hydrogen and a low-pressure stream of tail gas. The tail gas is accounted for in terms of its lower heating value (LHV). The PSA system is modeled according to industrial practice using a hydrogen recovery of 76.2% and a hydrogen purity (HP) of 99.9996 mol%.

4.5.1.5 Compression and Condensation

Depending on the separation technology of choice, the carbon dioxide product is delivered either at atmospheric pressure (MDEA, SEWGS) or at elevated pressure. The installation of any carbon dioxide capture technology presupposes a destination for the carbon dioxide, other than emission to the atmosphere. The carbon dioxide stream is therefore compressed to a pressure of 110 bar(a), creating a common basis for comparing the technologies. Compression stages and interstage cooling are modeled according to EBTF recommendations, except for Selexol where a literature value is used (discussed above).

4.5.1.6 Flowsheeting and Equation of State

Flowsheets for the cases outlined above are briefly summarized in Fig. 35. The flowsheets are modeled in Aspen Plus V8.4, using the Peng–Robinson cubic equation of state, which is particularly suitable for light gases and at high temperature and pressure (Peng and Robinson, 1976).

4.5.1.7 Process Efficiency and Steam Consumption

The efficiency of the process is expressed in terms of the LHVs of the feed and product streams, i.e., cold gas efficiency. It is important to note that the cases for electricity production have also been modeled based on LHV efficiency for the conversion of fuel to power. Thus, specific modeling of a cooled gas turbine, which has been presented in literature (Chiesa and Macchi, 2004; Gazzani et al., 2014) and is outside the scope of the current work, can be circumvented. Three efficiencies are defined accordingly:

$$\eta_0 = \frac{LHV_{\mathrm{H_2}\,product} F_{m,\mathrm{H_2}\,product}}{LHV_{feed} F_{m,feed}} \tag{17}$$

$$\eta_1 = \frac{LHV_{\mathrm{H_2}\,product} F_{m,\mathrm{H_2}\,product} + \eta_{byproduct} LHV_{byproduct} F_{m,byproduct}}{LHV_{feed} F_{m,feed}} \tag{18}$$

$$\eta_2 = \frac{LHV_{\mathrm{H_2}\,product} F_{m,\mathrm{H_2}\,product} + \eta_{byproduct} LHV_{byproduct} F_{m,byproduct} - \Delta W}{LHV_{feed} F_{m,feed}} \tag{19}$$

Thus, η_0 is defined as the cold gas efficiency, based on LHV of the hydrogen-rich product stream divided by the LHV of the syngas input. It is a measure to what extent the energy content in the syngas is preserved in the hydrogen-rich product. Similarly, η_1 represents the cold gas efficiency while taking into account the energy content of the byproduct that is produced, which can be utilized as fuel. Finally, η_2 is the theoretical net efficiency, accounting for the net work performed by the system in the process (Boon et al., 2016), e.g., the consumption or production of steam and electricity. (Note that the theoretical net efficiency does not represent the efficiency according to the second law of thermodynamics.) The net efficiency thereby provides a measure of the required work for the conversion of carbon monoxide and the separation of carbon dioxide from the syngas.

Apart from the efficiencies, the performance of the processes is quantified through a number of key metrics: the hydrogen recovery factor (HRF), HP, carbon capture ratio (CCR), and carbon purity (CP),

$$\mathrm{HRF} = \frac{y_{\mathrm{H_2},\mathrm{H_2}\,product} F_{mol,\mathrm{H_2}\,product}}{(y_{\mathrm{H_2}} + y_{\mathrm{CO}}) F_{mol,feed}} \tag{20}$$

$$\mathrm{HP} = y_{\mathrm{H_2},\mathrm{H_2}\,product} \tag{21}$$

$$\mathrm{CCR} = \frac{(y_{\mathrm{CO_2}} + y_{\mathrm{CO}} + y_{\mathrm{CH_4}})_{\mathrm{CO_2}\,product} F_{mol,\mathrm{CO_2}\,product}}{(y_{\mathrm{CO_2}} + y_{\mathrm{CO}} + y_{\mathrm{CH_4}})_{feed} F_{mol,feed}} \tag{22}$$

$$\mathrm{CO}_x\mathrm{CR} = \frac{(y_{\mathrm{CO_2}} + y_{\mathrm{CO}})_{\mathrm{CO_2}\,product} F_{mol,\mathrm{CO_2}\,product}}{(y_{\mathrm{CO_2}} + y_{\mathrm{CO}})_{feed} F_{mol,feed}} \tag{23}$$

$$\mathrm{CP} = y_{\mathrm{CO_2},\mathrm{CO_2}\,product} \tag{24}$$

Since all remaining carbon monoxide and methane will be oxidized to carbon dioxide in downstream processes, they are accounted for in the carbon capture ratio (CCR). The capture ratio of carbon oxides (CO_xCR) serves to assess the capture of carbon dioxide and carbon monoxide, without accounting for the unconverted methane in the natural gas cases.

4.5.1.8 Work of Separation

The work of separation is performed by equipment (electricity) and streams (steam) in the process. The work is done by the pressure of the feed, by the compression, and by the additional process streams involved. With Selexol, MDEA, and SEWGS, pressure swing is used to capture and release the carbon dioxide from the process stream, which means that the carbon dioxide is produced at low pressure. Additional input can be steam (SEWGS rinse gas, SEWGS purge gas) and side product can be PSA tail gas which has significant calorific value.

Heat to work efficiencies are calculated according to basic Carnot and Lorentz cycles for heat engines, as discussed by Boon et al. (2016). The Carnot cycle operates between a fixed high-temperature stream and a fixed low-temperature sink, and its efficiency represents the maximum efficiency for converting heat into power. The Lorentz cycle operates between a cooled high-temperature stream and a constant low-temperature sink (Lee and Kim, 1992). With the basic thermodynamic cycles, the efficiencies for converting heat into work are calculated for high temperature (>350°C, Figs. 36 and 37) and similarly for medium temperature heat exchangers (150–350°C) in the process. For all cases, the stream with the highest temperature is cooled and consequently the resulting heat engine corresponds to the Lorentz cycle. The lower temperature stages typically have an efficiency in between the maximum Carnot efficiency and the Lorentz efficiency.

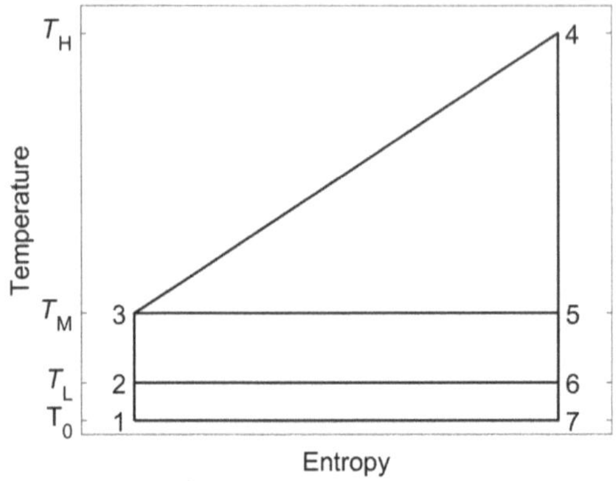

Fig. 36 Temperature–entropy diagram for heat recovery by high-temperature heat exchanger: Lorentz cycle (3–4–5) and two Carnot cycles (2–3–5–6 and 1–2–6–7). *Reprinted from Boon J, Spallina V, van Delft Y, van Sint Annaland M: Comparison of the efficiency of carbon dioxide capture by sorption-enhanced water–gas shift and palladium-based membranes for power and hydrogen production,* Int J Greenhouse Gas Control *50:121–134, 2016, Copyright (2016) with permission from Elsevier.*

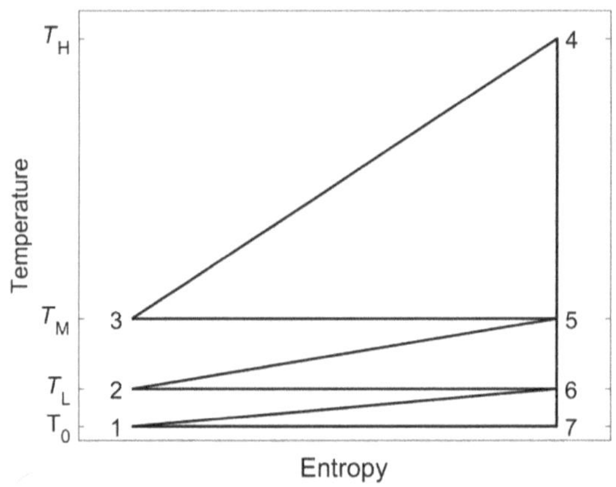

Fig. 37 Temperature–entropy diagram for heat recovery by high-temperature heat exchanger: three Lorentz cycles (3–4–5, 2–5–6, and 1–6–7). *Reprinted from Boon J, Spallina V, van Delft Y, van Sint Annaland M: Comparison of the efficiency of carbon dioxide capture by sorption-enhanced water–gas shift and palladium-based membranes for power and hydrogen production,* Int J Greenhouse Gas Control *50:121–134, 2016, Copyright (2016) with permission from Elsevier.*

The efficiency for converting heat duty to work is therefore calculated by taking the arithmetic mean between Lorentz–Carnot and Lorentz–Lorentz cycles efficiencies. This way, the detailed simulation of the entire plant (including concomitant choices and assumptions) is avoided while a realistic performance of a derived steam cycle operated with the heat sources of the plant is taken into account: (i) different steam streams at high, intermediate, and low pressure are generated during heat recovery; (ii) the Lorentz cycle does not take into account that steam production occurs through evaporation and super heating; and (iii) for example a stream at 650°C which is cooled down to ambient temperature would achieve a Carnot efficiency of 69%, a Lorentz efficiency of 44% and based on our methodology will achieve an overall efficiency of 56% which is in line with state-of-the-art steam cycle performance (Franco and Russo, 2002). The cases and typical efficiencies have been discussed in more detail in Boon et al. (2016).

Added steam, as opposed to steam used for heating, is accounted for in terms of lost work (which could otherwise have been performed by the steam in a steam turbine). The amount of work that steam of T_M can perform depends on its temperature and pressure (through the dew point T_d). As shown in Fig. 38, the lost work by steam addition is quantified from the heat duty to raise the steam (Q_s) through Lorentz cycle efficiency,

$$\eta_s = \frac{(T_s - T_0)\left(1 - \frac{T_0}{\sqrt{T_s T_0}}\right) - (T_d - T_0)\left(1 - \frac{T_0}{\sqrt{T_d T_0}}\right)}{T_s - T_d} \tag{25}$$

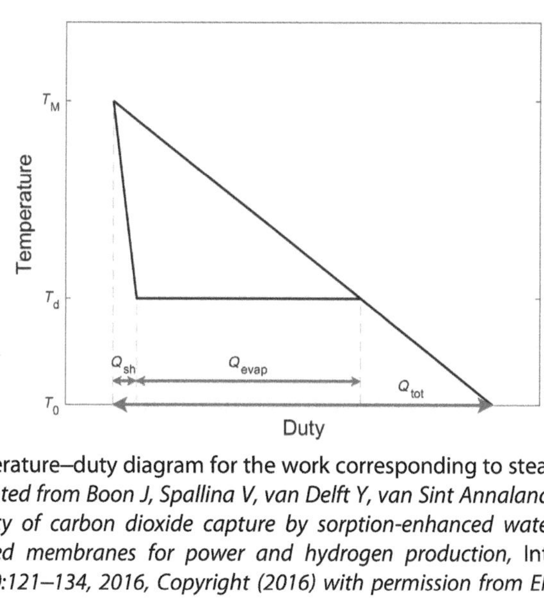

Fig. 38 Temperature–duty diagram for the work corresponding to steam added to the process. *Reprinted from Boon J, Spallina V, van Delft Y, van Sint Annaland M: Comparison of the efficiency of carbon dioxide capture by sorption-enhanced water–gas shift and palladium-based membranes for power and hydrogen production,* Int J Greenhouse Gas Control *50:121–134, 2016, Copyright (2016) with permission from Elsevier.*

The remaining work is electric. The used compressors and expanders were discussed above.

4.5.1.9 Utilization of Byproduct

In several of the cases, a byproduct stream of significant calorific value is produced. It is assumed that such a stream can be converted into work via a steam boiler, using a state-of-the-art efficiency of 45% LHV (Cau et al., 2014).

4.5.1.10 Work of Separation

As stated, this study aims to develop a framework for the evaluation of different technologies for precombustion carbon dioxide capture. The common denominator in the evaluation is the work of separation. Using the efficiencies derived above in the flowsheeting work, the heat demand or production of unit operations (Q), the steam added to the process (Q_s), and the utilization of byproducts (LHV) can all be converted to work.

4.5.2 Efficiencies of Carbon Capture: Results

Flowsheets for the eight cases have been modeled, all resulting in meeting the product specifications of hydrogen product and carbon dioxide product (Table 11). The major results are summarized in Table 12. First, the results for the different cases are discussed.

4.5.2.1 Conventional Absorption Technology

In Cases I and II, the Selexol process achieves a CCR of 92%. In Cases VII and VIII, at lower carbon dioxide partial pressure, and with a relatively high methane content, a CCR of 76% is reached with MDEA (see discussion below).

4.5.2.2 Sorption-Enhanced Water–Gas Shift (SEWGS)

The SEWGS process in this study has been designed for a specified carbon dioxide purity. It has been shown and discussed above that SEWGS can readily achieve very high carbon capture ratios. The present work is based on the performance of the Alkasorb sorbent as developed by ECN; adopting the novel Alkasorb + would further improve the SEWGS performance (Jansen et al., 2013, 2015).

Table 11 Product Stream Specifications (Franco, 2009; Gazzani et al., 2013; Manzolini et al., 2013a)

Species		Hydrogen	Hydrogen-Rich Fuel	Carbon Dioxide for Sequestration
H_2	mol%	99.9	<60	<4
CO	mol%			<0.2
CO_2	mol%			>90
H_2O	mol%			
N_2	mol%			<4
CH_4	mol%			<2
H_2S	ppmv		<20	<200
Temperature	°C		200–350	25

Reprinted from Boon J, Cobden PD, van Dijk HAJ, van Sint Annaland M: High-temperature pressure swing adsorption cycle design for sorption-enhanced water–gas shift, *Chem Eng Sci* 122:219–231, 2015, Copyright (2015) with permission from Elsevier.

Results of the carbon dioxide purity from SEWGS cycle simulations are shown in Fig. 39. Clearly, both the rinse steam amount and the number of pressure equalizations strongly affect the carbon dioxide purity. The target purity of 96 mol% can be achieved with eight columns using $S/C_r=0.1$ and $S/C_p=0.1$. The corresponding schedule is shown in Fig. 40.

SEWGS simulations for the natural gas reformate have recently been presented by Boon et al. (2016) and Walspurger et al. (2015) and are briefly summarized in Fig. 41. The lower partial pressure of carbon oxides in the natural gas reformate translates into a higher steam consumption. The required carbon dioxide purity is reached at $S/C_r=0.5$ and $S/C_p=0.4$, but only six columns are required. Note that this is not only a significant decrease in steam consumption but also in the number of columns required compared to previous cases (cf. the study by Gazzani et al. (Manzolini et al., 2013a) using a nine columns SEWGS train) (Fig. 42).

For the production of hydrogen fuel (Cases III and IX), additional dilution was required. Note that SEWGS captures hydrogen sulfide together with carbon dioxide and the concentration of hydrogen sulfide in the carbon dioxide product reaches 3600 ppmv (Cases III and IV), which exceeds the specification of 200 ppmv (Table 11). The sulfur could be recovered as

Table 12 Major Results Summarized for the Selected Cases

Case	Feed	Main Technology	Product	η_0	η_1	η_2	HRF	HP	CCR	CO_xCR	CP
I	Coal	Selexol	Fuel	0.89	0.89	0.76	0.966	0.600	0.918	0.918	0.985
II			H_2	0.66	0.76	0.67	0.736	1.000	0.918	0.918	0.985
III		SEWGS	Fuel	0.87	0.87	0.75	0.964	0.600	0.978	0.978	0.961
IV			H_2	0.65	0.75	0.65	0.735	1.000	0.978	0.978	0.961
VII	NG	MDEA	Fuel	0.98	0.98	0.94	0.989	0.600	0.759	0.939	0.969
VIII			H_2	0.62	0.78	0.78	0.754	1.000	0.759	0.939	0.969
IX		SEWGS	Fuel	0.98	0.98	0.95	0.987	0.600	0.765	0.946	0.963
X			H_2	0.62	0.78	0.79	0.752	1.000	0.765	0.946	0.963

Reprinted from Boon J, Cobden PD, van Dijk HAJ, van Sint Annaland M: High-temperature pressure swing adsorption cycle design for sorption-enhanced water–gas shift, *Chem Eng Sci* 122:219–231, 2015, Copyright (2015) with permission from Elsevier.

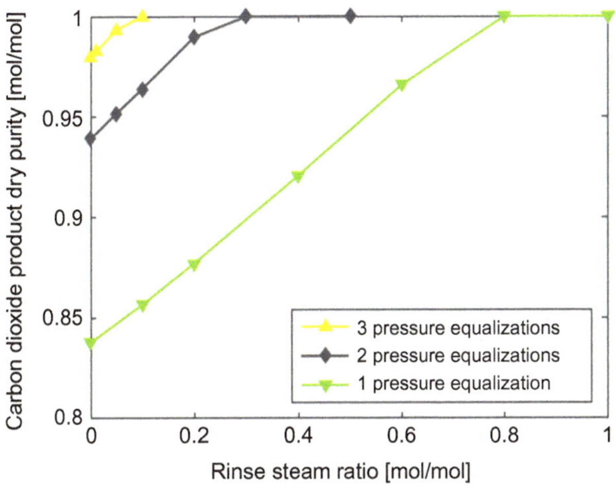

Fig. 39 SEWGS performance, carbon dioxide product dry purity vs rinse steam consumption for 3, 2, and 1 pressure equalizations with coal syngas after preshift and $S/C_p = 0.1$. *Reprinted from Boon J, Spallina V, van Delft Y, van Sint Annaland M: Comparison of the efficiency of carbon dioxide capture by sorption-enhanced water–gas shift and palladium-based membranes for power and hydrogen production,* Int J Greenhouse Gas Control *50:121–134, 2016, Copyright (2016) with permission from Elsevier.*

Column	1/4				1/8		1/16	1/16	1/16	1/4				1/16	1/16	1/16
1	Adsorption				Rinse		PEQ1 ↓	PEQ2 ↓	Blowdn	Purge				PEQ2 ↑	PEQ1 ↑	Rep
2	PEQ1 ↑	Rep	Adsorption				Rinse		PEQ1 ↓	PEQ2 ↓	Blowdn	Purge				PEQ2 ↑
3	Blowdn	Purge			Adsorption				Rinse		PEQ1 ↓	PEQ2 ↓	Blowdn	Purge		
4	Rinse		PEQ1 ↓	PEQ2 ↓	Blowdn	Purge	Adsorption				Rinse		PEQ1 ↓	PEQ2 ↓	Blowdn	Purge
5	Blowdn	Purge				PEQ2 ↑	PEQ1 ↑	Rep	Adsorption				Rinse		PEQ1 ↓	PEQ2 ↓
6	PEQ1 ↓	PEQ2 ↓	Blowdn	Purge				PEQ2 ↑	PEQ1 ↑	Rep	Adsorption				Rinse	
7	Rinse		PEQ1 ↓	PEQ2 ↓	Blowdn	Purge				PEQ2 ↑	PEQ1 ↑	Rep	Adsorption			
8	Adsorption		Rinse		PEQ1 ↓	PEQ2 ↓	Blowdn	Purge				PEQ2 ↑	PEQ1 ↑	Rep	Adsorption	

Fig. 40 SEWGS cycle for coal syngas with two pressure equalizations, using eight columns (*PEQ*: pressure equalization between two columns; *Blowdn*: blowdown/depressurization to atmospheric pressure; *Rep*: repressurization with hydrogen product). *Reprinted from Boon J, Spallina V, van Delft Y, van Sint Annaland M: Comparison of the efficiency of carbon dioxide capture by sorption-enhanced water–gas shift and palladium-based membranes for power and hydrogen production,* Int J Greenhouse Gas Control *50:121–134, 2016, Copyright (2016) with permission from Elsevier.*

elemental sulfur through the Claus process, but no technology is currently industrially available with sufficient selectivity to produce concentrated hydrogen sulfide from a hydrogen sulfide–carbon dioxide mixture: Selexol or Rectisol are not sufficiently selective to produce concentrated hydrogen sulfide stream required by a Claus process (Kohl and Nielsen, 1997). Gazzani et al. (2013) have proposed the production of calcium sulfate though stoichiometric combustion of hydrogen sulfide with oxygen. This was predicted to have an energy efficiency penalty of a mere 0.2%-points.

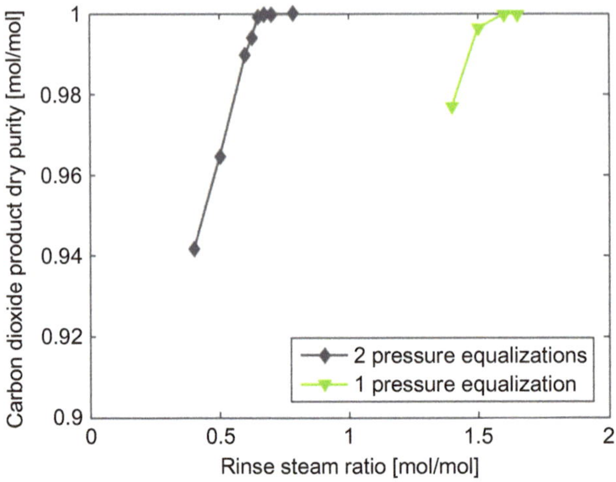

Fig. 41 SEWGS performance, carbon dioxide product dry purity vs rinse steam consumption for natural gas syngas after preshift. *Reprinted from Boon J, Spallina V, van Delft Y, van Sint Annaland M: Comparison of the efficiency of carbon dioxide capture by sorption-enhanced water–gas shift and palladium-based membranes for power and hydrogen production,* Int J Greenhouse Gas Control *50:121–134, 2016, Copyright (2016) with permission from Elsevier.*

Column	1/6		1/6		1/12	1/12	1/12	1/6		1/12	1/12	1/12
1	Adsorption		Rinse		PEQ1 ↓	PEQ2 ↓	Blowdn	Purge		PEQ2 ↑	PEQ1 ↑	Rep
2	PEQ1 ↑	Rep	Adsorption		Rinse		PEQ1 ↓	PEQ2 ↓	Blowdn	Purge		PEQ2 ↑
3	Purge	PEQ2 ↑	PEQ1 ↑	Rep	Adsorption		Rinse		PEQ1 ↓	PEQ2 ↓	Blowdh	Purge
4	Blowdn	Purge		PEQ2 ↑	PEQ1 ↑	Rep	Adsorption		Rinse		PEQ1 ↓	PEQ2 ↓
5	PEQ1 ↓	PEQ2 ↓	Blowdh	Purge		PEQ2 ↑	PEQ1 ↑	Rep	Adsorption		Rinse	

Fig. 42 SEWGS cycle for natural gas reformate with two pressure equalizations, using six columns (*PEQ*: pressure equalization between two columns; *Blowdn*: blowdown/depressurization to atmospheric pressure; *Rep*: repressurization with hydrogen product). *Reprinted from Boon J, Spallina V, van Delft Y, van Sint Annaland M: Comparison of the efficiency of carbon dioxide capture by sorption-enhanced water–gas shift and palladium-based membranes for power and hydrogen production,* Int J Greenhouse Gas Control *50:121–134, 2016, Copyright (2016) with permission from Elsevier.*

Since the current sequestration requirements are subject to ongoing debate, the role of hydrogen sulfide in the carbon dioxide product is not further considered in this work.

4.5.2.3 System Efficiencies and Work of Separation

The results were summarized in Table 12 are detailed in terms of the relative contributions to the efficiency penalty in Figs. 43–46. The LHV (net heating value) multiplied by the feed mass flow rate is scaled to 100, all other contributions are scaled on the same basis. The resulting values of η_0, η_1, and η_2 are also scaled to 100 and plotted in between.

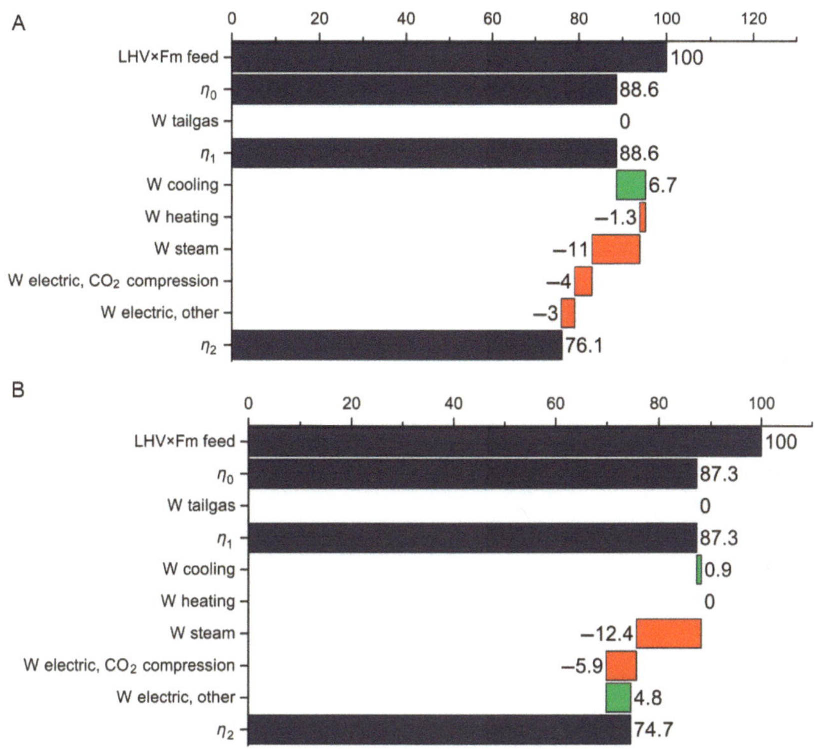

Fig. 43 Efficiency of carbon dioxide capture for coal syngas to decarbonized fuel. (A) Case I: Selexol and (B) Case III: SEWGS. *Reprinted from Boon J, Spallina V, van Delft Y, van Sint Annaland M: Comparison of the efficiency of carbon dioxide capture by sorption-enhanced water–gas shift and palladium-based membranes for power and hydrogen production,* Int J Greenhouse Gas Control *50:121–134, 2016, Copyright (2016) with permission from Elsevier.*

The carbon capture ratio, which is the translation of the rationale behind all these process schemes, is the single most important parameter not yet accounted for in the efficiency comparison in Figs. 43–46. As summarized in Table 12, the carbon capture ratios for the natural gas cases are similar. For the coal syngas cases, the carbon capture ratio for SEWGS (Cases III and IV) is significantly higher than for the other cases. The achieved carbon capture ratio is plotted in Fig. 47 vs the work of separation, as measured by the loss in theoretical net efficiency.

4.5.3 Discussion of the Efficiency of Carbon Capture

The results for the separation of carbon dioxide from hydrogen in the cases defined compare best in four subsets, for coal syngas and for natural gas

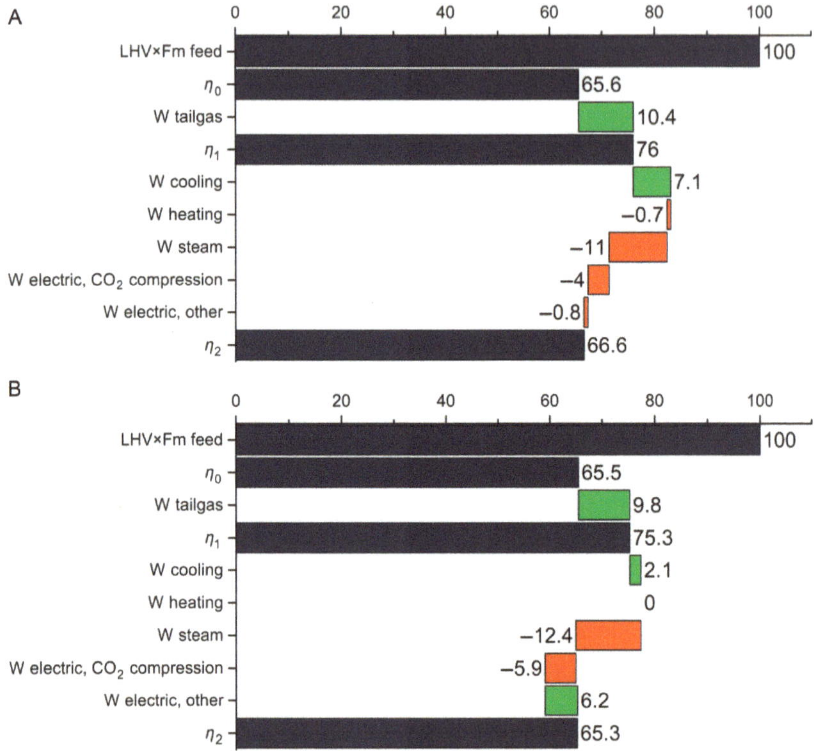

Fig. 44 Efficiency of carbon dioxide capture for coal syngas to hydrogen. (A) Case II: Selexol and (B) Case IV: SEWGS. *Reprinted from Boon J, Spallina V, van Delft Y, van Sint Annaland M: Comparison of the efficiency of carbon dioxide capture by sorption-enhanced water–gas shift and palladium-based membranes for power and hydrogen production,* Int J Greenhouse Gas Control *50:121–134, 2016, Copyright (2016) with permission from Elsevier.*

reformate, for decarbonized fuel and for pure hydrogen product. The current study focuses on the work for carbon dioxide separation, producing hydrogen-rich products from syngas. Therefore, no assessment is made of the efficiency of producing syngas, or of the efficiency with which power can be produced. Regarding the last aspect, it should be noted that the decarbonized fuel product is defined to be used in combined cycle power plants. Alternatively, concentrated hydrogen product streams can be used in fuel cells for power generation (Song, 2002), creating an alternative route for syngas to power with a potentially high efficiency.

4.5.3.1 Coal Syngas to Decarbonized Fuel (Cases I and III)

For the conversion of coal-derived syngas to hydrogen fuel, conventional Selexol and SEWGS end up with similar net efficiencies η_2, 76% and

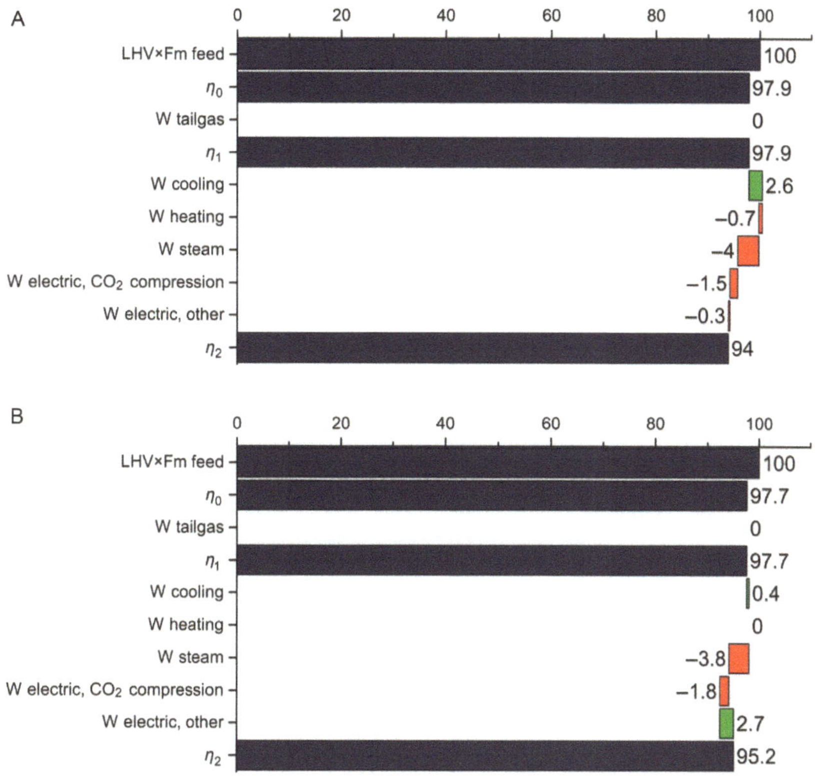

Fig. 45 Efficiency of carbon dioxide capture for natural gas reformate to decarbonized fuel. (A) Case VII: MDEA and (B) Case IX: SEWGS. *Reprinted from Boon J, Spallina V, van Delft Y, van Sint Annaland M: Comparison of the efficiency of carbon dioxide capture by sorption-enhanced water–gas shift and palladium-based membranes for power and hydrogen production,* Int J Greenhouse Gas Control *50:121–134, 2016, Copyright (2016) with permission from Elsevier.*

75%, respectively (Fig. 43). The reduced cooling and reheating duties that result from the high-temperature operation of SEWGS are clear from the corresponding heating and cooling duties, but this aspect is actually favorable for Selexol (more heat is recovered during cooling than required for reheating). Compared to Selexol, SEWGS requires only 1.4%-points of additional efficiency loss for the addition of steam. Note that the steam requirement of the SEWGS process is calculated using improved understanding of the interaction of carbon dioxide and steam with the sorbent, as discussed above. The steam consumption has been decreased significantly compared to previous assessments (Gazzani et al., 2013; Manzolini et al., 2013a). In addition, a remarkable reduction in capital expenditure has been made with the reduction from 9 to 6 columns, which leads to a decrease in

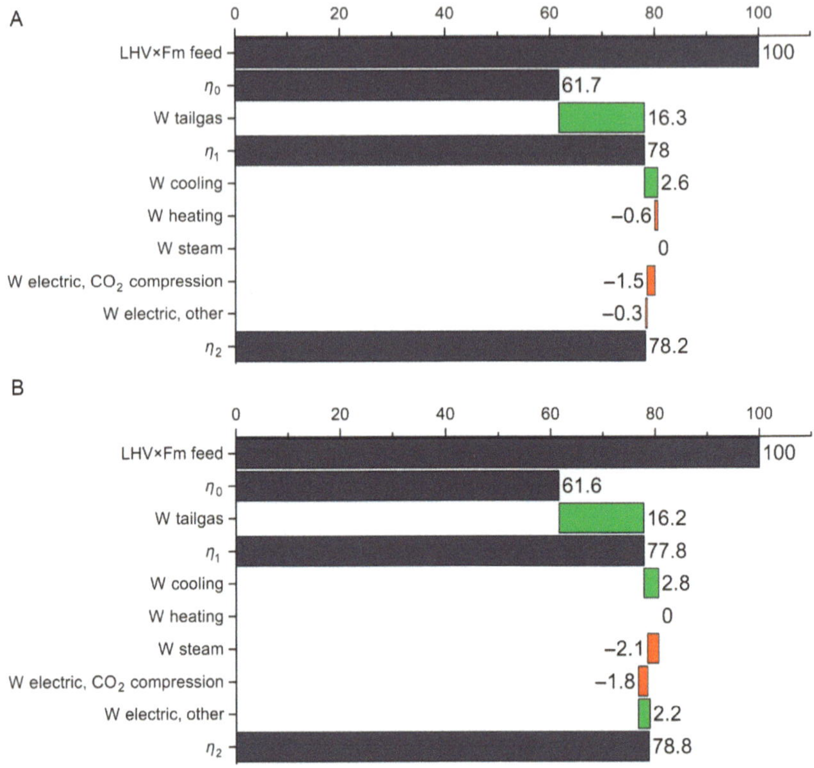

Fig. 46 Efficiency of carbon dioxide capture for natural gas reformate to hydrogen. (A) Case VIII: MDEA and (B) Case X: SEWGS. *Reprinted from Boon J, Spallina V, van Delft Y, van Sint Annaland M: Comparison of the efficiency of carbon dioxide capture by sorption-enhanced water–gas shift and palladium-based membranes for power and hydrogen production,* Int J Greenhouse Gas Control *50:121–134, 2016, Copyright (2016) with permission from Elsevier.*

capital expenditure and will further improve the already interesting economic prospects for SEWGS (Manzolini et al., 2013b). Consequently, further reduction of the SEWGS steam demand (e.g., by increasing the number of pressure equalizations as discussed above) is expected to be of little consequence in terms of the overall efficiency penalty. The main advantage of SEWGS, other than the higher carbon capture ratio, appears in the reduced net electric work, resulting from the produced work by the two expanders and from avoiding electricity consumption for carbon dioxide capture (in contrast to Selexol).

While Selexol has a slightly higher net efficiency ($\eta_2=0.76$) than SEWGS ($\eta_2=0.75$), accounting for the strongly enhanced carbon capture rate yields SEWGS as the technology with the lowest efficiency penalty

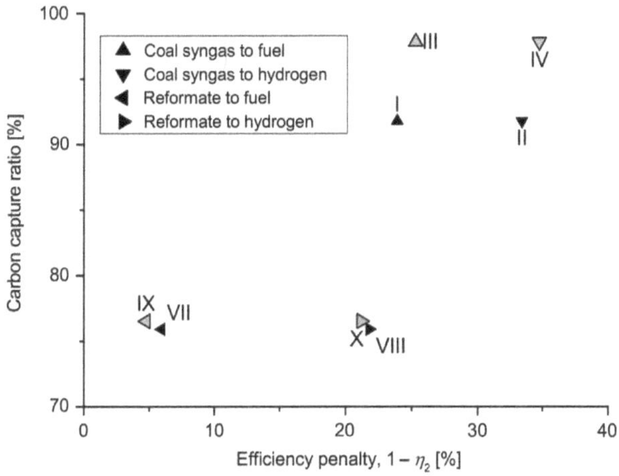

Fig. 47 Comparison of the 12 cases in terms of carbon capture ratio (CCR) vs required efficiency penalty ($1-\eta_2$). *Reprinted from Boon J, Spallina V, van Delft Y, van Sint Annaland M: Comparison of the efficiency of carbon dioxide capture by sorption-enhanced water–gas shift and palladium-based membranes for power and hydrogen production,* Int J Greenhouse Gas Control *50:121–134, 2016, Copyright (2016) with permission from Elsevier.*

per unit carbon dioxide captured to produce decarbonized fuel from coal syngas. Cormos (2012) predicted a lower efficiency penalty for Selexol, but with a lower carbon capture ratio and a significantly higher efficiency assumption for compression of the carbon dioxide product. The efficiency of Selexol in this work is identical to the efficiency of the Selexol case while for SEWGS the efficiency is slightly lower than the values reported by Gazzani et al. (2013) ($\eta_2=0.803$), which is related to the SEWGS cycle design. The current cycle design, based on an improved understanding of the interaction of carbon dioxide and steam with the sorbent as discussed above, achieves a carbon capture ratio of 98% with $S/C_r=S/C_p=0.1$ whereas the cycle used by Gazzani et al. achieves the same capture ratio using $S/C_r=0.34$ and $S/C_p=1.4$. Note that the sorbent in this work is still the alpha-type sorbent and the potential further improvement by using the higher-capacity sorbent beta is not yet accounted for (Gazzani et al., 2013).

The precombustion capture schemes for IGCC presented here outperform postcombustion capture by absorption from pulverized coal power plants, reaching an efficiency of $\eta_2=0.6$ and a target value of $\eta_2=0.71$ (House et al., 2009). A recently presented design study reports higher efficiencies, approaching the efficiencies in this work for Selexol and SEWGS with a newly developed solvent, albeit at a lower carbon dioxide capture ratio (Manzolini et al., 2015; Sanchez Fernandez et al., 2014).

4.5.3.2 Coal Syngas to Pure Hydrogen (Cases II and IV)

For the conversion of coal-derived syngas to pure hydrogen, conventional Selexol and SEWGS again perform similar in terms of the net efficiency η_2 (Fig. 44). Their performance suffers equally from the relatively low product recovery in the hydrogen PSA unit, from which the tail gas is used but at an LHV efficiency of 45%. Again SEWGS outperforms Selexol in terms of the efficiency penalty per unit of carbon dioxide captured.

4.5.3.3 Natural Gas Reformate to Decarbonized Fuel (Cases VII and IX)

For the conversion of natural gas reformate to hydrogen fuel, SEWGS has a higher net efficiency η_2 than MDEA (Fig. 45). The theoretical net efficiency penalty for MDEA is 6%, while the overall system efficiency penalty of carbon dioxide capture by MDEA in NGCC typically reaches 16% (8%-points) (Jansen et al., 2015). As for the coal cases, the net effect of cooling down and reheating of the process gas is slightly favorable for MDEA, because more heat is recovered in the cooling than is required for reheating. Conversely, the SEWGS-based process requires 2.7%-points less electricity; in the SEWGS flowsheet, the expanders result in a net electricity production. Compared to the prediction by Cobden et al. (2007), the steam consumption has been reduced and the efficiency penalty $(1-\eta_2)$ for SEWGS has been reduced from up to 12%–5% which is much closer to the minimum value of 2.7%. Gazzani et al. also reported a slightly higher efficiency for SEWGS compared to MDEA (Manzolini et al., 2013a). Note that the low capture ratio is not due to the employed separation processes but instead due to the methane conversion assumed, as witnessed from the carbon oxides capture ratio in Table 12. Postcombustion capture would in principle allow for higher carbon capture ratios because it allows to also capture part of the carbon dioxide that is produced by the combustion of PSA offgas, but Iyer et al. found 66.5% CCR because carbon dioxide was not captured from the capture unit boiler (Iyer et al., 2015a). Of the carbon (carbon monoxide, carbon dioxide, and methane) in the reformate, 19 mol% is methane which can be used to underfire the reformer and is thus not captured. In the MDEA and SEWGS cases for fuel, the methane ends up in the hydrogen fuel to be combusted. Based on carbon oxides only, MDEA achieves a CCR of 94% while SEWGS would achieve 95%.

Outside the scope of the current study, in the case of natural gas reformate to power, postcombustion carbon dioxide capture could be preferred from a thermodynamic efficiency point of view (Manzolini et al., 2013a), although a recent study found an efficiency penalty of 12% for carbon dioxide capture in NGCC (Iyer et al., 2015b), which is higher than the currently predicted efficiency penalties for MDEA (Case VII) and SEWGS (Case IX). Still, newly developed amine solvents could reduce the efficiency penalty for postcombustion carbon dioxide capture (Manzolini et al., 2015).

Because in the current study it yields the highest efficiency as well as the highest carbon capture ratio, SEWGS is the most efficient process to produce decarbonized fuel from natural gas reformate.

4.5.3.4 Natural Gas Reformate to Pure Hydrogen (Cases VIII and X)

For the conversion of natural gas reformate to pure hydrogen (Fig. 46), conventional MDEA and SEWGS have a similar net efficiency with an advantage for SEWGS. The efficiencies are largely determined by the performance of the PSA system for the production of pure hydrogen, as in the corresponding coal syngas cases. In the MDEA and SEWGS cases for hydrogen, the methane is recovered in the PSA tail gas as accounted for in the models and similar performances are achieved in terms of carbon capture ratio and process efficiency (Fig. 47).

4.5.4 SEWGS Improves Efficiency of Precombustion Carbon Capture

SEWGS has been compared to conventional technologies for hydrogen–carbon dioxide separation based on the carbon capture ratio and energy efficiency penalty. Four major process routes were assessed, each leading to different conclusions. For the production of decarbonized fuel from coal syngas, SEWGS yields the lowest efficiency penalty per unit of carbon dioxide captured. For the production of pure hydrogen from coal syngas, SEWGS has a significantly higher carbon capture ratio than Selexol. For the production of decarbonized fuel from natural gas reformate, SEWGS is more efficient than MDEA in terms of efficiency penalty. For the production of pure hydrogen from natural gas syngas, SEWGS has a slightly better performance than MDEA: higher carbon capture ratio and lower efficiency penalty.

5. CONCLUSIONS

SEWGS combines the WGS reaction with in situ carbon dioxide capture at high temperature, also enhancing the carbon monoxide conversion. Its inherent advantages overcompeting technologies can be summarized as:

- High-temperature hydrogen product (typically around 400°C),
- High-pressure hydrogen product (typically 30bar),

- Medium purity hydrogen product (typically 90–95 mol%),
- High-temperature carbon dioxide product (typically around 400°C),
- Carbon dioxide product at ambient pressure, and
- High carbon monoxide conversion.

Given these characteristics, it is not surprising that SEWGS has a long-standing history in developing precombustion carbon dioxide capture. It has also been suggested in applications related to distributed refinery hydrogen, refinery process hydrogen, refinery fuel gas, ammonia and urea production, production of methanol, processing of steel plant gas streams, and coal to chemicals or fuels.

The sorbent of choice for state-of-the-art SEWGS is K-HTC. K-HTC has a layered structure and contains hydroxides, carbonates, and water groups. Between 300 and 500°C, the material goes through several dehydroxylation and decarbonation steps. The layered structure collapses above approximately 350°C. Therefore, under SEWGS operating conditions, the K-HTC sorbent is in transition to mixed-metal oxide, while still containing carbonate and hydroxide groups, and consequently difficult to fully define. Several experimental studies have elucidated part of the physical and chemical interactions of carbon dioxide and steam with the sorbent. Various measurements proved that temperature and time during the desorption step determine the measured cyclic working capacity for CO_2. The adsorption of CO_2, which shows very fast reaction kinetics in the beginning followed by a somewhat slower adsorption, does not seem to be influenced significantly by the temperature and by the adsorption time. Depending on the desorption conditions the decrease in the adsorption rate and the final CO_2 cyclic working capacity change. The desorption kinetics can be increased with an increase in temperature but are in general much slower than the adsorption kinetics. CO_2 adsorption seems to be a mixture between physical adsorption and chemical reaction. In contrast, the adsorption mechanism for H_2O seems to be mainly physical adsorption, which has been elaborated with temperature dependency of H_2O adsorption.

Smart designed TGA cycle in combination with PBR experiments was used to develop a model capable of describing the interactions of CO_2 and H_2O on K-HTC at 400°C. It was found that at least four different adsorption sites are required to describe the cyclic working capacity of both CO_2 and H_2O. Site A (H_2O) and site B (CO_2) can be regenerated with a stream of N_2. One additional site (site C) which can be either occupied by CO_2 or by H_2O showed a strong dependency on the feed gas composition. If a high partial pressure of CO_2 is used H_2O can be desorbed from this site and vice versa. This site can be explained by an equilibrium reaction where a metal oxide is transformed in either a hydroxide or a carbonate. Therefore, the cyclic working capacity of the adsorbent can be significantly improved if H_2O is used for the regeneration step. During adsorption of CO_2 a lower H_2O partial pressure is beneficial because more CO_2 can be adsorbed on site C. The in general increased adsorption capacity if the adsorbent is exposed to both CO_2 and H_2O in one step was described with an additional adsorption site D.

High-pressure cyclic breakthrough tests on a SEWGS bench scale unit have yielded the capacity and kinetics of the adsorption on K-HTC over a wide range of partial pressures, up to 24 bar(a) and at 400°C. Capacities of up to 1.5 mol kg^{-1} have been measured, both for CO_2 and for H_2O. This dataset has been used to fit a multicomponent adsorption isotherm. The isotherm consists of a lower partial pressure part where surface adsorption dominates and a high partial pressure nanopore adsorption part.

Based on the experimentally determined isotherm, a SEWGS reactor model has been developed. The model uses a heterogeneous linear driving force approximation for the intraparticle mass transfer resistance, reflecting the porous character of the particle. With the SEWGS model, a pressure swing cycle is proposed for the SEWGS process. It has been shown that coadsorption of steam during the SEWGS cycle is beneficial for the efficiency of the rinse step. A parameter study based on numerical simulations has yielded a SEWGS cycle that consumes significantly less steam than previously reported cycle designs in literature. With the results from the SEWGS cycle simulations, process efficiencies have been calculated for the capture of CO_2, and these have been compared to the efficiencies for conventional absorption technology. It is shown that SEWGS outperforms conventional technologies for hydrogen–carbon dioxide separation, based on the carbon capture ratio and energy efficiency penalty.

Future work on the sorbents will focus on increasing the sorption capacity and stability of the sorbents. From a research point of view, a better understanding of the phenomena occurring on the sorbent surface could be achieved by applying in situ measurements such as FTIR and/or operando computer tomography. Additionally, the same sorbents can be used with syngas containing impurities, and a better study on the effect of other molecules on the sorption capacity for CO_2 should be studied.

On the process level, the new models for sorption of CO_2, clearly change the process scheme and operation, so that advanced a SEWGS cycles can be designed which have much higher efficiency.

ACKNOWLEDGMENTS

These projects have received funding from and by the Dutch Ministry of Economic affairs through the CATO and CATO-2 programs (http://www.co2-cato.nl/) and through the ADEM innovation lab program (project number TUEP05) and from the European Union's Sixth and Seventh Framework Programs under grant agreements nos. 019972 and 213206.

REFERENCES

Abanades JC, Alvarez D: Conversion limits in the reaction of CO_2 with lime, *Energy Fuels* 17:308–315, 2003. https://doi.org/10.1021/ef020152a.

Alpay E, Chadwick D, Kershenbaum LS, Barrie PJ, Sivadinarayana C, Gladden LF: TEOM analysis of the equilibria and kinetics of n-hexane and n-heptane adsorption on FCC catalyst/silicalite, *Chem Eng Sci* 58:2777–2784, 2003.

Beavis R: The EU FP6 CACHET project—final results, *Energy Procedia* 4:1074–1081, 2011. https://doi.org/10.1016/j.egypro.2011.01.157.

Bird RB, Stewart WE, Lightfoot EN: *Transport phenomena,* New York, 1960, Wiley.

Birol F: *Redrawing the energy-climate MAP: World energy outlook special report*, 2013World Energy Outlook Special Report Paris 134.

Boll W, Hochgesand G, Higman C, et al: Gas production, 3. Gas treating. In *Ullmann's encyclopedia of industrial chemistry*, 2000, Wiley-VCH-Verlag. https://doi.org/10.1002/14356007.o12_o02.

Boon J, Van Dijk E, Pirgon-Galin Ö, Haije W, Van Den Brink R: Water–gas shift kinetics over FeCr-based catalyst: effect of hydrogen sulphide, *Catal Lett* 131:406–412, 2009.

Boon J, Cobden PD, van Dijk HAJ, Hoogland C, van Selow ER, van Sint Annaland M: Isotherm model for high-temperature, high-pressure adsorption of CO_2 and H_2O on K-promoted hydrotalcite, *Chem Eng J* 248:406–414, 2014. https://doi.org/10.1016/j.cej.2014.03.056.

Boon J, Cobden PD, van Dijk HAJ, van Sint Annaland M: High-temperature pressure swing adsorption cycle design for sorption-enhanced water–gas shift, *Chem Eng Sci* 122:219–231, 2015. https://doi.org/10.1016/j.ces.2014.09.034.

Boon J, Spallina V, van Delft Y, van Sint Annaland M: Comparison of the efficiency of carbon dioxide capture by sorption-enhanced water–gas shift and palladium-based membranes for power and hydrogen production, *Int J Greenhouse Gas Control* 50:121–134, 2016. https://doi.org/10.1016/j.ijggc.2016.04.033.

Caesar: *The SEWGS process*, 2011(WWW Document).

Carbo MC, Boon J, Jansen D, et al: Steam demand reduction of water–gas shift reaction in IGCC power plants with pre-combustion CO_2 capture, *Int J Greenhouse Gas Control* 3:712–719, 2009.

Carbo MC, Smit R, van der Drift B, Jansen D: Bio energy with CCS (BECCS): large potential for BioSNG at low CO_2 avoidance cost, *Energy Procedia* 4:2950–2954, 2011. https://doi.org/10.1016/j.egypro.2011.02.203.

Cau G, Tola V, Deiana P: Comparative performance assessment of USC and IGCC power plants integrated with CO_2 capture systems, *Fuel* 116:820–833, 2014. https://doi.org/10.1016/j.fuel.2013.06.005.

Cavani F, Trifiro F, Vaccari A: Hydrotalcite-type anionic clays: preparation, properties and applications, *Catal Today* 11:173–301, 1991.

Chiesa P, Macchi E: A thermodynamic analysis of different options to break 60% electric efficiency in combined cycle power plants, *J Eng Gas Turbines Power* 126:770–785, 2004. https://doi.org/10.1115/1.1771684.

Chiesa P, Romano MC, Spallina V, Turi DM, Mancuso L: Efficient low CO_2 emissions power generation by mixed conducting membranes, *Energy Procedia* 37:905–913, 2013. https://doi.org/10.1016/j.egypro.2013.05.185.

Choi S, Drese JH, Jones CW: Adsorbent materials for carbon dioxide capture from large anthropogenic point sources, *ChemSusChem* 2:796–854, 2009. https://doi.org/10.1002/cssc.200900036.

Cobden PD, van Beurden P, Reijers HTJ, et al: Sorption-enhanced hydrogen production for pre-combustion CO_2 capture: thermodynamic analysis and experimental results, *Int J Greenhouse Gas Control* 1:170–179, 2007. https://doi.org/10.1016/S1750-5836(07)00021-7.

Coenen K, Gallucci F, Cobden P, Dijk Ev, Hensen E, Annaland MvS: Chemisorption working capacity and kinetics of CO_2 and H_2O of hydrotalcite-based adsorbents for sorption-enhanced water–gas-shift applications, *Chem Eng J* 293:9–23, 2016a. https://doi.org/10.1016/j.cej.2016.02.050.

Coenen K, Gallucci F, Pio G, Cobden P, Dijk Ev, Annaland MvS: On the influence of steam on the CO_2 chemisorption capacity of a hydrotalcite-based adsorbent for SEWGS applications, *Chem Eng J* 314:554–569, 2016b. https://doi.org/10.1016/j.cej.2016.12.013.

Cormos C-C: Integrated assessment of IGCC power generation technology with carbon capture and storage (CCS), *Energy* 42:434–445, 2012. https://doi.org/10.1016/j.energy.2012.03.025.

Debecker DP, Gaigneaux EM, Busca G: Exploring, tuning, and exploiting the basicity of hydrotalcites for applications in heterogeneous catalysis, *Chemistry* 15:3920–3935, 2009. https://doi.org/10.1002/chem.200900060.

Di Cosimo JI, Diez VK, Xu M, Iglesia E, Apestegu CR: Structure and surface and catalytic properties of Mg–Al basic oxides, *J Catal* 510:499–510, 1998a.

Ding Y, Alpay E: Equilibria and kinetics of CO_2 adsorption on hydrotalcite adsorbent, *Chem Eng Sci* 55:3461–3474, 2000. https://doi.org/10.1016/S0009-2509(99)00596-5.

Do DD: *Adsorption analysis: equilibria and kinetics,* London, 1998, Imperial College Press.

Doong SJ, Yang RT: A simple potential-theory model for predicting mixed-gas adsorption, *Ind Eng Chem Res* 27:630–635, 1988. https://doi.org/10.1021/ie00076a017.

Du H, Ebner AD, Ritter Ja: Temperature dependence of the nonequilibrium kinetic model that describes the adsorption and desorption behavior of CO_2 in K-promoted HTlc, *Ind Eng Chem Res* 49:3328–3336, 2010. https://doi.org/10.1021/ie901210y.

Ebner AD, Reynolds SP, Ritter JA: Understanding the adsorption and desorption behavior of CO_2 on a K-promoted hydrotalcite-like compound (HTlc) through nonequilibrium dynamic isotherms, *Ind Eng Chem Res* 45:6387–6392, 2006. https://doi.org/10.1021/ie060389k.

Ebner AD, Reynolds SP, Ritter JA: Nonequilibrium kinetic model that describes the reversible adsorption and desorption behavior of CO_2 in a K-promoted hydrotalcite-like compound, *Ind Eng Chem Res* 46:1737–1744, 2007.

Franco F: *Common framework definition document*, 2009. Technical Report D4.6.

Franco A, Russo A: Combined cycle plant efficiency increase based on the optimization of the heat recovery steam generator operating parameters, *Int J Therm Sci* 41:843–859, 2002. https://doi.org/10.1016/S1290-0729(02)01378-9.

Gazzani M, MacChi E, Manzolini G: CO_2 capture in integrated gasification combined cycle with SEWGS—part A: thermodynamic performances, *Fuel* 105:206–219, 2013. https://doi.org/10.1016/j.fuel.2012.07.048.

Gazzani M, Chiesa P, Martelli E, Sigali S, Brunetti I: Using hydrogen as gas turbine fuel: premixed versus diffusive flame combustors, *J Eng Gas Turbines Power* 136:51504–51510, 2014.

Gazzani M, Romano M, Manzolini G: CO_2 capture in integrated steelworks by commercial-ready technology and SEWGS process, *Int J Greenhouse Gas Control* 41:249–267, 2015. https://doi.org/10.1016/j.ijggc.2015.07.012.

Gielen D: CO_2 removal in the iron and steel industry, *Energ Conver Manage* 44:1027–1037, 2003. https://doi.org/10.1016/S0196-8904(02)00111-5.

Gluud, W., Keller, K., Schönfelder, R., Klempt, W., 1931. Production of hydrogen. 1816523.

Gülker, F., 1927. Herstellung von Wasserstoff. 446488.

Halabi MH, de Croon MHJM, van der Schaaf J, Cobden PD, Schouten JC: High capacity potassium-promoted hydrotalcite for CO_2 capture in H_2 production, *Int J Hydrogen Energy* 37:4516–4525, 2012. https://doi.org/10.1016/j.ijhydene.2011.12.003.

Hanif A, Dasgupta S, Divekar S, Arya A, Garg MO, Nanoti A: A study on high temperature CO_2 capture by improved hydrotalcite sorbents, *Chem Eng J* 236:91–99, 2014. https://doi.org/10.1016/j.cej.2013.09.076.

Hatch CD, Wiese JS, Crane CC, Harris KJ, Kloss HG, Baltrusaitis J: Water adsorption on clay minerals as a function of relative humidity: application of BET and Freundlich adsorption models, *Langmuir* 28:1790–1803, 2012. https://doi.org/10.1021/la2042873.

Hedin N, Chen L, Laaksonen A: Sorbents for CO_2 capture from flue gas-aspects from materials and theoretical chemistry, *Nanoscale* 2:1819–1841, 2010. https://doi.org/10.1039/c0nr00042f.

House KZ, Harvey CF, Aziz MJ, Schrag DP: The energy penalty of post-combustion CO_2 capture & storage and its implications for retrofitting the US installed base, *Energ Environ Sci* 2:193–205, 2009. https://doi.org/10.1039/B811608C.

Hufton JR, Mayorga S, Sircar S: Sorption-enhanced reaction process for hydrogen production, *AIChE J* 45:248–256, 1999. https://doi.org/10.1002/aic.690450205.

Hutson ND, Attwood BC: High temperature adsorption of CO_2 on various hydrotalcite-like compounds, *Adsorption* 14:781–789, 2008. https://doi.org/10.1007/s10450-007-9085-6.

Iyer M, Beavis R, Tarrant T, Bullen T, Butler D: Techno-economic assessment of deep CO_2 capture from hydrogen plants. In Gerdes KF, editor: *Carbon dioxide capture for storage in deep geological formations*, Thatcham, UK, 2015a, CPL, pp 153–164.

Iyer M, Beavis R, Tarrant T, Bullen T, Butler D: Techno-economic assessment of power generation from natural gas in combined cycle ({NGCC}) systems with integrated capture. In Gerdes KF, editor: *Carbon dioxide capture for storage in deep geological formations*, Thatcham, UK, 2015b, CPL, pp 317–328.

Jang HM, Lee KB, Caram HS, Sircar S: High-purity hydrogen production through sorption enhanced water gas shift reaction using K_2CO_3-promoted hydrotalcite, *Chem Eng Sci* 73:431–438, 2012. https://doi.org/10.1016/j.ces.2012.02.015.

Jansen D, Van Selow E, Cobden P, et al: SEWGS technology is now ready for scale-up! *Energy Procedia* 37:2265–2273, 2013. https://doi.org/10.1016/j.egypro.2013.06.107.

Jansen D, Gazzani M, Manzolini G, Dijk EV: Pre-combustion CO_2 capture, *Int J Greenhouse Gas Control* 40:167–187, 2015. https://doi.org/10.1016/j.ijggc.2015.05.028.

Kanezaki E: Thermal behavior of the hydrotalcite-like layered structure of Mg and Al-layered double hydroxides with interlayer carbonate by means of in situ powder HTXRD and DTA/TG, *Solid State Ion* 106:279–284, 1998. https://doi.org/10.1016/S0167-2738(97)00494-3.

Kohl AL, Nielsen RB: *Gas purification*, ed 5, 1997, ISBN: 978-0-88415-220-0. https://www.elsevier.com/books/gas-purification/kohl/978-0-88415-220-0.

Lee W-Y, Kim S-S: The maximum power from a finite reservoir for a Lorentz cycle, *Energy* 17:275–281, 1992. https://doi.org/10.1016/0360-5442(92)90055-5.

Lee KB, Verdooren a, Caram HS, Sircar S: Chemisorption of carbon dioxide on potassium-carbonate-promoted hydrotalcite, *J Colloid Interface Sci* 308:30–39, 2007. https://doi.org/10.1016/j.jcis.2006.11.011.

Lee KB, Beaver MG, Caram HS, Sircar S: Reversible chemisorbents for carbon dioxide and their potential applications, *Ind Eng Chem Res* 47:8048–8062, 2008.

Liu Y, Lotero E, Goodwin JG, Mo X: Transesterification of poultry fat with methanol using Mg–Al hydrotalcite derived catalysts, *Appl Catal Gen* 331:138–148, 2007. https://doi.org/10.1016/j.apcata.2007.07.038.

Manzolini G, Macchi E, Gazzani M: CO_2 capture in natural gas combined cycle with SEWGS. Part B: economic assessment, *Int J Greenhouse Gas Control* 12:502–509, 2013a. https://doi.org/10.1016/j.ijggc.2012.06.021.

Manzolini G, MacChi E, Gazzani M: CO_2 capture in integrated gasification combined cycle with SEWGS. Part B: economic assessment, *Fuel* 105:220–227, 2013b. https://doi.org/10.1016/j.fuel.2012.07.043.

Manzolini G, Sanchez Fernandez E, Rezvani S, Macchi E, Goetheer ELV, Vlugt TJH: Economic assessment of novel amine based CO_2 capture technologies integrated in power plants based on European benchmarking task force methodology, *Appl Energy* 138:546–558, 2015. https://doi.org/10.1016/j.apenergy.2014.04.066.

Maroño M, Torreiro Y, Gutierrez L: Influence of steam partial pressures in the CO_2 capture capacity of K-doped hydrotalcite-based sorbents for their application to SEWGS processes, *Int J Greenhouse Gas Control* 14:183–192, 2013. https://doi.org/10.1016/j.ijggc.2013.01.024.

Maroño M, Torreiro Y, Montenegro L, Sánchez J: Lab-scale tests of different materials for the selection of suitable sorbents for CO_2 capture with H_2 production in IGCC processes, *Fuel* 116:861–870, 2014. https://doi.org/10.1016/j.fuel.2013.03.067.

Meis NNAH, Bitter JH, Jong KPD: On the influence and role of alkali metals on supported and unsupported activated hydrotalcites for CO_2 sorption, *Ind Eng Chem Res* 49:8086–8093, 2010.

Miguel CV, Trujillano R, Rives V, et al: High temperature CO_2 sorption with gallium-substituted and promoted hydrotalcites, *Sep Purif Technol* 127:202–211, 2014. https://doi.org/10.1016/j.seppur.2014.03.007.

Newsome DS: The water–gas shift reaction, *Catal Rev Sci Eng* 21:275–318, 1980.

Ochoa-Fernández E, Zhao T, Rønning M, Chen D: Effects of steam addition on the properties of high temperature ceramic CO_2 acceptors, *J Environ Eng* 135:397–403, 2009.

Oliveira ELG, Grande CA, Rodrigues AE: CO_2 sorption on hydrotalcite and alkali-modified (K and Cs) hydrotalcites at high temperatures, *Sep Purif Technol* 62:137–147, 2008. https://doi.org/10.1016/j.seppur.2008.01.011.

Ozawa S, Kusumi S, Ogino Y: Physical adsorption of gases at high pressure. IV. An improvement of the Dubinin—Astakhov adsorption equation, *J Colloid Interface Sci* 56:83–91, 1976.

Peng D-Y, Robinson DB: A new two-constant equation of state, *Ind Eng Chem Fundamen* 15:59–64, 1976.

Pérez-Ramírez J, Abelló S: Thermal decomposition of hydrotalcite-like compounds studied by a novel tapered element oscillating microbalance (TEOM): comparison with TGA and DTA, *Thermochim Acta* 444:75–82, 2006.

Pevida C, Plaza MG, Arias B, Fermoso J, Rubiera F, Pis JJ: Surface modification of activated carbons for CO_2 capture, *Appl Surf Sci* 254:7165–7172, 2008. https://doi.org/10.1016/j.apsusc.2008.05.239.

Poling BE, Prausnitz JM, O'Connell JP: *The properties of gases and liquids,* New York, 2001, McGraw-Hill.

Ram Reddy MK, Xu ZP, (Max) Lu GQ, Diniz da Costa JC: Influence of water on high-temperature CO_2 capture using layered double hydroxide derivatives, *Ind Eng Chem Res* 47:2630–2635, 2008.

Reichle WT: Catalytic reactions by thermally activated, synthetic, anionic, *Clay Miner* 557:547–557, 1985.

Reichle W: The nature of the thermal decomposition of a catalytically active anionic clay mineral, *J Catal* 101:352–359, 1986a. https://doi.org/10.1016/0021-9517(86)90262-9.

Reichle WT: Synthesis of anionic clay minerals (mixed metal hydroxides, hydrotalcite), *Solid State Ion* 22:135–141, 1986b.

Reichle W, Kang S, Everhardt D: The nature of the thermal decomposition of a catalytically active anionic clay mineral, *J Catal* 352–359, 1986.

Reijers HTJ, Valster-Schiermeier SEa, Cobden PD, Van Den Brink RW: Hydrotalcite as CO_2 sorbent for sorption-enhanced steam reforming of methane, *Ind Eng Chem Res* 45:2522–2530, 2006. https://doi.org/10.1021/ie050563p.

Reijers HT, Boon J, Elzinga GD, Cobden PD, Haije WG, Brink RW: Modeling study of the sorption-enhanced reaction process for CO_2 capture. II. Application to steam-methane reforming, *Ind Eng Chem Res* 48:6975–6982, 2009a.

Reijers HT, Boon J, Elzinga GD, Cobden PD, Haije WG, Van den Brink RW: Modeling study of the sorption-enhanced reaction process for CO_2 capture. I. Model development and validation, *Ind Eng Chem Res* 48:6966–6974, 2009b.

Reijers R, Van Selow E, Cobden P, Boon J, Van Den Brink R: SEWGS process cycle optimization, *Energy Procedia* 4:1155–1161, 2011.

Rg UK, Hudson MJ, Carlino S, Apperleyb DC: Thermal conversion of a layered (Mg/Al) double hydroxide to the oxide, *J Mater Chem* 5:323–329, 1995.

Ruthven DM: *Principles of adsorption and adsorption processes*, 464, 1984, ISBN: 978-0-471-86606-0. http://eu.wiley.com/WileyCDA/WileyTitle/productCd-0471866067.html.

Sanchez Fernandez E, Goetheer ELV, Manzolini G, Macchi E, Rezvani S, Vlugt TJH: Thermodynamic assessment of amine based CO_2 capture technologies in power plants based on European benchmarking task force methodology, *Fuel* 129:318–329, 2014. https://doi.org/10.1016/j.fuel.2014.03.042.

Soares JL, Moreira RFPM, José HJ, Grande CA, Rodrigues AE: Hydrotalcite materials for carbon dioxide adsorption at high temperatures: characterization and diffusivity measurements, *Sep Sci Technol* 39:1989–2010, 2005. https://doi.org/10.1081/SS-120039307.

Song C: Fuel processing for low-temperature and high-temperature fuel cells: challenges, and opportunities for sustainable development in the 21st century, *Catal Today* 77:17–49, 2002.

Spallina V, Romano MC, Chiesa P, Gallucci F, van Sint Annaland M, Lozza G: Integration of coal gasification and packed bed CLC for high efficiency and near-zero emission power generation, *Int J Greenhouse Gas Control* 27:28–41, 2014. https://doi.org/10.1016/j.ijggc.2014.04.029.

STEPWISE [WWW Document], 2015.

Talu O: Measurement and analysis of mixture adsorption equilibrium in porous solids, *Chemie Ing Tech* 83:67–82, 2011.

Toops TJ, Crocker M: New sulfur adsorbents derived from layered double hydroxides: II. DRIFTS study of COS and H_2S adsorption, *Appl Catal Environ* 82:199–207, 2008. https://doi.org/10.1016/j.apcatb.2008.01.013.

Vaccari A: Preparation and catalytic properties of cationic and anionic clays, *Catal Today* 41:53–71, 1998. https://doi.org/10.1016/S0920-5861(98)00038-8.

van Dijk E, Walspurger S, Cobden PD, van den Brink RW, de Vos FG: Testing of hydrotalcite-based sorbents for CO_2 and H_2S capture for use in sorption enhanced water gas shift, *Int J Greenhouse Gas Control* 5:505–511, 2011.

van Dijk HAJ, Cohen D, Hakeem AA, Makkee M, Damen K: Validation of a water–gas shift reactor model based on a commercial FeCr catalyst for pre-combustion CO_2 capture in an IGCC power plant, *Int J Greenhouse Gas Control* 29:82–91, 2014. https://doi.org/10.1016/j.ijggc.2014.07.005.

van Selow ER, Cobden PD, van den Brink RW, Hufton JR, Wright a: Performance of sorption-enhanced water–gas shift as a pre-combustion CO_2 capture technology, *Energy Procedia* 1:689–696, 2009a. https://doi.org/10.1016/j.egypro.2009.01.091.

van Selow ER, Cobden PD, Verbraeken Pa, Hufton JR, Brink RWVD: Carbon capture by sorption-enhanced water–gas shift reaction process using hydrotalcite-based material, *Ind Eng Chem Res* 48:4184–4193, 2009b. https://doi.org/10.1021/ie801713a.

Van Selow ER, Cobden PD, Van den Brink RW, et al: Pilot-scale development of the sorption enhanced water gas shift process. In *Carbon dioxide capture for storage in deep geologic formations*, Berks, 2009c, CPL Press, pp 157–180.

Van Selow ER, Cobden PD, Wright aD, Van Den Brink RW, Jansen D: Improved sorbent for the sorption-enhanced water–gas shift process, *Energy Procedia* 4:1090–1095, 2011. https://doi.org/10.1016/j.egypro.2011.01.159.

van Selow ER, Cobden PD, van Dijk HAJ, Walspurger S, Verbraeken PA, Jansen D: Qualification of the ALKASORB sorbent for the sorption-enhanced water–gas shift process, *Energy Procedia* 37:180–189, 2013. https://doi.org/10.1016/j.egypro.2013.05.100.

Verfahren zur Herstellung von Wasserstoff, 1915. 284816.

Walspurger S, Boels L, Cobden PD, Elzinga GD, Haije WG, Brink RWvd: The crucial role of the K+-aluminium oxide interaction in K+-promoted alumina- and hydrotalcite-based materials for CO_2 sorption at high temperatures, *ChemSusChem* 1:643–650, 2008. https://doi.org/10.1002/cssc.200800085.

Walspurger S, Cobden PD, Safonova OV, Wu Y, Anthony EJ: High CO(2) storage capacity in alkali-promoted hydrotalcite-based material: in situ detection of reversible formation of magnesium carbonate, *Chemistry* 16:12694–12700, 2010. https://doi.org/10.1002/chem.201000687.

Walspurger S, de Munck S, Cobden PD, Haije WG, van den Brink RW, Safonova OV: Correlation between structural rearrangement of hydrotalcite-type materials and CO_2 sorption processes under pre-combustion decarbonisation conditions, *Energy Procedia* 4:1162–1167, 2011. https://doi.org/10.1016/j.egypro.2011.01.169.

Walspurger S, Boon J, Peper S, Cobden PD, van Dijk HAJ, Overwater K: *Process intensification is the key to CO_2 emission abatement: options for steam reforming to hydrogen*, 2015. https://www.ecn.nl/fileadmin/ecn/corp/Beurspagina/ECCE_2015/Abstract_Submission_Process_intensification_is_the_key_to_CO2_emissions_abatement333333333.pdf.

Wang Q, Tay HH, Guo Z, et al: Morphology and composition controllable synthesis of Mg–Al–CO_3 hydrotalcites by tuning the synthesis pH and the CO_2 capture capacity, *Appl Clay Sci* 55:18–26, 2012. https://doi.org/10.1016/j.clay.2011.07.024.

Wang J, Huang L, Yang R, et al: Recent advances in solid sorbents for CO_2 capture and new development trends, *Energ Environ Sci* 7:3478–3518, 2014. https://doi.org/10.1039/C4EE01647E.

Westerterp KR, Van Swaaij WPM, Beenackers AACM: *Chemical reactor design and operation,* Chichester, 1987, Wiley.

Wright aD, White V, Hufton JR, Quinn R, Cobden PD, van Selow ER: CAESAR: development of a SEWGS model for IGCC, *Energy Procedia* 4:1147–1154, 2011. https://doi.org/10.1016/j.egypro.2011.01.167.

Wu YJ, Li P, Yu JG, Cunha aF, Rodrigues aE: K-promoted hydrotalcites for CO_2 capture in sorption enhanced reactions, *Chem Eng Technol* 36:567–574, 2013. https://doi.org/10.1002/ceat.201200694.

Yang RT: Gas separation by adsorption processes, In *Butterworth-Heinemann*, ed 1, ISBN: 9781483162669, 1987, p 632. https://www.elsevier.com/books/gas-separation-by-adsorption-processes/yang/978-0-409-90004-0.

Yang RT: *Gas separation by adsorption processes, series on chemical engineering,* London, 1997, Imperial College Press.
Yong Z, Mata V, Rodrigues E: Adsorption of carbon dioxide onto hydrotalcite-like compounds (HTlcs) at high temperatures, *Ind Eng Chem Res* 40:204–209, 2001. https://doi.org/10.1021/ie000238w.
Yong Z, Mata V, Rodrigues AE: Adsorption of carbon dioxide at high temperature—a review, *Sep Purif Technol* 26:195–205, 2002.
Zhang J, Singh R, Webley PA: Alkali and alkaline-earth cation exchanged chabazite zeolites for adsorption based CO_2 capture, *Microporous Mesoporous Mater* 111:478–487, 2008. https://doi.org/10.1016/j.micromeso.2007.08.022.

FURTHER READING

Di Cosimo JI, Dı´ez VK, Xu M, Iglesia E, Apesteguı´a CR: Structure and surface and catalytic properties of Mg–Al basic oxides, *J Catal* 178:499–510, 1998b. https://doi.org/10.1006/jcat.1998.2161.

CHAPTER TWO

CO_2 Sorbents for Sorption-Enhanced Steam Reforming

Maria C. Iliuta[1]
Université Laval, Ville de Québec, QC, Canada
[1]Corresponding author: e-mail address: maria-cornelia.iliuta@gch.ulaval.ca

Contents

Abstract

Sorption-enhanced steam reforming (SESR) is an integrated process which combines reforming reaction (H_2 formation) and selective separation (in situ CO_2 sorption by solid sorbent materials) in a single reactor, in order to shift thermodynamically the reforming reaction and increase the hydrogen production. The key element in the SESR processes is the selection of appropriate sorbent materials. The choice of suitable sorbents for selective in situ CO_2 removal from the reaction medium is a great challenge due to the severe operating conditions, especially the high temperature and the presence of steam during the sorption process. Special attention has therefore been devoted to the development of efficient materials as well as the evaluation of their performance in different SESR processes. High sorption capacity and kinetics, together with a good stability in the long-term cyclic process under severe reaction conditions, are the main expected characteristics. Many promising materials have been developed as potential sorbents, including CaO-based, hydrotalcite-based, and alkaline ceramic sorbents.

Advances in Chemical Engineering, Volume 51
ISSN 0065-2377
https://doi.org/10.1016/bs.ache.2017.08.001

Particular emphasis has been given to the development of different strategies aimed at improving the sorbents' stability and performance in cyclic SESR process. This chapter provides an overview of different aspects of high-temperature CO_2 sorbents and their application in sorption-enhanced reforming processes.

ABBREVIATIONS

$Al(NO_3)_3 \cdot 9H_2O$	aluminum nitrate nonahydrate
Al_2O_3	aluminum oxide
$Al(O^iPr)_3$	aluminum isopropoxide
BET	Brunauer–Emmett–Teller (nitrogen adsorption–desorption isotherms)
$CaAl_2O_4$	monocalcium aluminate
$Ca_3Al_2O_6$	calcium aluminate (β form)
$Ca_{12}Al_{14}O_{33}$	mayenite
$Ca_3(C_6H_3O_6)_2$	calcium aconitate
$CaCO_3$	calcium carbonate
$CaMg(CO_3)_2$	dolomite
$CaMg_3(CO_3)_4$	huntite
$Ca(NO_3)_2 \cdot 4H_2O$	calcium nitrate tetrahydrate
CaO	calcium oxide
CaO-Gm	Graymont limestone-based CaO
$Ca(OH)_2$	calcium hydroxide
CaO-Nf	Newfoundland limestone-based CaO
$Ca_2(SiO_4)$	lamite
$Ca_5(SiO_4)_2CO_3$	spurrite
$CaTiO_3$	calcium titanate
$CaZrO_3$	calcium zirconate
CMC	critical micelle concentration
CO_2	carbon dioxide
CRHA	rice husk ash
CTAB	cetyl trimethylammonium bromide or hexadecyltrimethylammonium bromide (synonyms)
ESR	ethanol steam reforming
FSP	flame spray pyrolysis
GSR	glycerol steam reforming
HTlc	hydrotalcite
K	temperature-dependent rate constant, cf. Eq. (2)
$K\text{-}Li_2ZrO_3$	potassium-doped lithium zirconate
La_2O_3	lanthanum oxide
LDHs	layered double hydroxides (hydrotalcite and hydrotalcite-like compounds)
Li_2CO_3	lithium carbonate
Li_2SiO_3	lithium metasilicate
Li_4SiO_4	lithium orthosilicate
Li_2ZrO_3	lithium zirconate
MgO	magnesium oxide

MMT	lamellar montmorillonite
MSR	methane steam reforming
Na_2CO_3	sodium carbonate
Na_2ZrO_3	sodium zirconate
NiO	nickel oxide
ReaxFF	reactive force field
S_0	initial specific surface area, cf. Eq. (2)
S	specific surface at the sintering time t, cf. Eq. (2)
S/C	steam to carbon ratio
SCVR	sorbent/catalyst volume ratio
SEESR	sorption-enhanced ethanol steam reforming
SEGSR	sorption-enhanced glycerol steam reforming
SEM	scanning electron microscope
SEMSR	sorption-enhanced methane steam reforming
SERP	sorption-enhanced reaction process
SESR	sorption-enhanced steam reforming
SiO_2	silicon dioxide
SrO	strontium oxide
TEA	triethanolamine
T_m	melting temperature
TiO_2	titanium dioxide
TPO	temperature program oxidation
USP	ultrasonic spray pyrolysis
WGS	water gas shift
XRD	X-ray diffraction
Y_2O_3	yttrium oxide
$Zr(OH)_4$	zirconium hydroxide
ZrO_2	zirconium oxide
γ	specific surface area reduction kinetic exponent, cf. Eq. (2)

1. INTRODUCTION

Energy and environmental issues are today major concerns; fossil fuels use increased rapidly to solve huge energy needs, leading to major negative effect on the environment. As sustainable alternatives, one can mention (i) substitute fuels (e.g., H_2 or renewable sources (e.g., solar)) and (ii) carbon capture and storage.

On one hand, carbon dioxide is considered by experts and climate researchers to have a major contribution to the global climate change. Most anthropogenic CO_2 emissions contributing to climate change are issued from fossil fuels utilization for energy production in power plants, car engines, and heating processes. Coal, oil, and natural gas provide over 80% of total industrial energy needs. In addition, various industrial processes (oil refineries, cement works, iron and steel sector, and aluminum production) annually emit huge amounts of CO_2 into the atmosphere. The concept of CO_2 capture and sequestration has recently attracted considerable interest in the field of CO_2 emission reduction. The aim of CO_2 capture is to separate the pure gas from different gaseous mixtures in order to produce a concentrated stream ready for sequestration or further use.

On the other hand, the significant reduction of fossil fuel sources and their harmful influence on the environment and human health justify the necessity of alternative energy source development with the advantages of low emission of pollutants and more energy per mass. As a promising alternative energy carrier, hydrogen has attracted considerable interest in recent research, in connection with its production, storage, and utilization. Because it does not exist freely in nature, it is generated from conventional (fossil fuels) and renewable (hydroelectric, wind, and solar power, biomass, electrolysis or photocatalysis of water) sources. Nowadays it is still mainly (~95%) produced by reforming processes from nonrenewable resources, namely coal, petroleum, and natural gas. Among them, methane steam reforming is the most common process, due to the highest thermal efficiency and the lowest capital investment, despite its high energy consumption. However, the synthesis of hydrogen is accompanied by the generation of important amounts of CO_2 as by-product along with some impurities like CO and unconverted methane (99 wt% CO_2 of the total air emissions). 84% of these emissions are released at the hydrogen plant, corresponding to 10.62 kg CO_2 (vs 59.8 g CH_4 and 5.7 g CO) for each kg of H_2 produced in a typical reforming plant of 1.5 million Nm^3/day (Spath and Mann, 2000). Hydrogen purification can increase significantly the cost and complexity of the global process, depending on the application requirements. Considerable efforts have therefore been focussed in the last decades on the reduction of the production cost of highly pure hydrogen and the improvement of production efficiency.

Coupling the reforming reaction system (hydrocarbon reforming and water gas shift) with in situ CO_2 separation using sorbents (sorption-enhanced reaction process, SERP) results in an intensified process whose compelling advantages have been widely reported in the literature (Harrison, 2008; Wu et al., 2016). These hybrid configurations based on the Le Chatelier's principle may lead to significant improvement of the reactant conversion close to equilibrium, as well as the product selectivity. The saturated sorbent can be regenerated by heat treatment (calcination) to produce very pure CO_2 for sequestration or further use. Catalytic conversion of CO_2 into high value-added products like hydrocarbons, urea, salicylic acid, polycarbonates, methanol, and dimethyl ether is a promising and efficient alternative to CO_2 sequestration by chemical recycling (Centi and Perathoner, 2009; Iliuta et al., 2011).

The key element for a successful SESR process is the appropriate selection of high-temperature sorbents. The choice of suitable sorbents for selective in situ CO_2 removal from the reaction medium is a great challenge due to the severe operating conditions, especially the high temperature and the presence of steam during the sorption process. Special attention has been given to the development of efficient high-temperature CO_2 sorbents as well as the evaluation of their performance in different SESR processes (Barelli et al., 2008; Harrison, 2008; Shokrollahi Yancheshmeh et al., 2016a). High sorption capacity and kinetics (at relatively low CO_2 partial pressures), together with a good stability in long-term cyclic process under severe reaction conditions, and a reasonable production cost are the main expected characteristics.

Many promising materials have been developed as potential sorbents, including CaO-based, hydrotalcite-based, and alkaline ceramic sorbents (Dou et al., 2016b; Shokrollahi Yancheshmeh et al., 2016a; Wu et al., 2016). A particular emphasis has been given to the development of different strategies aimed at improving the sorbents' stability and performance in cyclic SESR process. To provide high and stable CO_2 capture capacity and appropriate sorption kinetics, the sorbent materials have to present a meso- and macroporous structure with high specific surface area, small particle size, and large pore volume. However, keeping these characteristics in long-term cyclic sorption/desorption (regeneration) operation, as required in industrial applications, is a huge challenge. As demonstrated in all available investigations, the deactivation of sorbents (synthetic or derived from natural sources) is inevitable. Depending on the category of sorbents, many different strategies have therefore been proposed in hundreds of articles to improve the sorbents' performance such as (i) the use of various precursors, (ii) development of improved synthesis techniques, (iii) incorporation of stable inert materials into the sorbent structure, (iv) reactivation and thermal treatment, and (v) addition of dopants. Particular attention has also been devoted to the evaluation of their performance in different SESR processes and the optimization of the operation parameters.

This chapter focusses on various aspects of high-temperature CO_2 sorbents and their application in different sorption-enhanced reforming processes for hydrogen production.

2. HIGH-TEMPERATURE CO_2 SOLID SORBENTS

2.1 CaO-Based Materials

Among the high-temperature CO_2 sorbent materials, CaO-based sorbents have been the most investigated due to the high sorption capacity (theoretical: 0.786 g of CO_2/g of CaO), adequate kinetics, and low cost:

$$\mathrm{CaO(s) + CO_2(g) \leftrightarrow CaCO_3(s)} \quad \Delta H^\circ_{298} = -175.7\,\mathrm{kJ/mol} \tag{1}$$

CaO-based sorbents have been widely applied for CO_2 capture in different processes. However, the weakness of thermal stability of CaO results in the reduction of sorption capacity in time. This loss of CO_2 sorption capacity during cyclic operation mainly results from an increase in sintering with each performed cycle, which leads to a progressive reduction of sorbent porosity and specific surface area due to the coalescence of micrograins produced by thermal decomposition of particles of calcium carbonate or hydroxide, changes in pore shapes, increases in pores size, and pore shrinkage. Because these modifications in the material surface area and porosity strongly affect its reactivity in gas–solid processes, the sintering mechanism and kinetics made the object of extensive studies. A two-sphere model of sintering particles (Fig. 1) was used by German and Munir (1976) to develop a kinetic equation for the surface area reduction during sintering, $\Delta S = S_0 - S$:

$$(\Delta S/S_0)^\gamma = K \cdot t \tag{2}$$

where, S_0 is the initial specific surface area, S is the specific surface area at the sintering time t, K is the temperature-dependent rate constant (time^{-1}), and the coefficient γ is related to the sintering mechanism through particle coordination for monosized spheres (parameter dependent on the mechanism of sintering and powder packing). Based on the distinctive value of γ for a specific mechanism, the proposed model offers useful tools in determining the

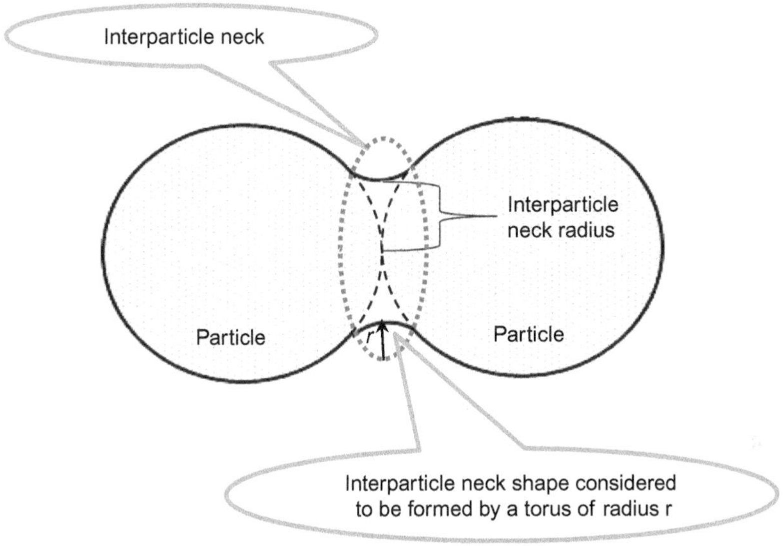

Fig. 1 Two-sphere model of sintering particles. *Adapted from German RM, Munir ZA: Surface-area reduction during isothermal sintering,* J Am Ceram Soc *59:379–383, 1976.*

dominant mass transport mechanism during sintering: lattice diffusion (2.7), grain boundary diffusion (3.3), and surface diffusion (3.5).

Conform to the model illustrated in Fig. 1, the connecting neck—assumed to be formed during the sintering process in each point of contact of adjacent particles—grows in diameter during the time (German and Munir, 1976). The approximation that the interparticle neck shape is formed by a torus is assumed valid for small neck sizes. ΔS (determined by BET analysis) depends on neck diameter and initial contact points between particles. The neck growth by mass transfer together with particle fusion results in a reduction of surface area.

By applying the model in a study on the sintering rate of CaO (produced from ultrapure $CaCO_3$, limestone, or $Ca(OH)_2$) in an inert (N_2) atmosphere (to exclude the catalytic effects of CO_2 and water) and temperatures of 700–1100°C, it was revealed that the rate of surface reduction was independent of particle size between 2 and 20 μm (Borgwardt, 1989b). However, it was found to be significantly influenced by temperature and the presence of impurities. The formation of micropores within grain clusters and a macropore network between clusters (in the early stage of the sintering process) is followed by a reduction of porosity with the coalescence of fused clusters and the elimination of macropores. The increase in temperature results in a rapid intensification of the sintering rate and the sintering rate constant (K) follows the Arrhenius relationship. For CaO prepared from pure $CaCO_3$, limestone or $Ca(OH)_2$ a mean γ value of 2.7 confirms the mechanism of lattice diffusion. Concerning the rates of sintering, the results were found to be different for the three precursors investigated. In comparison with limestone, the lower sintering rate and the higher activation energy of CaO prepared from pure $CaCO_3$ are consistent with the lattice diffusion mechanism. The increased sintering behavior of limestone-based CaO is attributed to the defects caused by impurities (especially aliovalent ions), which result in the acceleration of lattice diffusion. The behavior of $Ca(OH)_2$-based CaO was similar to that of limestone, partly on account of the presence of impurities, but to a greater extent due to the lower porosity, 0.40 compared to 0.48 (more points of contact between particles for denser structures). In a further study, the sintering phenomenon was found to be accelerated in the presence of both CO_2 and water, with corresponding increase of K and γ (Borgwardt, 1989a). As pointed out above, higher γ values suggest that other diffusion mechanisms prevail. Besides, Lysikov et al. (2007) reported that temperature, type of precursor, and carbonation duration in cyclic operation strongly affect the decline of CaO uptake capacity in cyclic operation. Through a reactive molecular dynamics study with the reactive force field (ReaxFF) performed in NVE ensembles, Zhang et al. (2012) revealed that the expansion occurring during CO_2 sorption is the source of CaO sintering and the process is strongly affected by temperature and particle separation distance (the higher the sorption temperature and the shorter the distance between two sorbent particles, the faster the sintering rate during the sorption process).

During CaO carbonation by CO_2 chemisorption on CaO particles, a first step (fast regime) controlled by chemical reaction at the CaO–$CaCO_3$ interface is followed by a second one (slow regime) controlled by ion diffusion through the carbonate ($CaCO_3$) product layer which covers most of the surface of CaO particles (Bhatia and Perlmutter, 1983; Dou et al., 2010b; Li et al., 2012, 2014; Mohammadi et al., 2014; Rout et al., 2014; Sun et al., 2012). The fast step lasts until completing the carbonate layer surrounding the unreacted CaO core. Then, the slow kinetic step that is controlled by gas diffusion starts. During this step, the product layer restricts the access of CO_2 molecules to reactive sites. Alvarez and Abanades (2005a) reported that the gas diffusion and thus the sorption rate can be limited above a critical carbonate layer thickness of 50 nm.

CaO can be derived from natural and abundant sources (such as limestone, dolomite, and huntite) or synthesized from different laboratory-grade Ca-derivative precursors, like $CaCO_3$ (Gupta and Fan, 2002; Hsu et al., 2014; Nieto-Sanchez et al., 2013), $Ca(OH)_2$ (Hsu et al., 2014; Materic et al., 2014b; Wu et al., 2007), organometallic (Grasa et al., 2007; Liu et al., 2010b; Lu et al., 2006, 2008, 2009b; Silaban et al., 1992; Yang et al., 2013a), and nanosized $CaO/CaCO_3$ (Barker, 1974; Florin and Harris, 2008, 2009; Liu et al., 2010b; Lu et al., 2009a,b; Luo et al., 2013; Santos et al., 2012).

2.1.1 Limestone: CaO Natural Sorbent

Limestone, a sedimentary rock, is the most abundant and cheapest natural source of CaO. Except for CaO which is the major component of the natural rock, depending on the source the limestone may usually contain MgO, Na_2O, K_2O, SiO_2, Al_2O_3, Fe_2O_3, and TiO_2. The composition of limestones from various sources is an essential issue to be considered in sorbent synthesis. On one hand, limestone rocks used for producing CaO sorbents can contain different mineral impurities with a negative effect on both CO_2 sorption capacity and thermal stability of prepared CaO absorbent. For example, the presence of sodium ion (as main impurity) in the high-purity La Blanca (Spanish) limestone (1.07% Na_2O) resulted in poor thermal stability compared with other kinds of natural limestones, even after thermal pretreatment (Section 2.1.3.1.2; Alvarez and Abanades, 2005b; Grasa and Abanades, 2006; Manovic et al., 2008). The presence of NaCl in another high-purity natural limestone from Newfoundland (Canada), even in a relatively small amount (3.36 wt.%), results in a much smaller uptake capacity than expected, even at a sorption temperature of 700°C (Radfarnia and Iliuta, 2013a). It was already demonstrated that the presence of sodium ions into the CaO lattice resulted in the enhancement of mass transfer and sintering phenomenon (Borgwardt, 1989b). SiO_2 is another impurity present in some limestone rocks (Havelock (Canada), 1.07 wt.% and Katowice (Poland), 0.85 wt.%) which can negatively affect the sorbent activity in cyclic operation, on account of local melting on the particle surface and change in particle morphology, which become more pronounced in realistic calcination conditions (higher temperature (950°C) and CO_2 concentration (100%) (Manovic et al., 2009). On the other hand, the almost complete absence of Al (<0.1 wt.% Al_2O_3) in La Blanca limestone has a negative effect during the thermal pretreatment for sorbent self-reactivation during the cyclic operation (Manovic and Anthony, 2008).

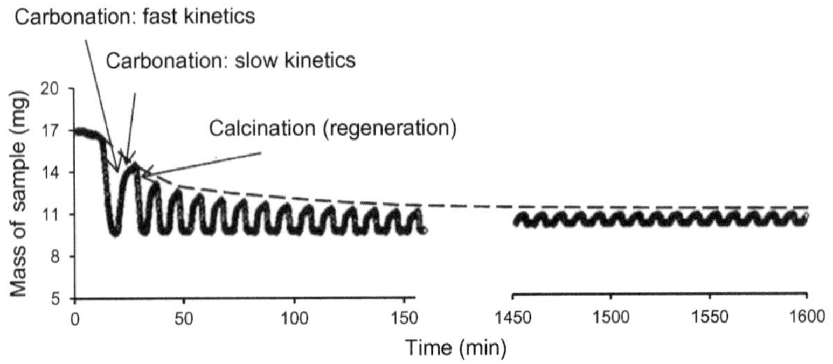

Fig. 2 Effect of sintering on CaO sorbent in cyclic (160 cycles) carbonation/calcination process. *Adapted from Grasa GS, Abanades JC: CO_2 capture capacity of CaO in long series of carbonation/calcination cycles,* Ind Eng Chem Res *45:8846–8851, 2006.*

Despite the appropriate initial capacity and kinetics of fresh material, this sorbent is affected by loss of CO_2 uptake capacity in cyclic operation, which is an important drawback in industrial applications (Abanades, 2002; Sun et al., 2007, 2008). It is therefore not surprising that the CO_2 sorption behavior of limestone was intensively studied in long-term cyclic operation. Capture tests on Piaseck (Polish) limestone during 500 carbonation/calcination cycles resulted in a significant loss of CO_2 uptake capacity during the first 20 cycles followed by a stabilization to around 0.075–0.08 residual conversion (Grasa and Abanades, 2006; Fig. 2). Similarly, Sun et al. (2008) studied the behavior of limestone over more than 1000 cycles. After 150 carbonation/calcination cycles, the conversion can decrease up to 4%–17%, depending on the carbonation time. Silaban et al. (1996) also noticed a reduction of limestone capacity from 61% to 35% of its theoretical value after six cycles only. The improvement of limestone sorption efficiency in cyclic operation to increase its reversibility in carbonation/decarbonation reaction and reduce the sintering process is not only a significant but a very difficult challenge. In this appealing research area, different promising approaches can be applied, including hydration, thermal pretreatment, recarbonation, and chemical pretreatment (Table 1).

2.1.2 Synthetic CaO Sorbents

2.1.2.1 Organometallic Precursors

The possibility to synthesize CaO from organometallic precursors like calcium acetate hydrate, calcium citrate tetrahydrate, calcium D-gluconate monohydrate, calcium formate, and calcium L-lactate hydrate has been

Table 1 Brief Summary of Some Investigations on Limestone Treatment

Treatment Type	Treatment Conditions	CO_2 Capacity Test: Carbonation/Regeneration Conditions	Carbonation/Regeneration Duration (min)	Cycle Number	CO_2 Uptake at Last Cycle (g CO_2/g ads)	References
Hydration	Steam	700°C, 15% CO_2/850°C, 100% N_2	30/10	10	0.55	Manovic and Anthony (2007)
		650°C, 15% CO_2/940°C, 70% CO_2	15/20	5	0.37	Coppola et al. (2014)
		650°C, 15% CO_2/925°C, 60% CO_2, 15% H_2O	20/5	15	~0.11	Champagne et al. (2013)
		650°C, 15% CO_2, 15% H_2O/925°C, 60% CO2, 40% H_2O	20/5	15	~0.16	Champagne et al. (2013)
		650°C, 15% CO_2/925°C, 60% CO_2, 15% H_2O	20/5	15	~0.11	Champagne et al. (2013)
Thermal pretreatment	Preheat[a]	800°C, 50% CO_2/800°C, 100% N_2	30/10	30	0.39	Manovic and Anthony (2008)
		650°C, 15% CO_2/850°C, 100% air	5/5	50	0.27	Valverde et al. (2014b)
Chemical pretreatment	Acetic acid	650°C, 15% CO_2/920°C, 100% CO_2	20/10	20	0.09	Ridha et al. (2013b)
		650°C, 15% CO_2/850°C, 100% N_2	20/5	20	0.23	Ridha et al. (2013a)
	Vinegar	650°C, 15% CO_2/850°C, 100% N_2	20/5	20	0.15	Ridha et al. (2013a)

Continued

Table 1 Brief Summary of Some Investigations on Limestone Treatment—cont'd

		CO_2 Capacity Test				
		Carbonation/Regeneration				
Treatment Type	Treatment Conditions	Conditions	Duration (min)	Cycle Number	CO_2 Uptake at Last Cycle (g CO_2/g ads)	References
	Formic acid	650°C, 15% CO_2/850°C, 100% N_2	20/5	20	0.22	Ridha et al. (2013a)
	Oxalic acid	650°C, 15% CO_2/850°C, 100% N_2	20/5	20	0.25	Ridha et al. (2013a)
	10 wt.% aluminate cement, 10% starch	700°C, 15% CO_2 in air/ 900°C, 100% air	30/20	10	~0.65	Ridha et al. (2013a)

[a]Preheat (950°C, 100% air, 12 h)/recarbonation (850°C, 90% CO_2, 3 min).

investigated with the aim of developing sorbents possessing meso- and macroporous structures with large surface area which could lead to the improvement of sorption performance (Grasa et al., 2007; Liu et al., 2010b; Lu et al., 2006, 2008, 2009b; Silaban et al., 1992). In general, materials prepared using calcium acetate and calcium D-gluconate give the best results for CO_2 sorption in cyclic operation (Grasa et al., 2007; Liu et al., 2010b; Lu et al., 2006). The corresponding synthetic sorbents possess a fluffy structure with high specific surface area and large pore volume. While the synthetic sorbents perform well under mild reaction conditions, a dramatic decay in the capture capacity occurs when the cyclic CO_2 sorption is performed in realistic calcination conditions (regeneration under CO_2 atmosphere at around 900°C) (Grasa et al., 2007), a similar behavior noticed in the case of limestones. Even though the sorbent derived from calcium acetate exhibits a slightly higher sorption capacity at the end of cyclic operation compared to limestone, it does not justify the use of costly chemical precursors.

2.1.2.2 Nanosized CaO and $CaCO_3$ Precursors

The use of nanosized CaO and $CaCO_3$ as calcium precursors is worth investigating with the aim to produce nanosized sorbents with critical particle size less than 44 nm or large single CaO crystal with a critical crystal size of less than 220 nm for which the carbonation reaction completes within the fast kinetically controlled regime (no slow solid-state diffusion-controlled regime) (Abanades and Alvarez, 2003; Barker, 1974). The capture performance of CaO nanosized sorbents is highly enhanced compared to microsized CaO. As an example, the structure of CaO nanosized sorbents developed by the sol–gel method from calcium nitrate tetrahydrate and citric acid monohydrate precursors presents well-dispersed and uniform particles (200 nm). The sorption capacity over 20 cycles in both mild and severe calcination conditions (0.51 and 0.20 g of CO_2/g of sorbent) is significantly higher compared to CaO sorbents fabricated from commercial micro- and nanosized $CaCO_3$. Moreover, the material shows very high carbonation rate (60% calcium conversion ratio within 20 s) and better resistance to sintering.

However, as drawbacks can practically never be avoided, the application of these nanosized sorbents mainly reveals the manifestation of sintering because of their very high surface area (Florin and Harris, 2009). The cost and complexity of the methods applied for their synthesis (sol–gel and flame spray pyrolysis (FSP)) is a matter of concern as well.

2.1.3 Improvement of the Performance of CaO-Based Sorbents

The improvement of material stability which is the general main drawback of this kind of high-temperature solid sorbents remains a significant challenge. Their cyclic stability in long-time operation still needs to be much improved to make them appropriate candidates for different SESR processes. Different strategies have therefore been investigated to improve the stability of CaO-based sorbents.

2.1.3.1 Treatment of Natural and Synthetic Sorbents

2.1.3.1.1 Hydration Sorbent reactivation by hydration (steam or water addition) can be performed during carbonation, calcination, or as a separate treatment on $CaO/CaCO_3$, depending on the stage where steam or water is introduced in the system.

Steam addition during carbonation is known to affect positively the carbonation conversion. The role of steam and the mechanism by which the steam affects the carbonation reaction raise several questions. The increase of CO_2 sorption capacity can, for example, be attributed to the increase in the carbonation rate of the fast kinetically controlled regime (Dou et al., 2010b; Symonds et al., 2009; Wang et al., 2014c; Yang and Xiao, 2008). Different possible explanations tried to elucidate this behavior: (1) the formation of $Ca(OH)_2$ as transient intermediate whose carbonation is thermodynamically more favorable compared to CaO (Dou et al., 2010b; Symonds et al., 2009; Wang et al., 2014c); (2) the catalytic effect of steam by which hydroxyl groups and bicarbonates can play the main role (Yang and Xiao, 2008); this is based on the observation that the steam promotes CaO carbonation conversion remarkably well with or without the formation of $Ca(OH)_2$ as intermediate in the reaction medium. Another theory trying to elucidate the positive effect of steam is based on the improvement of solid-state diffusion into the product ($CaCO_3$) layer. Given the observation that the presence of steam during carbonation does not affect significantly the CO_2 capture rate in the fast kinetic regime, modifications in the porous structure and/or the acceleration of solid-state diffusion in the product layer could explain the enhancement in the carbonation conversion (Arias et al., 2012a; Blamey et al., 2016; Broda et al., 2014; Gonzalez et al., 2016; Li et al., 2014; Linden et al., 2011; Manovic and Anthony, 2010; Qin et al., 2013; Rong et al., 2013; Wang et al., 2013; Wu et al., 2007; Yang et al., 2013b). This enhancement of solid-state diffusion was found to be more pronounced at lower carbonation temperature and for more sintered sorbents (Broda et al., 2014; Manovic and Anthony, 2010), which confirms the ability of steam to promote the carbonation by the enhancement of solid-state diffusion in the product layer. According to Broda et al. (2014), the enhancement of the solid-state diffusion in the product layer could be due to the formation of OH^- groups. Based on the hypothesis that the carbonation can be controlled by the (i) CO_3^{2-} diffusion from the interface gas/$CaCO_3$ to $CaCO_3$/CaO interface and (ii) O^{2-} diffusion from the interface CaO/$CaCO_3$ to gas/$CaCO_3$ interface, OH^- groups can be formed in the presence of steam, through H^+–O^{2-} interactions. Due to the higher diffusion rate of OH^- from $CaCO_3$/CaO interface to $CaCO_3$/gas interface compared to O^{2-}, the steam helped promote the solid-state diffusion in the product layer, thus enhancing the overall carbonation rate. A combination of steam catalysis and solid-state diffusion enhancement in the $CaCO_3$ layer could also explain the positive effect of steam injection during the carbonation step (Donat et al., 2012).

The *effect of steam addition during calcination* did not reach a general consensus yet. Some investigations concluded that the effect was positive. For example, a study on the reactivation of two Canadian limestones (Cadomin and Havelock) by steam injection (up to 40 vol.%) during the calcination step over 15 carbonation/calcination cycles (Champagne et al., 2013) revealed an increase of the carbonation reactivity of sorbents at all steam concentrations investigated. The explanation of the increase of carbonation conversion in these conditions is founded on the formation of larger pore diameter and lower specific surface area via enhancing particle sintering in the presence of steam. It is known that the reduction of the specific surface area of sorbents results in a decrease of the capture capacity; however, the diffusional resistance can be reduced due to the formation of larger pores during the process of sintering. On the contrary, Rong et al. (2013) concluded in cyclic experiments on CaO sorbent prepared by calcining reagent-grade $CaCO_3$ that the injection of steam of different concentrations between 20 and 40 vol.% during calcination resulted in a decrease in CaO conversion over 10 carbonation/calcination cycles of 2%–3% with respect to the case without steam reactivation. The negative effect of the hydration treatment during calcination on the sorbent activity is probably due to a loss of porosity in the presence of steam during the calcination step. A comparison between the hydration treatment throughout carbonation or calcination revealed that the use of steam during the carbonation step results in a much larger increase in the sorbent carbonation capacity compared to the treatment performed during calcination (Champagne et al., 2013).

Steam reactivation of spent CaO-based sorbents by *separate hydration after calcination* is another possible approach which can lead to excellent performance (Coppola et al., 2014; Manovic and Anthony, 2007; Martinez et al., 2011; Materic et al., 2014a; Rong et al., 2013; Sun et al., 2013b; Wang et al., 2008; Zeman, 2008). Compared to the steam addition during carbonation or calcination, a separate hydration treatment under low temperature performed after calcination can recover to a great extend the porosity (pore volume) and specific surface area of the spent sorbent. This is due to the increase of molar volume associated with the production of calcium hydroxide ($CaO \rightarrow Ca(OH)_2$) in the hydration process which results in the breaking of the agglomerates present in the spent sorbent structure. As an example, the steam reactivation of spent limestone at 200°C can result in a considerable intensification of CO_2 capture from 0.27–0.31 g of CO_2/g of original (fresh) sorbent to about 0.55 g of CO_2/g of sorbent after 10 cycles carbonation/calcination (Manovic and Anthony, 2007). Moreover, both reactivity and durability of CaO-based spent sorbents can be recovered significantly by separate hydration after calcination (Rong et al., 2013).

Several operational parameters including the concentration of steam and the hydration temperature and frequency have an influence on the reactivation of spent sorbent (Rong et al., 2013):

1. High steam concentrations result in a better reactivation performance by increasing the ultimate conversion of the fast kinetically controlled step of the carbonation process.
2. The efficiency is more significant at lower hydration temperatures. Due to severe $Ca(OH)_2$ sintering, performing the hydration at high temperatures results in a decrease in the carbonation conversion in time (cyclic operation).
3. Because of the difficulty to limit the sintering of $Ca(OH)_2$-derived CaO sorbent, periodic steam hydration (every three carbonation/calcination cycles) or one single treatment does not result in an adequate reactivation performance during cyclic CO_2 capture. However, performing the hydration after every calcination step limits significantly the fast decay of the capture capacity of CaO sorbent during 10 carbonation/calcination cycles.

Although the hydration of spent limestone recovers the capture capacity, it leads to a considerably increase of the attrition tendency. Partial hydration and pelletization have been proposed as alternatives to decrease the negative impact of steam on the mechanical strength of reactivated material, improve the sorbent mechanical strength, and reduce the cost of the process (Broda et al., 2014; Martinez et al., 2011; Materic et al., 2014a).

Although most hydration treatments concern different kinds of limestone, they can equally be applied to synthetic CaO materials. The influence of the addition of various concentrations of steam (up to 9.5 vol.%) on the sorption efficiency of an Al-stabilized CaO-based synthetic sorbent containing 78 wt.% CaO and 22 wt.% $Ca_9Al_6O_{18}$ (sorption at 550–650°C using 15 vol.% CO_2; calcination at 800°C using 100 vol.% N_2) was investigated in either carbonation or calcination steps (Shokrollahi Yancheshmeh et al., 2016b).

As in the case of the investigations on CaO-based materials (mostly limestone) discussed earlier, the steam addition during the carbonation step has a positive effect. The sorbent reactivity increases with the increase of concentrations of steam injected (2.3–9.5 vol.%) (Fig. 3A). However, a significant increase occurs between 0 and 2.3 vol.%. Similar observations were reported for both limestones (different sources) and synthetic sorbent (Manovic and Anthony, 2010). The presence of steam during carbonation promotes the CO_2 sorption by steam catalysis and improvement of solid-state diffusion into the $CaCO_3$ layer (extension of the fast reaction step). First, the increase in the rate of the fast kinetically controlled step by steam addition (Fig. 3B) can explain a catalytic effect of the steam, rather than the formation of $Ca(OH)_2$ which is insignificant in the process conditions (Symonds et al., 2009; Yang and Xiao, 2008). Indeed, the formation of hydroxyl groups by water dissociation on the basic CaO is accelerated in the presence of steam and can lead to the production of bicarbonates by the reaction with CO_2 (Bandi et al., 2002; Yang and Xiao, 2008). Consequently, the formation of bicarbonate as intermediates enhances the CO_2 capture by their conversion into $CaCO_3$ through the carbonation reaction. Second, the extension of the fast reaction step in the presence of steam clearly shows an acceleration of the solid-state diffusion into the product ($CaCO_3$) layer. This hypothesis is also supported by BET and SEM analysis. The sorbent-specific surface area and pore volume show a significant increase when the carbonation is performed in the presence of 9.5 vol.% steam, due to the ability of steam to increase the solid-state diffusion in the product layer, which results in a structural rearrangement of the porous structure. This behavior is in agreement with the changes in the pore size distribution; the increase in pore volume after 15 carbonation/calcination cycles due

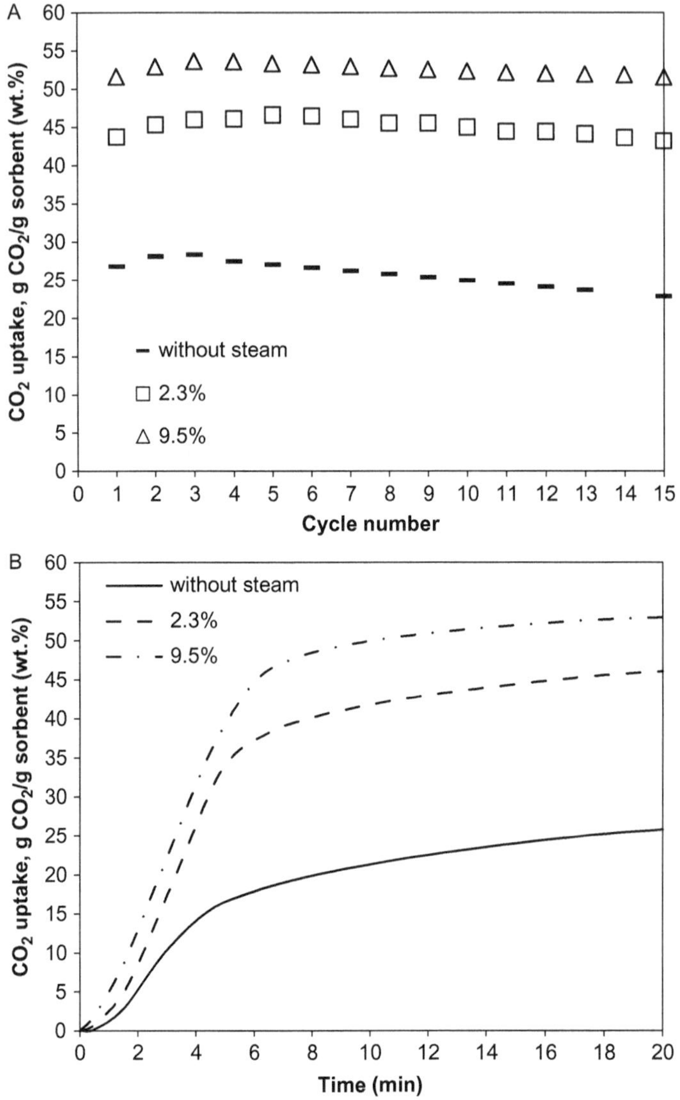

Fig. 3 Effect of steam addition during carbonation (sorption at 650°C/15 vol.% CO_2; desorption at 800°C/100 vol.% N_2): (A) cyclic behavior; (B) example for the eighth cycle. *Data from Shokrollahi Yancheshmeh M, Radfarnia HR, Iliuta MC: Influence of steam addition during carbonation or calcination on the CO_2 capture performance of $Ca_9Al_6O_{18}$-CaO sorbent,* J Nat Gas Sci Eng *36:1062–1069, 2016.*

to the material structural rearrangement after cyclic operation is more evident in the presence of steam during the carbonation step. Similar conclusions on the influence of steam during carbonation were reported by Manovic and Anthony (2010) and Donat et al. (2012).

By contrast with the effect of steam on the carbonation process, the steam concentration in the feed influences significantly the CO_2 capture capacity when steam is introduced in the calcination step. Thus, while 2.3 vol.% has a negative effect all over the 15 cycles, an increase of the capture capacity occurs for a concentration of steam of 9.5 vol.% during the nine initial cycles (Fig. 4).

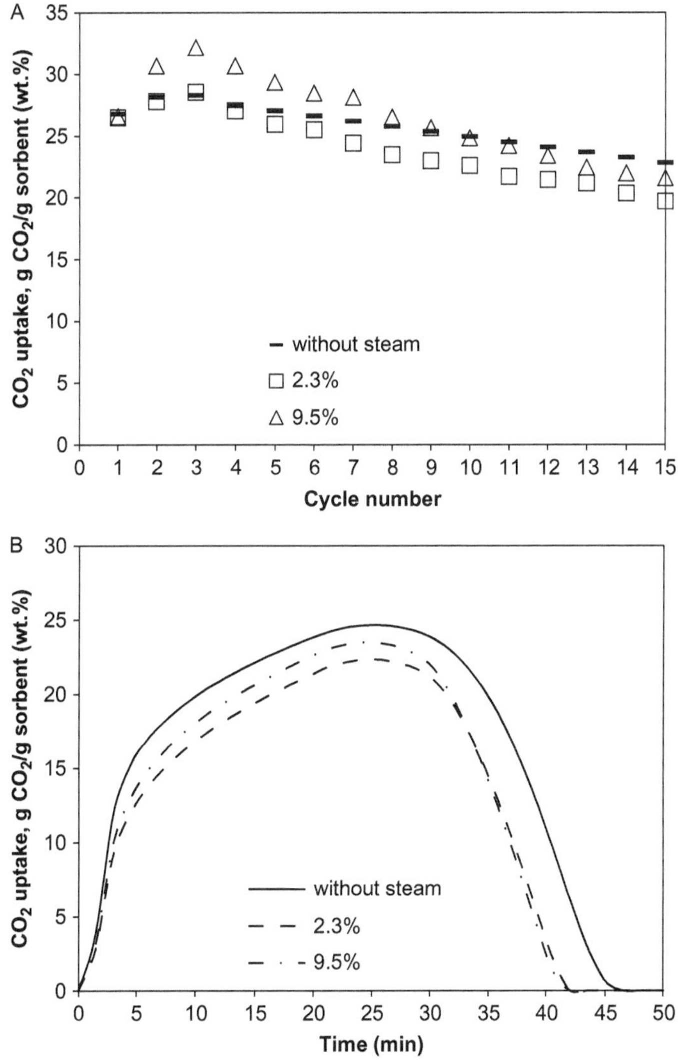

Fig. 4 Effect of steam addition during calcination (sorption at 650°C/15 vol.% CO_2; desorption at 800°C; steam balanced with N_2): (A) cyclic behavior; (B) example for the 13th cycle. *Data from Shokrollahi Yancheshmeh M, Radfarnia HR, Iliuta MC: Influence of steam addition during carbonation or calcination on the CO_2 capture performance of $Ca_9Al_6O_{18}$-CaO sorbent, J Nat Gas Sci Eng 36:1062–1069, 2016.*

In fact, the presence of steam in the calcination process does not influence the carbonation rate of the fast kinetically controlled step, but the carbonation rate for the slow diffusion-limited step. Morphological changes occurring throughout the calcination process in the presence of steam can explain this behavior. As the presence of steam in calcination intensifies the particle sintering and produces larger pore diameter and lower specific surface area (Borgwardt, 1989a; Donat et al., 2012), the uptake capacity can be enhanced as long as the positive effect of larger pores can compensate the negative effect of lower surface area. Indeed, a low steam concentration of 2.3 vol.% results in a reduction of the specific surface area (localized sintering), but seems not to be sufficient to develop a porous structure with larger pores, and consequently, a reduction of the capture capacity can be observed. A larger steam concentration of 9.5 vol.% seems appropriate to promote the formation of larger pores during the first nine cycles, after which the gradual increase of sintering reaches a limit where the positive effect of the increase of CO_2 diffusion through large pores could not compensate the negative effect of the decrease of specific surface area. Overall, regardless of the steam concentration, the presence of steam during calcination enhances the CaO sintering, accelerating the deactivation of sorbents in cyclic carbonation–calcination operation.

2.1.3.1.2 Thermal Pretreatment and Recarbonation The aim of the *thermal pretreatment* is the stabilization of sorbent structure and its activation in cyclic carbonation/calcination operation. Manovic and Anthony (2008) investigated the thermal activation of four Canadian limestones (Kelly Rock, Cadomin (formerly called Luscar), Graymont, and Havelock) by performing different pretreatments at 800–1300°C between 6 and 48 h. After thermal pretreatment, the sorbents' performance was improved, showing better conversion stability in cyclic operation compared to natural limestones. An increase of sorbent conversion with the increase of the number of calcination/carbonation cycles was also observed in some conditions, especially for powdered sorbents pretreated at 1000°C.

The proposed pore-skeleton model describes this self-reactivation effect (Fig. 5) by morphologic changes in the material (modification of pore size distribution and transformation of smaller pores into large ones) as a result of expanding and shrinking of material structure in cyclic operation (continuous formation and decomposition of $CaCO_3$).

Also, the model suggests that the changes in the pore skeleton produced during the thermal pretreatment strongly influence the subsequent

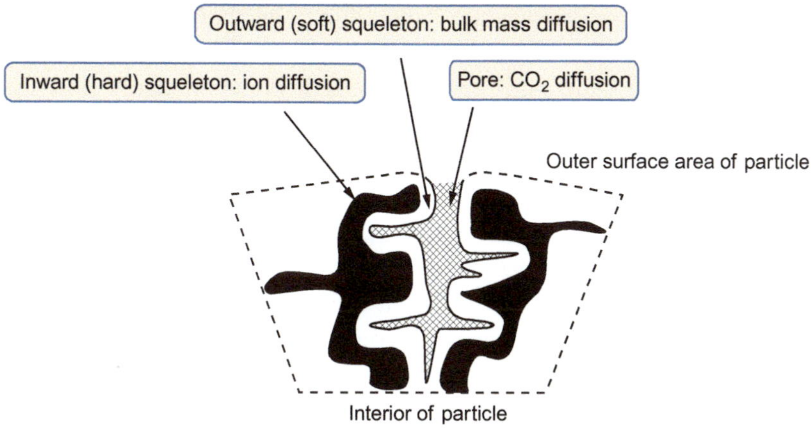

Fig. 5 Pore-skeleton model. *Adapted from Manovic V, Anthony EJ: Thermal activation of CaO-based sorbent and self-reactivation during CO_2 capture looping cycles,* Environ Sci Technol *42:4170–4174, 2008.*

carbonation/calcination cycle. During carbonation/calcination cycling, a competition between the bulk mass transfer associated to the formation/decomposition of $CaCO_3$ and ion diffusion in the crystal structure of CaO occurs and two types of structure are formed: an internal unreacted structure (inward (hard) skeleton) which stabilizes and protects the particle morphology and an external structure (outward (soft) skeleton) that changes constantly during the formation/decomposition of $CaCO_3$. In the period of thermal pretreatment, the bulk diffusion occurs only for the duration of calcination and ion diffusion continues, when the calcination of $CaCO_3$ is finished, and leads to the stabilization of the skeletal structure and the formation of a hard skeleton. In the initial carbonation/calcination cycles, CO_2 sorption capacity is reduced because the hard skeleton is less reactive and the ion diffusion is more difficult in this stable hard skeleton structure. In the succeeding carbonation/calcination cycles, the soft skeleton is formed and the carbonation rate is accelerated. The continuous growth of the soft skeleton and the decline of the hard structure lead to the enhancement of sorbent activity until the stabilization takes place between skeleton changeovers. Conversion of the hard skeleton formed during the thermal pretreatment into a soft skeleton during the carbonation/calcination cyclic operation results in an increase of CO_2 sorption of sorbent by facilitating the mass transfer.

The self-activation phenomenon was later reported in several works. It was, for example, noticed in the behavior of two kinds of CaO natural sources (Strassburg limestone and Arctic dolomite) thermally pretreated at 1000°C and 1100°C, respectively, in cyclic operation (1000 cycles) (Chen et al., 2009b) and on the treated Newfoundland limestone (Radfarnia and Iliuta, 2013a). However, this behavior may not occur under typical conditions of a circulating fluidized-bed carbonator, where the reaction time is limited to few minutes at low CO_2 partial pressures (below 10 kPa) and around 650°C (Arias et al., 2011).

The conversion of thermal pretreated CaO sorbent can also be explained by a balance between the enhancement of the specific surface area due to improved solid-state diffusion carbonation and the reduction of surface area of restored CaO structure due to sintering during calcination (Sanchez-Jimenez et al., 2014; Valverde et al., 2013). The application of harsh calcination conditions of temperature and duration (e.g., above 900°C for 5 min or 850°C for 15 min) prevents the self-reactivation of thermally pretreated sorbent by accelerating the sintering of soft skeleton and reducing CaO conversion in the fast kinetically controlled phase. A slower decline of the carbonation rate in the diffusion-controlled step and an increase of conversion in the fast kinetically controlled step can be obtained by grinding the sorbent prior to thermal pretreatment (Sanchez-Jimenez et al., 2014). A study on raw and preground limestone (without thermal pretreatment) demonstrated that the conversion of carbonation in the stage controlled by solid-state diffusion can be significantly improved by pregrinding attributed to local high stresses induced by grinding which surpass the cohesion forces between the lattice atoms and lead to crystal cracking and enhancement of structural defects. This process facilitates the CO_2 diffusion through the solid. The limestone heat pretreatment performed in a CO_2 atmosphere resulted in a fast calcination along with the formation of a CaO skeleton with a significantly high reactivity in the kinetically controlled phase and high resistance to solid-state diffusion; this behavior was attributed to annealing of the crystal structure (Liu and Evans, 1997).

Another option is to include *recarbonation* between the carbonation and calcination steps at high temperature and CO_2 partial pressure to increase the CO_2 capture capacity of CaO-based sorbents in cyclic operation (Arias et al., 2012b), based on a forced increase of CaO conversion slightly above the "maximum carrying capacity" usually obtained by the end of the fast carbonation step under normal conditions. For example, short recarbonation periods of 100–200 s performed on partially carbonated sorbents resulted in the stabilization of capture capacities at 0.15–0.20 molar conversion (Grasa et al., 2014). The carbonation reaction in the recarbonation step mainly occurs in the slow diffusion-controlled reaction phase and is strongly favored by high recarbonation temperatures of 750–800°C and CO_2 partial pressures higher than 60 kPa, in the presence of steam.

Combining thermal pretreatment and recarbonation processes was also found to result in an improvement of the conversion of limestone-based CaO sorbent in cyclic operation (Valverde et al., 2014b). The synergy between these two processes is reflected in a high and stable conversion of CaO sorbent in the cyclic sorption/desorption process due to a significant enhancement of solid-state diffusion during recarbonation of the pretreated limestone and consequently that leads to a thermally stable and reactive CaO structure after calcination. However, when this combining procedure is applied under realistic calcination conditions (high CO_2 partial pressure in the calcination step) (Valverde et al., 2014a), the results are in contradiction with those obtained by performing the regeneration under low CO_2 partial pressures (Valverde et al., 2014b). This behavior suggests that either recarbonation or the combination between thermal pretreatment and recarbonation results in an adverse effect under high CO_2 concentrations when a drastic reduction occurs in the reactivity of the fast carbonation process of the regenerated sorbent. The effect of recarbonation is therefore highly dependent on the CO_2 partial pressure in the calciner.

2.1.3.1.3 Chemical Pretreatment Along with hydration, thermal pretreatment, and recarbonation, the use of chemical solutions is another promising treatment approach to enhance the capture efficiency. Treatment by aqueous ethanol solutions (50%–90%) increases the uptake capacity compared to the sorbent directly prepared from limestone or treated by hydration with water, due to the tendency of ethanol to increase the affinity and penetrability of water molecule into CaO, which results in higher specific surface area and larger pore volume after calcination (Li et al., 2008). Using higher ethanol concentrations results in higher uptake capacity and better antisintering performance. A similar treatment applied to sorbent derived from calcium acetate results in a higher capacity and better stability on account of the formation of smaller size particles, and larger surface area and pore volume (Wang et al., 2013). The use of pure ethanol or water shows a negative effect.

Organic acids are another class of chemicals offering improved sintering resistance (Li et al., 2009b; Radfarnia and Iliuta, 2013a; Ridha et al., 2013a; Sun et al., 2013a). Limestone treatment using 50% acetic acid solution offers a significant increase of sorption capacity (0.39 g of CO_2/g of sorbent) after

20 cycles, compared to natural limestone (0.12 g of CO_2/g of sorbent) and a better stability in cyclic operation, attributed to the formation of structures with smaller particle size, higher surface area, and pore volume (Li et al., 2009b). Similarly, an enhancement of cyclic stability can also be obtained by treatment with pyroligneous acid whose price is lower compared to acetic acid (Li et al., 2011). Treatment by organic acids like acetic acid, vinegar, formic acid, and oxalic acid offers a certain improvement of cyclic capacity as well (Ridha et al., 2013a). The acidification widens the pores of the modified sorbents and improves their resistance to sintering. The activity of the modified sorbents was however found to deteriorate like for untreated limestone. In brief, the limestone capture capacity can be increased by organic acid treatment, but not in a high extent, and the process is not economically attractive. Moreover, this chemical pretreatment resulted in an improved sorbent reactivity toward SO_2, thus accelerating the decay of CO_2 capture capacity compared to untreated limestone.

To reduce the cost of sorbent production, natural limestone (Canadian Newfoundland) was also considered by Radfarnia and Iliuta (2013a) instead of synthetic Ca precursors to prepare highly porous CaO with better stability in high-temperature operation compared to raw (untreated) limestone. A novel synthesis technique was developed consisting in the acidification of limestone by citric acid (natural-based product, >1 million tons/year produced for multiple applications in industry) coupled with a two-step calcination procedure initially in argon and followed by air. The acid treatment results first in pure calcium citrate precipitation, followed by controlled thermal decomposition. In the first step (under Ar (inert) atmosphere), (i) calcium citrate decomposes initially into intermediate compounds (calcium aconitate ($Ca_3(C_6H_3O_6)_2$), $CaCO_3$, and some gases (CO and CH_4)) between 325°C and 570°C, and (ii) at temperatures >570°C, calcium aconitate and carbonate are further converted into CaO, CO, H_2O, and carbon. During this step, the in situ carbon formed by calcium citrate decomposition covered effectively the CaO particles and prevented their growth. In the second step (air atmosphere, 850°C), the carbon combustion helps to control the thermal sintering, promote a good dispersion of particle agglomerates, and improve the sorbent porosity by releasing a high volume of gas (CO_2, H_2O, and NO_2). The sorbent was found to present a highly fluffy and porous structure compared to natural limestone. The combination of acid treatment and two-step calcination involving consecutive treatments in Ar and air streams resulted in remarkable improvement of material morphology, porosity, and sorption efficiency. A study of the influence of primary calcination temperature revealed that performing the calcination at higher temperatures (950°C) resulted in an initial increase in uptake capacity (self-reactivation) followed by further reduction during the cyclic operation.

Despite the improvement of capture efficiency of CaO-based sorbents, hydration and acidification have the disadvantage of reducing the material mechanical strength. Pelletization might represent an interesting option to improve the sorbent mechanical strength, as required in industry. Among different kinds of binders for pelletization like Na_2CO_3, Na- and Ca-bentonite (Na^+- or Ca^{2+}-hydrated aluminosilicate clays), and calcium aluminate cements (with monocalcium aluminate ($CaAl_2O_4$) as the main active component), calcium aluminate cements show very promising results in long-time operation over 30–35 cycles, while Na_2CO_3 and bentonites are not suitable (Manovic and Anthony, 2009). The enhanced sintering observed in the case of Na_2CO_3 and bentonites, which results in a faster decay in activity, is on account of the formation of calcium-silica compounds with low melting points like $Ca_2(SiO_4)$ and $Ca_5(SiO_4)_2CO_3$. In comparison, no enhanced sintering occurs for calcium aluminate cements; moreover, these low-cost materials provide fast setting and good refractory properties. The behavior of Na_2CO_3 and bentonites leads to the conclusion that an appropriate binder should not contain Si, Na, and K. An examination of the capture behavior of CaO-based pellets from raw and acidified (10 vol.% acetic acid) limestone, obtained with calcium aluminate cement binder (10–14 wt.%), showed 41% higher capacity of acidified pellets after 20 cycles carbonation (650°C, 15% CO_2)/calcination (920°C, 100% CO_2) (Ridha et al., 2013b). The positive performance of acidified pellets is on account of their improved morphology consisting of large pores (20–200 nm), which allows hosting more product ($CaCO_3$). The addition of starch along with aluminate cement (58 wt.% Al_2O_3) in the pelletization process of limestone (5–10 wt.% starch and 10 wt.% aluminate cement) can improve significantly the mechanical properties of pellets due to the high mechanical strength of aluminate and the performance of cyclic CO_2 capture (capacity and stability) due to the increase of pore volume of pellets through starch decomposition during calcination (Chen et al., 2013). Nonetheless, being an expensive process, for economic considerations its application can be suitable when it demonstrates considerable improvement of the sorption performance. Even with such an increase of 41% in CO_2 capture capacity, the applied treatments are not justified because of the high price of acetic acid (around \$900/ton). Moreover, as the acidified pellets show an improved reactivity toward SO_2, the removal of SO_2 from the flue gases is required as well.

2.1.3.2 Metal-Stabilized CaO Sorbents

One of the efficient methods to improve the stability of synthetic sorbents in cyclic operation is the incorporation of inert support materials with a high Tammann temperature (Table 2) like aluminum oxide (Al_2O_3), magnesium oxide (MgO), zirconium oxide (ZrO_2), titanium oxide (TiO_2), silica (SiO_2), yttrium oxide (Y_2O_3), and cobalt oxide (CoO) into the sorbent structure. Dispersing inert support materials among the CaO particles during the synthesis helps reduce the sintering phenomenon during the carbonation/calcination cyclic operation in order to enhance the performance of synthesized sorbents. The optimum ratio between inert and active material is a matter of concern for the majority of research. Indeed, the presence of a high amount of inert material in an industrial absorbent candidate is undesirable as it results in reduced sorption capacity. Table 3 presents a brief summary of different examined stabilizers which offer promising results.

Table 2 Tammann Temperature of Different Oxides (Approximated as $0.5 \cdot T_m$ (K)[a])

Oxide	$T_{Tammann}$ (°C)
Al_2O_3 (corundum)	890
CaO	1313
$CaCO_3$ (calcite)	529
$CaZrO_3$	1139
$Ca_3Al_2O_6$	631
$Ca_{12}Al_{14}O_{33}$ (mayenite)	572
CeO_2	1064
CoO	779
Cr_2O_3	1028
MgO	1276
SiO_2 (cristobalite)	725
SrO	1129
TiO_2	785
Y_2O_3	1083
ZrO_2	1218

[a]Sources: CRC Handbook of Chemistry and Physics (Lide, 2003), except for mayenite (Lourenço et al., 2013).

Table 3 Brief Summary of Different Investigations on Metal-Stabilized CaO Sorbents

Stabilizer		Carbonation/Regeneration				
Type	Content (wt.%)	Conditions	Duration (min)	Cycle Number	CO_2 Uptake at Last Cycle (g CO_2/g ads)	References
$Ca_{12}Al_{14}O_{33}$	25	690°C, 14% CO_2/850°C, 100% N_2	30/10	13	0.45	Li et al. (2005)
	25	690°C, 14% CO_2/950°C, 20% CO_2	30/10	13	0.333	Li et al. (2005)
	25	700°C, 20% CO_2/850°C, 100% N_2	30/5	50	0.41	Li et al. (2006a)
	25	700°C, 20% CO_2/980°C, 100% CO_2	30/5	56	0.22	Li et al. (2006a)
	25	690°C, 15% CO_2/850°C, 100% N_2	30/10	45	0.26	Martavaltzi and Lemonidou (2008a)
	7.7	690°C, 15% CO_2/850°C, 100% N_2	30/5	30	~0.37	Martavaltzi and Lemonidou (2008b)
	7.7	650°C, 15% CO_2/900°C, 15% CO_2	10/5	30	~0.22	Florin et al. (2010)
	9.2	750°C, 40%CO_2/750°C, 100% N_2	20/20	30	~0.55	Broda and Müller (2012)
	~42	700°C, 30% CO_2/700°C, 100% H_e	10/10	100	0.4	Koirala et al. (2012)
	~42	850°C, 100% CO_2/950°C, 30% CO_2	10/10	100	0.25	Koirala et al. (2012)
	7.7	650°C, 15% CO_2/900°C, 100% N_2	30/10	30	~0.6	Liu et al. (2013)

Continued

Table 3 Brief Summary of Different Investigations on Metal-Stabilized CaO Sorbents—cont'd

Stabilizer		Carbonation/Regeneration				
Type	Content (wt.%)	Conditions	Duration (min)	Cycle Number	CO_2 Uptake at Last Cycle (g CO_2/g ads)	References
$Ca_9Al_6O_{18}$	20	650°C, 15% CO_2/800°C, 100% N_2	30/10	28	0.51	Zhou et al. (2012)
	22.1	650°C, 15% CO_2/750°C, 100% Ar	30/30	25	0.33	Radfarnia and Iliuta (2013b)
	22.1	650°C, 15% CO_2/930°C, 50% CO_2	30/10	25	0.23	Radfarnia and Iliuta (2013b)
	20	650°C, 15% CO_2/800°C, 100% N_2	30/10	35	0.52	Xu et al. (2013)
	20	650°C, 15% CO_2/1000°C, 80% CO_2	30/10	35	0.2	Xu et al. (2013)
	7.5	650°C, 100% CO_2/800°C, 100% N_2	30/10	31	0.57	Radfarnia and Sayari (2015)
	7.5	650°C, 100% CO_2/930°C, 100% CO_2	30/5	31	0.33	Radfarnia and Sayari (2015)
$Ca_3Al_2O_6$	9	650°C, 20% CO_2/850°C, 100% N_2	30/10	100	0.35	Zhang et al. (2013a)
	34	690°C, 15% CO_2/800°C, 100% N_2	30/5	45	0.45	Angeli et al. (2014)
$CaZrO_3$	26.2	650°C, 15% CO_2/750°C, 100% Ar	30/30	25	0.29	Radfarnia and Iliuta (2013b)

	26.2	650°C, 15% CO_2/930°C, 50% CO_2	30/10	25	0.3	Radfarnia and Iliuta (2013b)
	10	650°C, 15% CO_2/800°C, 100% air	n.a.	30	0.37	Zhao et al. (2014)
	30	650°C, 100% CO_2/950°C, 100% CO_2	n.a.	30	0.31	Zhao et al. (2014)
	30	700°C, 100% CO_2/700°C, 100% He	30/30	1050	0.3	Reddy et al. (2014)
MgO	25	750°C, 25% CO_2/750°C, 100% N_2	20/30	1250	0.17	Albrecht et al. (2008)
	26	758°C, 100% CO_2/758°C, 100% He	30/30	50	0.53	Li et al. (2009a)
	25	650°C, 15% CO_2/900°C, 100% N_2	30/10	24	0.56	Liu et al. (2010a)
	25	650°C, 15% CO_2/900°C, 100% N_2	30/10	44	0.46	Liu et al. (2012)
	25	700°C, 20% CO_2/730°C, 100% N_2	10/10	50	0.54	Lan and Wu (2014)
Al_2O_3	19	750°C, 40% CO_2/750°C, 100% N_2	20/20	30	0.36	Kierzkowska et al. (2013)
	19	700°C, 40% CO_2/925°C, 100% CO_2	20/20	10	0.31	Kierzkowska et al. (2013)
	8	710°C, 100% CO_2/950°C, 100% N_2	30/5	15	0.62	Sayyah et al. (2013)

A comparison between different metal oxide (M: Al, Zr, Mg, and Y)-stabilized CaO sorbents derived from two Canadian limestones (Newfoundland (Nf) and Graymont (Gr)) by means of single-step acidification (citric acid) followed by a two-step calcination method (Ar and air atmospheres) (Radfarnia and Iliuta, 2013b) is illustrated in Fig. 6.

Production of CaO from limestone treatment by citric acid coupled with a heat treatment in argon and air streams, respectively (two-step calcination) (Radfarnia and Iliuta, 2013a) was found to produce considerable porosity to make the CaO matrix very suitable for incorporating stabilizers and improve even more its stability in cyclic operation. Special attention was focussed on the investigation of the effect of type of precursors (soluble or insoluble) and metal/Ca ratios (Table 4), as well as the calcination atmosphere, on the performance of prepared materials. As Newfoundland is a high-purity source compared to Graymont, the two types of limestones led to different results, indicating the importance of the composition of chosen limestone in material preparation. The effect of metal/Ca ratio and type of precursors is illustrated in Figs. 6–11, together with the behavior of CaO (no metal-stabilized) prepared by using the same acidification technique from both limestone sources (CaO-Nf and CaO-Gm).

The type of precursor is an essential parameter affecting the chemical nature and properties of sorbents. The stabilizers can act directly as self-stabilizers and/or through the products that can be formed during the synthesis by the reaction of different precursors, depending on their affinity and conditions (synthesis parameters). In the case of insoluble precursors, the attainment of a suitable inert dispersion which can affect considerably the sorbent properties is a matter of concern. XRD analysis revealed stabilizer formation in most investigated samples, through the interaction between CaO and the metal oxide precursor, like, for example, $Ca_9Al_6O_{18}$ (from soluble aluminum nitrate, as well as by solid–solid reaction involving insoluble aluminum acetate and calcium citrate) and $CaZrO_3$ (from soluble zirconoxy nitrate). However, no reaction occurred between insoluble Al_2O_3 and CaO to form calcium aluminate, and insoluble $Zr(OH)_4$ did not confirm complete reaction with CaO to form $CaZrO_3$. The consequences of these behaviors are well reflected in the results presented in Figs. 6, 9, and 10.

Under mild conditions, $Ca_{0.9}ZrN_{0.1}$ and $Ca_{0.9}AlN_{0.1}$ exhibit the best activity among the Zr- and Al-stabilized sorbents produced from soluble precursors (Fig. 6). Mg-stabilized sorbents produced from the soluble stabilizer precursor as well as from carbon-containing Mg and Ca precursors (Radfarnia and Iliuta, 2013a) were also expected to present good sorption

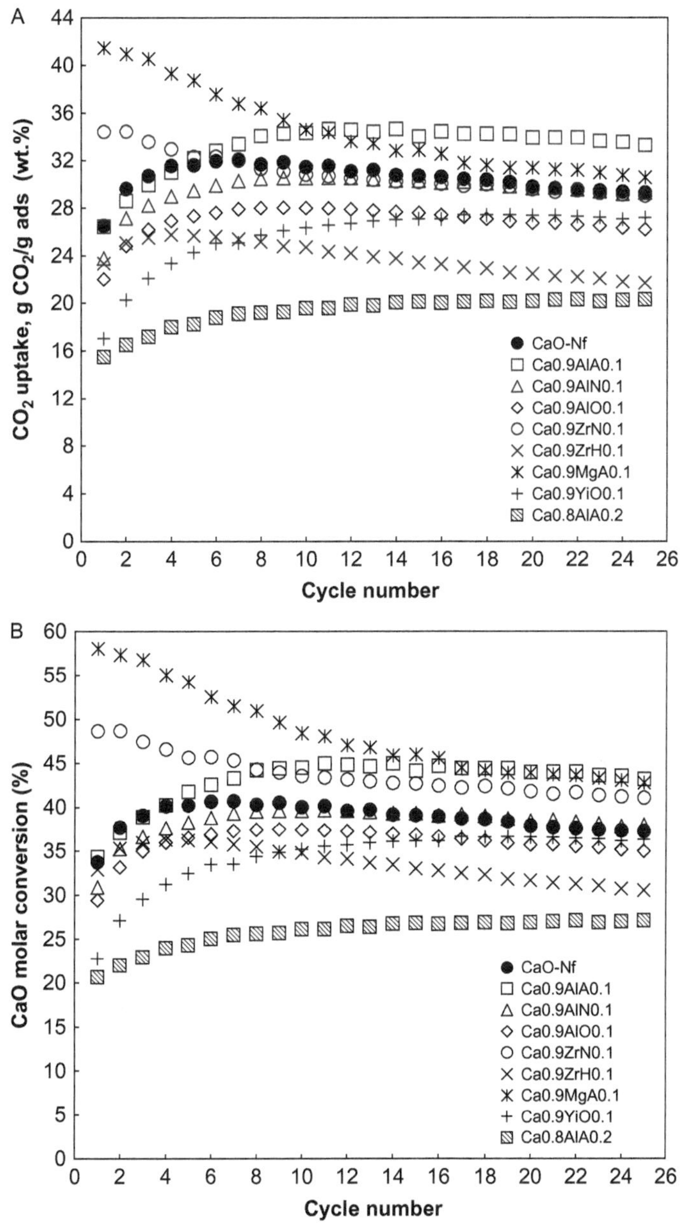

Fig. 6 CO_2 uptake (A) and molar conversion (B) of different metal-doped CaO sorbents derived from Newfoundland limestone—application in mild conditions (sorption: 650°C/15 vol.% CO_2; desorption: 750°C/100 vol.% Ar). *Data from Radfarnia HR, Iliuta MC: Metal oxide-stabilized calcium oxide CO_2 sorbent for multicycle operation,* Chem Eng J *232:280–289, 2013.*

Table 4 Specifications of Different Metal Oxide-Stabilized Sorbents Prepared Using Soluble and Insoluble Precursors.

Sorbent	Soluble Precursor	Insoluble Precursor	Molar Ratio Metal/Calcium	Stabilizer Type
$Ca_{0.9}AlN_{0.1}$	Aluminum nitrate nanohydrate		0.1	$Ca_9Al_6O_{18}$
$Ca_{0.8}AlN_{0.2}$			0.2	
$Ca_{0.7}AlN_{0.3}$			0.3	
$Ca_{0.6}AlN_{0.4}$			0.4	
$Ca_{0.9}MgA_{0.1}$	Magnesium acetate tetrahydrate		0.1	MgO
$Ca_{0.8}MgA_{0.2}$			0.2	
$Ca_{0.6}MgA_{0.4}$			0.4	
$Ca_{0.5}MgA_{0.5}$			0.5	
$Ca_{0.2}MgA_{0.8}$			0.8	
$Ca_{0.9}ZrN_{0.1}$	Zirconoxy nitrate		0.1	$CaZrO_3$
$Ca_{0.8}ZrN_{0.2}$			0.2	
$Ca_{0.7}ZrN_{0.3}$			0.3	
$Ca_{0.6}ZrN_{0.4}$			0.4	
$Ca_{0.9}ZrH_{0.1}$		Zirconium hydroxide	0.1	ZrO_2
$Ca_{0.9}AlA_{0.1}$		Aluminum acetate	0.1	$Ca_9Al_6O_{18}$
$Ca_{0.8}AlA_{0.2}$		Aluminium acetate	0.2	$Ca_9Al_6O_{18}$
$Ca_{0.9}YiO_{0.1}$		Yttrium oxide	0.1	Y_2O_3
$Ca_{0.9}AlO_{0.1}$		Aluminum oxide	0.1	Al_2O_3

Data from Radfarnia HR, Iliuta MC: Metal oxide-stabilized calcium oxide CO_2 sorbent for multicycle operation, Chem Eng J232:280–289, 2013.

activity. However, this was only observed during the initial cycles, after which their efficiency reduces significantly. Due to the smaller molar volume of the inert stabilizer (MgO) compared to $CaCO_3$, it does not seem able to prevent accurately the fusion of particles and pore structure failure during the sorption process (Zhang et al., 2012). Among the Mg-stabilized samples, $Ca_{0.6}MgA_{0.4}$ exhibits the best CO_2 uptake capacity

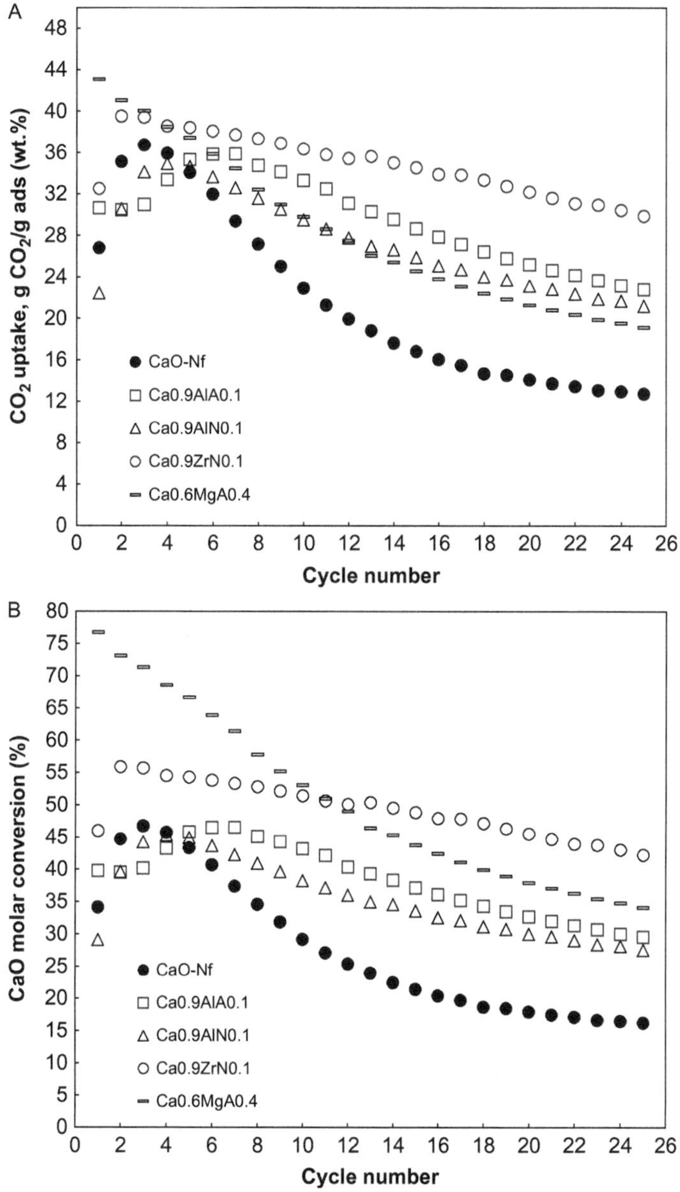

Fig. 7 CO_2 uptake (A) and molar conversion (B) of different metal-doped CaO sorbents derived from Newfoundland limestone—application in severe conditions (sorption: 650°C/15 vol.% CO_2; desorption: 930°C/50 vol.% CO_2). *Data Radfarnia HR, Iliuta MC: Metal oxide-stabilized calcium oxide CO_2 sorbent for multicycle operation,* Chem Eng J *232:280–289, 2013.*

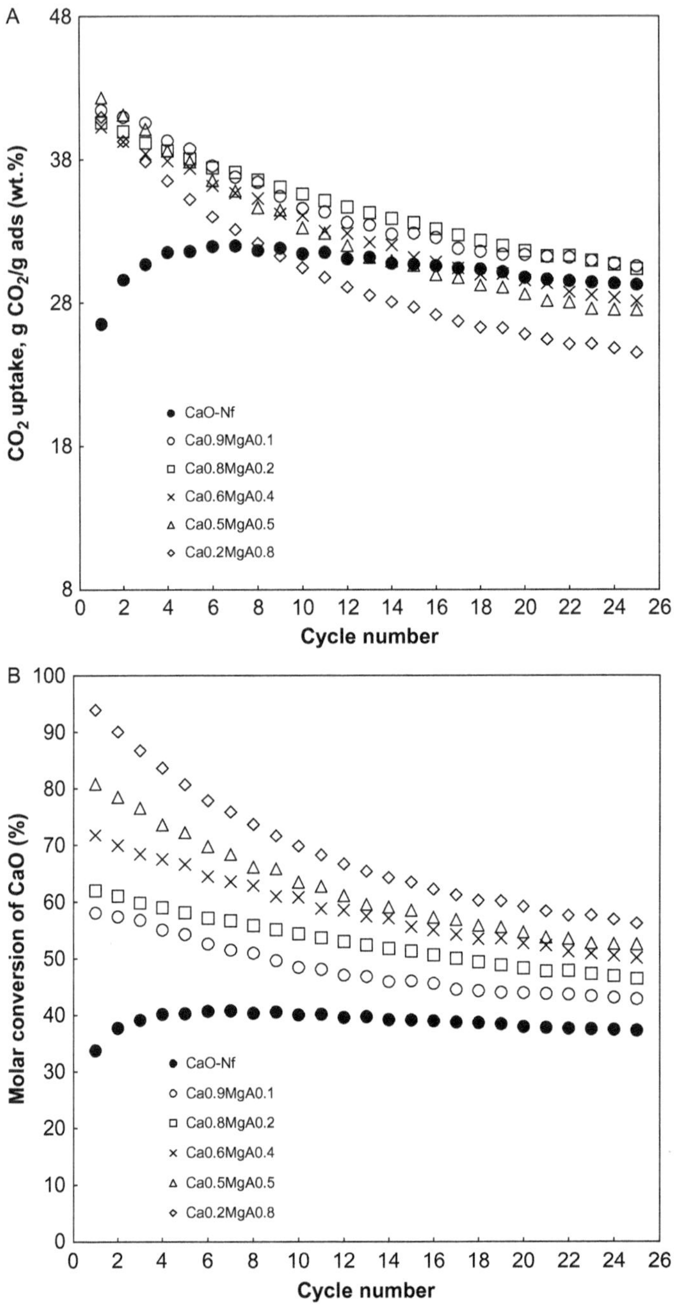

Fig. 8 Influence of metal/Ca ratio on CO_2 uptake (A) and molar conversion (B) of Mg-doped CaO sorbents derived from Newfoundland limestone—application in mild conditions (sorption: 650°C/15 vol.% CO_2; desorption: 750°C/100 vol.% Ar). *Data from Radfarnia HR, Iliuta MC: Metal oxide-stabilized calcium oxide CO_2 sorbent for multicycle operation,* Chem Eng J *232:280–289, 2013.*

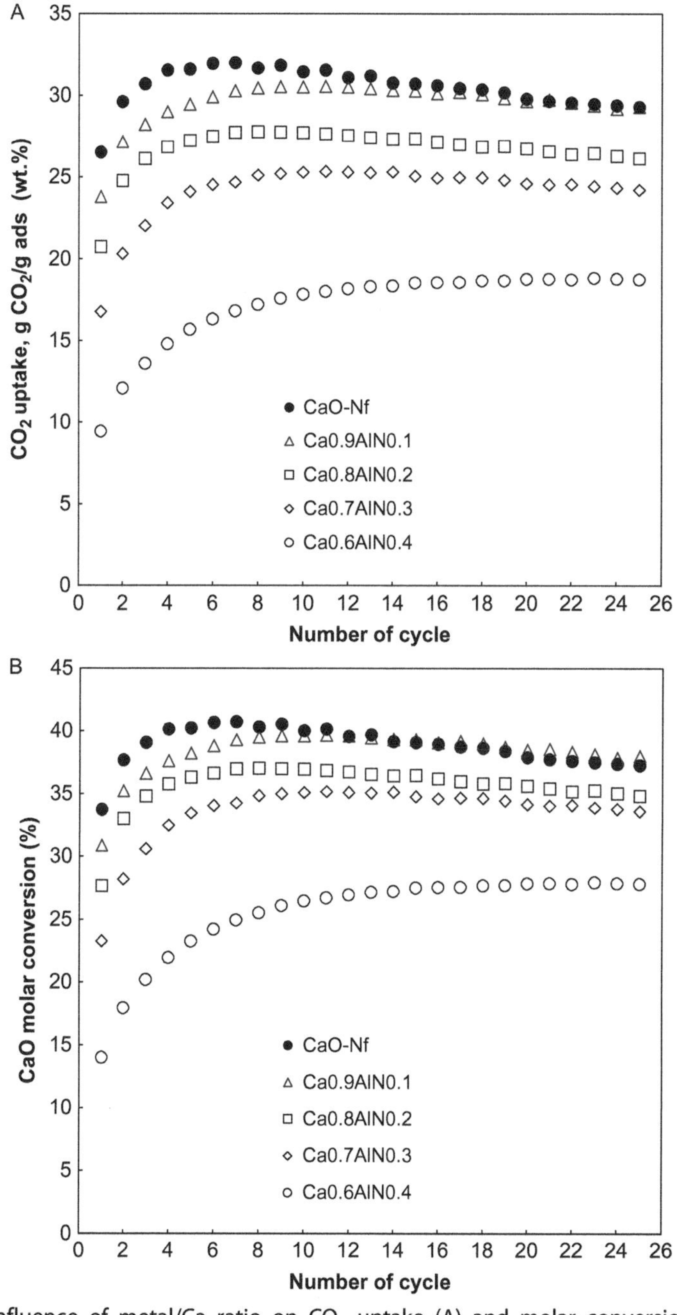

Fig. 9 Influence of metal/Ca ratio on CO_2 uptake (A) and molar conversion (B) of Al-doped CaO sorbents derived from Newfoundland limestone from soluble (aluminum nitrate) precursor—application in mild conditions (sorption: 650°C/15 vol.% CO_2; desorption: 750°C/100 vol.% Ar). *Data from Radfarnia HR, Iliuta MC: Metal oxide-stabilized calcium oxide CO_2 sorbent for multicycle operation,* Chem Eng J *232:280–289, 2013.*

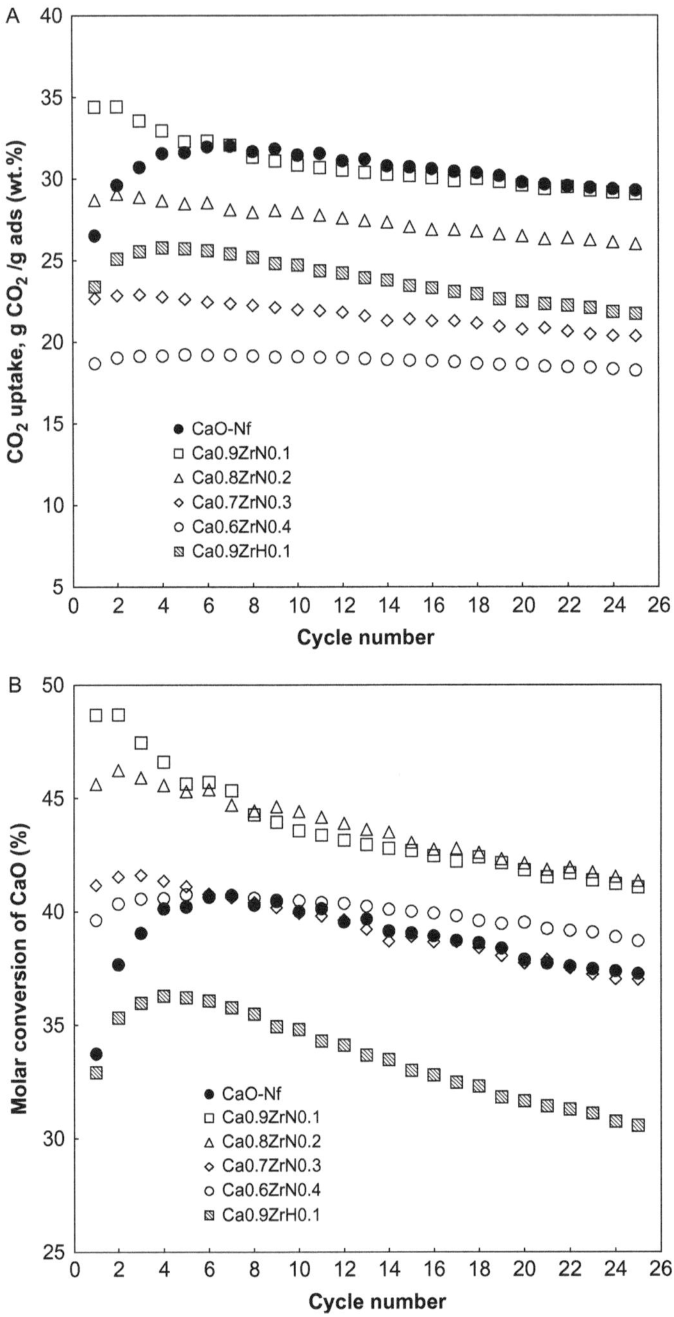

Fig. 10 Influence of metal/Ca ratio on CO_2 uptake (A) and molar conversion (B) of Zr-doped CaO sorbents derived from Newfoundland limestone from soluble (zirconoxy nitrate) and insoluble (zirconium hydroxide) precursor—application in mild conditions (sorption: 650°C/15 vol.% CO_2; desorption: 750°C/100 vol.% Ar). *Data from Radfarnia HR, Iliuta MC: Metal oxide-stabilized calcium oxide CO_2 sorbent for multicycle operation,* Chem Eng J *232:280–289, 2013.*

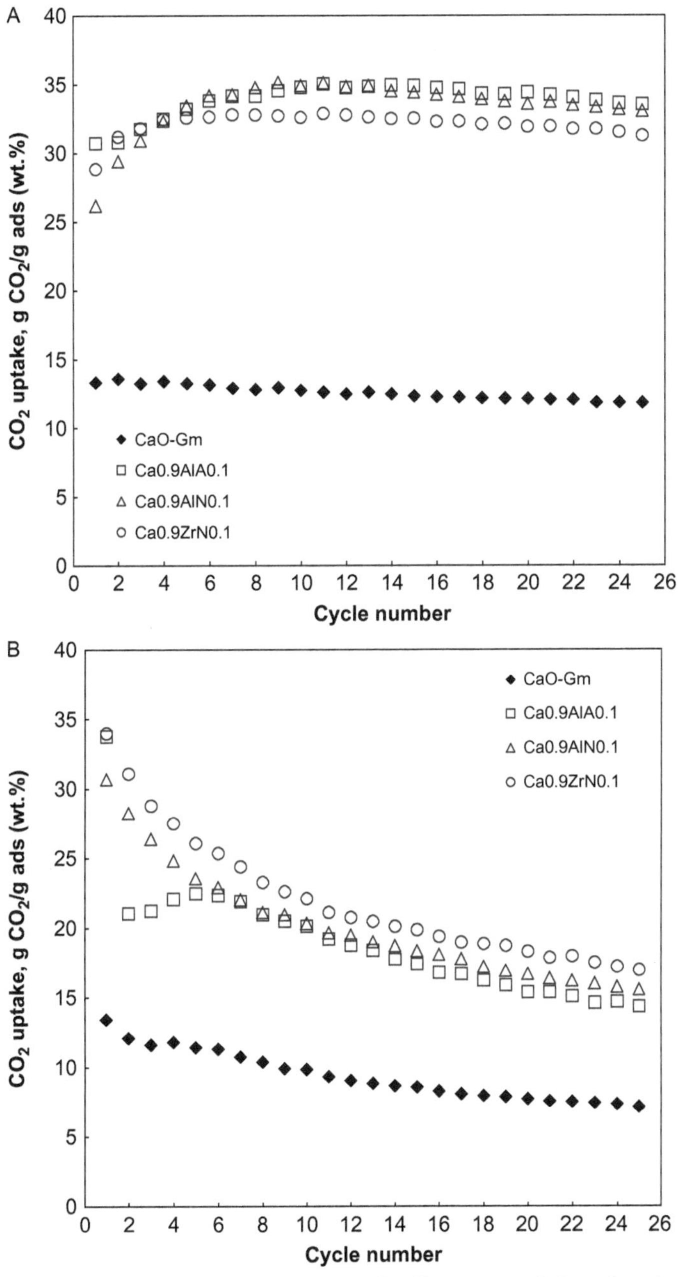

Fig. 11 CO_2 uptake and molar conversion of different metal-doped CaO sorbents derived from Graymont limestone—application in (A) mild conditions (sorption: 650°C/15 vol.% CO_2; desorption: 750 °C/100 vol.% Ar) and (B) severe conditions (sorption: 650°C/15 vol.% CO_2; desorption: 930°C/50 vol.% CO_2). *Data from Radfarnia HR, Iliuta MC: Metal oxide-stabilized calcium oxide CO_2 sorbent for multicycle operation,* Chem Eng J *232:280–289, 2013.*

and CaO molar conversion at the end of cyclic operation (Fig. 8). Among the sorbents obtained from insoluble stabilizer precursors, $Ca_{0.9}AlA_{0.1}$ has the best performance (Fig. 6). By burning off the in situ carbon previously formed from aluminum acetate in the Ar atmosphere, the evolved gas contributes to particle dispersion and increases the sorbent porosity (Radfarnia and Iliuta, 2013a). However, a larger content of insoluble aluminum has a negative effect on the sorbent activity, possibly due to the increase of particle agglomeration in the liquid solution. Accurate determination of the optimum mixing ratio of stabilizer/calcium is therefore a key parameter in material synthesis. The other sorbents prepared from insoluble zirconium, yttrium, and aluminum precursors, which are all free of carbon, did not demonstrate adequate performance. Compared to mild conditions where no CO_2 is present in the calcination atmosphere, the sorption operation under severe conditions reveals important deactivation of CaO-Nf but significant improvement of metal-stabilized sorbents. Fig. 7 illustrates the best performance of $Ca_{0.9}ZrN_{0.1}$ and $Ca_{0.9}AlA_{0.1}$ and an important deterioration of the activity of $Ca_{0.6}MgA_{0.4}$ (which proved the best efficiency in mild conditions).

Unlike the Newfoundland limestone-based sorbents, the activity of CaO-Gm material was found to be significantly less than that of CaO-Nf especially in severe operating conditions, due to structural sintering possibly resulted from the presence of impurities in the Graymont limestone which limits the CO_2 access to the active sites. Fig. 11 illustrates the role of stabilizers in improving the performance of natural limestone, by limiting the sorbent structural deterioration.

As Al_2O_3, ZrO_2, and MgO are considered the most promising supporting materials, the following discussion will be mostly focussed on these materials (Table 3). Among the proposed stabilizers, the use of ZrO_2 and Al_2O_3 has received more attention due to the appropriate thermal stability of these compounds.

2.1.3.2.1 Al-Stabilized CaO Most of the attention was directed at the development of CaO-based sorbents containing inert Al_2O_3. Depending on the synthesis method and both calcium and aluminum precursors, Al_2O_3, $Ca_{12}Al_{14}O_{33}$, $Ca_9Al_6O_{18}$, and $Ca_3Al_2O_6$ phases can be produced in the preparation process. Their incorporation into CaO structure during the sorbent synthesis aims to prevent sintering and enhance the sorbent stability in cyclic operation. Special consideration is given to the choice of precursors (Martavaltzi and Lemonidou, 2008a; Wu et al., 2008; Zhou et al., 2012) and synthesis technique (Angeli et al., 2014; Broda and Müller, 2012; Florin et al., 2010; Kierzkowska et al., 2013; Kim et al., 2013; Martavaltzi and Lemonidou, 2008b; Radfarnia and Iliuta, 2013b; Radfarnia and Sayari, 2015; Sayyah et al., 2013; Stendardo et al., 2013; Zhang et al., 2013a), as well as to the optimization of the ratio between Ca and Al (Dennis and Pacciani, 2009; Koirala et al., 2012; Pacciani et al., 2008; Yu and Chen, 2014).

In particular, the development of different preparation approaches like wet mixing (Dennis and Pacciani, 2009; Li et al., 2006a, 2005; Martavaltzi and Lemonidou, 2008a; Pacciani et al., 2008; Zhou et al., 2012), limestone acidification by citric acid followed by two-step calcination (Radfarnia and Iliuta, 2013b), solid-state reaction (Stendardo et al., 2013), ultrasonic spray pyrolysis (USP) (Sayyah et al., 2013), combination of precipitation and hydration (Kim et al., 2013), coprecipitation (Kierzkowska et al., 2013), citrate preparation (Zhang et al., 2013a), sol–gel (Angeli et al., 2014; Broda and Müller, 2012), citrate-assisted sol–gel technique followed by two-step calcination (Radfarnia and Sayari, 2015), single nozzle FSP (Koirala et al., 2012), and precipitation (Florin et al., 2010) aims to improve the sorbents' efficiency by increasing the material specific surface area, pore volume, and high dispersion of the supports within the sorbent structure.

For example, the incorporation of mayenite ($Ca_{12}Al_{14}O_{33}$) into CaO matrix helps to prevent the structural breakdown of CaO at high temperatures. Despite its relatively low melting temperature (Table 2), mayenite hydrates and hardens rapidly, thus finding an application in controlling the hardening rate of cements (Lee et al., 2001; Lourenço et al., 2013). Using aluminum nitrate nonahydrate ($Al(NO_3)_3 \cdot 9H_2O$) and calcium oxide precursors and applying the wet-mixing method, the synthesized $Ca_{12}Al_{14}O_{33}$-stabilized CaO sorbent containing 25 wt.% $Ca_{12}Al_{14}O_{33}$ shows a high sorption capacity of 0.41 g of CO_2/g of sorbent at the end of the 50th cycle under mild calcination conditions (850°C, 100% N_2) and 0.22 g of CO_2/g of sorbent after 56 cycles under more severe conditions (980°C, 100% CO_2) (Li et al., 2006a, 2005). The severe conditions result in important reduction of sorption capacity, which is however better in comparison with dolomite (0.16 g of CO_2/g of sorbent). Among the Al-stabilized CaO sorbents obtained by Zhou et al. (2012) from various calcium and aluminum precursors and by applying the same technique, $Ca_9Al_6O_{18}$-stabilized CaO material derived from calcium citrate and aluminum nitrate presented the best capture capacity after 28 cycles (0.51 g of CO_2/g of sorbent) attributed to the bimodal pore size distribution with an adequate number of small pores, the high specific surface area of sorbents, and the uniform distribution of inert support materials among CaO particles. A possible mechanism was proposed to explain the formation of Al-stabilized sorbent. During the synthesis of $CaO–Ca_{12}Al_{14}O_{33}$ (85 wt.% CaO) sorbent by using the same method, Martavaltzi and Lemonidou (2008b) inserted a hydration step between calcination steps and reported a good stability (21% deactivation only) during 45 cycles in mild conditions of carbonation (690°C, 30 min, 15 vol.% CO_2)/ calcination (850°C, 5 min, inert atmosphere).

Sorbents with higher Al/Ca ratio offer a better stability in both mild and severe conditions. A synthetic sorbent with the Al/Ca molar ratio of 3:10 prepared using the single nozzle FSP was shown to offer the best performance among different $Ca_{12}Al_{14}O_{33}$-stabilized CaO materials, with a stable uptake capacity over 100 cycles of 0.40 g of CO_2/g of sorbent in mild conditions and 0.25 g of CO_2/g of sorbent in severe conditions (values corresponding to the end of the 100th cycle) (Koirala et al., 2012). The uniform distribution of $Ca_{12}Al_{14}O_{33}$ among CaO particles can account for this enhanced durability. By offering homogenous dispersion of additives in the CaO matrix, high surface area, and high stability, the application of USP method can lead to materials with high sorption performance. It was reported that by using calcium nitrate tetrahydrate ($Ca(NO_3)_2{\cdot}4H_2O$) and aluminum nitrate nonahydrate ($Al(NO_3)_3{\cdot}9H_2O$) precursors and applying the USP method, synthesized Al_2O_3-stabilized CaO sorbents present good stability for an optimized Al/Ca ratio of around 0.08 (Sayyah et al., 2013). The sorption capacity shows a very low decrease from 0.65 to 0.62 g of CO_2/g of sorbent only over 15 cycles.

A $CaO–Al_2O_3$ (9 wt.% Al_2O_3) sorbent with larger pore volume (0.85 cm^3/g vs 0.03 cm^3/g for CaO) and higher specific surface area (958 m^2/g vs 5 m^2/g for CaO) was prepared from aluminum nitrate, citric acid, and $CaCO_3$ by applying the citrate method together with a four-step sorbent activation under the mild conditions (Zhang et al., 2013a). The existence of such a porous structure promoted by gas (CO_2, H_2O, and NO_2) release during the activation process along with the presence of the stabilizer formed in the system ($Ca_3Al_2O_6$) which prevents sintering of CaO particle offers a good CO_2 capture capacity and stability in long-term operation (~0.35 g of CO_2/g of sorbent after 100 cycles).

Sol–gel is an efficient and inexpensive method for synthesizing Ca-based sorbents with high surface area and large pore volume. Different modifications have been made to develop sorbents with improved morphological characteristics, especially high specific surface area and pore volume, pore size distribution, and good dispersion of inert materials into the CaO structure. A modified sol–gel method was applied by Angeli et al. (2014) to produce $CaO–Ca_3Al_2O_6$ sorbent by using triethanolamine (TEA) as complexing agent together with $Ca(NO_3)_2{\cdot}4H_2O$ and $Al(NO_3)_3{\cdot}9H_2O$ precursors. The high and stable CO_2 uptake capacity of 0.45 g of CO_2/g of sorbent (84% carbonation conversion) during cyclic operation (45 cycles) under mild conditions was on account of the formation of a coral-like structure promoted by a high amount of gas released during TEA decomposition in the calcination stage and the uniform dispersion of Ca and Al ions from TEA–ion complexes formed in the system. Moreover, under severe conditions (calcination at 950°C under 100% CO_2) the sorbent presented higher stability in comparison with other Ca-based sorbents (40% of the initial capacity at the end of the 100th cycle). More recently, a similar sorbent ($CaO/Ca_3Al_2O_6$) was synthesized by incorporating Al via a modified sol–gel method, by using calcium nitrate tetrahydrate, aluminum nitrate nonahydrate, and citric acid monohydrate as precursors (Jing et al., 2017). The ratio of CaO to Al_2O_3 was found to have an important effect on sorption capacity and stability of the sorbent. The best sample obtained for a ratio of 8:2 showed five times better capacity compared to pure synthesized CaO. Another modified (citrate-assisted) sol–gel method was applied to produce a stable Al-stabilized CaO sorbent (7.5 wt.% Al_2O_3) which showed no loss of activity for 31 cycles under mild conditions of operation, due to a uniform dispersion of $Ca_9Al_6O_{18}$ throughout the CaO matrix (Radfarnia and Sayari, 2015). The sol–gel procedure was coupled with the two-step calcination technique proposed by Radfarnia and Iliuta (2013a).

The sol–gel method can also be applied to produce Ca-based sorbents from Ca–Al layered double hydroxides (LDHs). A scaffold-like hybrid Ca–Al LDH-derived sorbent was synthesized by Chang et al. (2013) using $Al(O^iPr)_3$ and $Ca(NO_3)_2$ as precursors and hexadecyl trimethyl ammonium bromide (CTAB) as structure-directing agent. $Ca_{12}Al_{14}O_{33}$ phase (acting as inert support) was found to be formed by calcination at 600°C when the ratio Ca^{2+}/Al^{3+} was 1:1, while the increase of this ratio resulted in the formation of the CaO phase besides $Ca_{12}Al_{14}O_{33}$. By combining sintering-resistant metal-oxide supports and porous $CaO/CaCO_3$-based structure, these CaO-based Ca–Al–O mixed oxide (CAMO) with high specific surface areas (up to 191 m^2/g) and pore size distribution (3–6 nm) that allows a rapid CO_2 diffusion into the sorbent demonstrated high CO_2 uptake capacity (49 wt.%) and fast CO_2 sorption kinetics, besides being stable in long-term cyclic operation (2%–6% reduction of capacity after 50 cycles). Other methods like reverse microemulsion (Chang et al., 2014) and hydrothermal (Chang et al., 2011; Yu and Chen, 2014) have been employed to prepare Ca-based sorbents derived from Ca–Al LDHs. In general, a good stability in cyclic operation has been reported for these materials, due to the uniform distribution of calcium aluminates produced during the synthesis among CaO particles, thus limiting their contact and preventing sintering.

2.1.3.2.2 Zr-Stabilized CaO Along with Al_2O_3, ZrO_2 has also been attracting researchers to explore its capacity in stabilizing the CaO structure in the high-temperature capture process. Several works have reported the superior performance of Zr-stabilized CaO sorbents compared to other metal-stabilized CaO materials (Lu et al., 2009a; Radfarnia and Iliuta, 2013b). The good efficiency of this kind of metal-stabilized CaO sorbents is on account of the formation of $CaZrO_3$ (calcium zirconate) with a high Tammann temperature (1139°C) (Table 2) which helps to minimize the thermal sintering of CaO particles.

Operational parameters like the synthesis method (sol–gel (Broda and Müller, 2014; Reddy et al., 2014), surfactant template/sonication (Radfarnia and Iliuta, 2012b), deposition–precipitation (Reddy et al., 2014), coprecipitation (Reddy et al., 2014), FSP (Koirala et al., 2011; Lu et al., 2009a; Reddy et al., 2014), and wet chemical (Zhao et al., 2014)), calcium and zirconium precursors, Ca/Zr ratio, and calcination conditions were found to have a significant effect on CO_2 uptake performance of Zr-stabilized CaO sorbents.

Among several kinds of inert metal (Si, Ti, Cr, Co, Zr, and Ce)-doped CaO sorbents prepared by the FSP method, Zr-doped CaO demonstrated the best CO_2 sorption performance under the same operation conditions, due to the efficiency of ZrO_2 in improving the sorbent thermal stability (Lu et al., 2009a). As the application of FSP demonstrated the effectiveness in developing nanomaterials with high surface area and large pore volume, this method was further used to investigate the effect of Zr/Ca ratio on sorbent properties (Koirala et al., 2011). A molar ratio of 0.5 led to remarkable thermal stability up to 1200 carbonation/calcination cycles owing to a suitable dispersion of $CaZrO_3$ nanoparticles which helps to prevent the growth of CaO particles. Despite the FSP effectiveness, the complexity of this technique can be an important matter of concern in its application on a wider scale.

Radfarnia and Iliuta (2012b) developed Zr-stabilized CaO sorbents by applying the surfactant template-ultrasound synthesis method, an efficient way to produce nanoparticle porous materials (Radfarnia and Iliuta,

2011), from calcium acetate, zirconoxy nitrate, and CTAB as surfactant to control the morphology of Ca–Zr complex. The Zr/Ca molar ratio was found to strongly affect the sorption efficiency: a ratio of 0.114 resulted in a significant capacity reduction while 0.303 was the optimum value which led to a remarkable improvement in the sorbent cyclic stability and an adequate uptake capacity. The formation of $CaZrO_3$ by the reaction between ZrO_2 and CaO and its appropriate dispersion helped prevent structural sintering and consequently enhancing the sorbent stability in cyclic operation in comparison with pure CaO.

Reddy et al. (2014) compared the structure and sorption properties of Zr-stabilized CaO sorbents obtained by different methods (deposition–precipitation, coprecipitation, FSP, and sol–gel) and their behavior in the CO_2 capture. TEM measurements indicated that the sorbent prepared by applying the sol–gel procedure presented much smaller crystallite size (10–20 nm) compared to materials synthesized by deposition–precipitation and coprecipitation (more than 100 nm). In spite of the very high Ca/Zr atomic ratio for the sorbents prepared by deposition–precipitation (9.46) and coprecipitation (8.9) compared to those obtained by sol–gel (3.57), the much lower crystallite size of the sol–gel-synthesized material resulted in a better sorption performance in cyclic operation, attesting the important effect of this parameter. The sample prepared by sol–gel exhibited better capture capacity in comparison with the FSP-synthesized one (the corresponding Ca/Zr atomic ratio order is $3.57 > 1.5$). The sorbent synthesized by the sol–gel method also shows higher basicity compared to the other samples. Both sorbents prepared by sol–gel and FSP showed increased stability until 1200 carbonation/calcination cycles.

Among possible inorganic and organic precursors of Ca (hydroxide, nitrate, acetate) used together with zirconium(IV) propoxide to prepare ZrO_2-stabilized CaO sorbents by applying the sol–gel method, calcium hydroxide and calcium acetate led to high specific surface area and pore volume, as well as a good CO_2 capture capacity (Broda and Müller, 2014). On the contrary, the lower melting point of calcium nitrate influences negatively the capture capacity of the synthesized sorbents. The salt melting occurring during the calcination step results in the formation of very coarse grained material which presents low surface area and pore volume and consequently, reduced capture capacity. As reported by Reddy et al. (2014), a higher Ca/Zr ratio affects positively the sorbent capture capacity on account of the higher amount of active phase (CaO) in the material, but decreases its thermal stability. As example, with a uniform dispersion of $CaZrO_3$ into the CaO framework, the optimized sorbent produced from calcium hydroxide as Ca precursor and possessing a Ca/Zr ratio of 95:5 showed the highest capture capacity (0.34 g of CO_2/g of sorbent) at the end of the 90th cycle (carbonation at 800°C under 50% CO_2 and calcination at 800°C under 100% N_2), outperforming the reference limestone by 160%.

The self-reactivation behavior (Manovic and Anthony, 2008) was observed in the cyclic sorption process of a $CaZrO_3$-stabilized CaO sorbents prepared by the wet chemical method (Zhao et al., 2014). An increase of the capture capacity from 0.31 g of CO_2/g of sorbent was noticed until the 10th cycle, before becoming stabilized throughout the remaining 20 cycles (0.37 g of CO_2/g of sorbent). Based on the SEM analysis, this behavior was explained by a modification of a densely packed microstructure with little porosity in the as-prepared powder to a more porous structure with the pore size less than 100 nm after 10 carbonation/decarbonation cycles. A very high percentage of CaO (90 wt.%) in the sorbent composition offered the best performance in cyclic sorption under mild conditions (carbonation: 650°C/15% CO_2; calcination: 800°C/air). However, performing the process in severe carbonation (650°C/100% CO_2)/calcination (950°C/100% CO_2) conditions requests a higher Zr/Ca ratio to increase the sorbent thermal stability in cyclic operation. The best performance in respect to stability enhancement was noticed for a sorbent with 70 wt.% CaO, mainly due to the good dispersion of $CaZrO_3$ nanoparticles (20–80 nm) within the $\leq$200 nm CaO porous matrix.

2.1.3.2.3 Mg-Stabilized CaO Because of its high Tammann temperature (1276°C, Table 2), MgO is another inert support material that should be able to increase the performance of CaO in cyclic operation. However, Mg-stabilized CaO-based sorbents composed of microscopic $CaCO_3$ and $MgCO_3$ crystals showed a rapid loss of capture capacity similarly to natural limestone, indicating that the presence of an inert component is not sufficient to stabilize the pore structure (Filitz et al., 2011). The authors demonstrated that only mixing Ca and Mg ions on a molecular level similar to natural dolomite, a natural Ca/Mg-based sorbent, could lead to a synthesized Ca-rich dolomite sorbent possessing very high and stable cyclic capture capacity (Section 2.2). The application of the coprecipitation method revealed that aging time, precipitation agent, and pH at which the precipitation occurred strongly affected the mixing level (microscopic/molecular) between active (Ca) and inactive (Mg) compounds.

The main parameters that influence the Mg-stabilized CaO-based sorbents are the synthesis method (like coprecipitation, dry physical mixing, wet physical mixing, and solution mixing), Ca and Mg precursors, and Ca/Mg ratio.

Li et al. (2009a) synthesized several Mg-stabilized CaO sorbents using various techniques. A comparison between coprecipitation, solution mixing, dry physical mixing, and wet physical mixing methods revealed that the use of the two physical mixing options led to sorbents with the highest sorption capacity and stability in cyclic carbonation/calcination operation. The effect of Mg precursor (NanoActive MgO, $Mg(OH)_2$, acetate, and oxalate) was found less significant compared to the synthesis method; the best results were however obtained using MgO nanoparticles obtained from magnesium oxalate thermal decomposition. A study of the influence of different combinations of Ca and Mg precursors was also performed by Liu et al. (2010a). The application of a simple wet-mixing method involving several Ca precursors (D-gluconate monohydrate, acetate hydrate, L-lactate hydrate, and formate) and Mg precursors (D-gluconate, acetate tetrahydrate, and L-lactate hydrate) revealed that calcium and magnesium salts of D-gluconic acid offered the best performance in cyclic CO_2 capture due to the uniform distribution of MgO nanoparticles in the CaO structure. However, regardless of precursors, high capture capacity close to theoretical values and good stability in cyclic operation was observed. Another three-step wet-mixing method consists of the preparation of a magnesium sol from citric acid solution and MgO slurry, followed by mixing the sol to a nano-$CaCO_3$ slurry and finally calcination to produce nano-CaO/MgO-based sorbent (Lan and Wu, 2014). The later showed improved reaction rate and sorption capacity compared to nano-CaO/Al_2O_3-based sorbent. The gas (CO_2 and H_2O) emission during the decomposition of magnesium sol ($Mg(OH)_2 \times (C_6H_8O_7)_n$) in the precalcination step resulted in a more porous material (mesoporous and macroporous) with large specific surface area and average pore radius. Moreover, the increase of MgO content was found to increase both specific surface area and pore radius and has a positive effect on reaction rate and stability.

2.2 CaO–MgO Sorbents

Compared to limestone, natural CaO sorbents containing MgO, such as dolomite and huntite, possess suitable stability during cyclic operation. Dolomite ($CaMg(CO_3)_2$), an equimolar mixture of $CaCO_3$ and $MgCO_3$, is a natural material with affinity for CO_2. For its wide availability and low cost, it might be a proper choice as CO_2 sorbent. However, its uptake capacity is low compared to limestone due to the large amount of $MgCO_3$ which is not active as sorbent in the steam-reforming conditions (Filitz et al., 2011). Because of the much lower calcination temperature compared to $CaCO_3$, $MgCO_3$ acts as an inert during CO_2 capture. Silaban et al. (1992) noticed the stability of dolomite compared to limestone: after six cycles, while limestone uptake decreased from 61% to 35% of its theoretical value, dolomite kept a capacity of 40%. The presence of MgO offered to dolomite a better structural stability. Huntite is another carbonate mineral containing Mg and Ca ($Mg_3Ca(CO_3)_4$). A comparative study on the efficiency of calcite, dolomite, and huntite during 47 sorption/desorption cycles revealed the following respective order of their capacities at the end of the cyclic operation: 38%, 55%, and 84% of the initial capacity (Bandi et al., 2002). In spite of its good stability, huntite possesses a lower uptake capacity compared to the two others due to the higher content of inactive MgO.

2.3 Hydrotalcite-Based Sorbents

Hydrotalcites-like compounds (HTlc) belong to a large class of anionic and basic clays. They are also known as LDHs with a general formula $[M^{2+}_{1-x} M^{3+}_x(OH)_2]\text{–}[A^{n-}]_{x/n}\cdot nH_2O$, where M^{2+} and M^{3+} are, respectively, divalent (Mg^{2+}, Mn^{2+}, Ni^{2+}, Zn^{2+}, Cu^{2+}) and trivalent (Al^{3+}, Fe^{3+}, Cr^{3+}) cations, A^{n-} is an interlayer anion (OH^-, Cl^-, $NO_3{}^-$, $CO_3{}^{2-}$, $SO_4{}^{2-}$), and x is usually between 0.17 and 0.33 (Yong et al., 2001). For example, $Mg_6Al_2(CO_3)(OH)_{16}\cdot 4H_2O$ is a typical composition. They can be found in nature or synthesized (generally by coprecipitation) for modifying their properties by the substitution of cations and anions (Dewoolkar and Vaidya, 2016b), an important benefit of this kind of materials. Their alkalinity helps the CO_2 sorption, reduces carbon deposition, and increases H_2 selectivity in the reforming process (Section 3). Moreover, they can be easily regenerated. HTlc-based materials attracted therefore considerable attention due to their capacity to be used in a broad application range like precursors, catalysts, supports of catalysts, and sorbents (including CO_2 capture).

Yong et al. (2001) studied the effect of aluminum content, anion type, water content, and heat treatment temperature on the sorption capacity of six different commercial HTlc hydrotalcite-like compounds at high temperatures and drew several important conclusions: (i) $CO_3{}^{2-}$-containing HTlc exhibit higher capacity compared to those containing OH^-, (ii) lower water content improves the uptake capacity, (iii) the sorption capacity depends essentially on microporous volume, interlayer spacing, and layer charge density, and (iv) aluminum content and heat treatment temperature have a significant effect on sorption capacity and optimum values have to be considered in the sorption processes at high temperatures. A sorption capacity higher than 0.013 g CO_2/g sorbent at 300°C was reported for most samples. Similarly, Ding and Alpay (2000) studied the equilibrium and kinetics of CO_2 capture on potassium-promoted HTlc and reported a sorption capacity for K-HTlc of 0.029 g CO_2/g sorbent at 400°C in dry conditions (slightly higher in wet conditions, independent of water concentration in the feed). Similar results were reported by Yong and Rodrigues (2002).

The main drawback of these sorbents is therefore the low uptake capacity, even for those promoted by alkali metals like K and Cs (Oliveira et al., 2008), compared to CaO-based and ceramic-based materials. However, due to the abundance of basic sides in their structure and the high specific surface area, several works have focussed on the application of HTlc-like compounds in sorption-enhanced processes, especially in the preparation of catalysts and hybrid sorbent-catalyst materials for ethanol steam reforming (ESR) (Section 3.2.1).

2.4 Alkaline Ceramic Materials

Compared to CaO-based sorbents, the alkaline ceramic materials have the general drawback of low kinetics especially at low CO_2 partial pressures, thus limiting their application in the sorption-enhanced processes. They possess very low specific surface area and have a strong tendency to agglomerate in the reaction conditions, especially the high temperatures. Among the ceramic sorbents, lithium zirconate (Li_2ZrO_3), lithium orthosilicate (Li_4SiO_4), and sodium zirconate (Na_2ZrO_3) are the most investigated ones (Table 5). In order to improve the sorption performance, special attention has been given to the preparation approach, choice of precursors, and doping by different elements including Na, K, Al, Fe, and V.

2.4.1 *Lithium Zirconate (Li_2ZrO_3)*

Li_2ZrO_3 was introduced by Nakagawa and Ohashi (1998) as a promising CO_2 sorbent material:

$$Li_2ZrO_3(s) + CO_2(g) \leftrightarrow Li_2CO_3(s) + ZrO_2(s) \quad \Delta H^{\circ}_{298} = -160\,\mathrm{kJ/mol} \tag{3}$$

Table 5 Brief Summary of Different Investigations on Ceramic-Type Sorbents

Sorbent	Method[a]		Carbonation/Regeneration Conditions	Carbonation/Regeneration Duration (min)	Cycle Number	CO_2 Uptake at Last Cycle (g CO_2/g ads)	References
Li based	Li_2ZrO_3	CSG	500°C, 50% CO_2/650°C, 100% N_2	60/50	3	0.26	Xiao et al. (2011a)
	Li_2ZrO_3	CP	550°C, 100% CO_2/690°C, 100% N_2	20/10	3	0.23	Yi and Eriksen (2006)
	Li_2ZrO_3	LSSC	575°C, 100% CO_2/650°C, 100% Ar	12/n.a.	100	0.24	Ochoa-Fernandez et al. (2006)
	Li_2ZrO_3	SU	575°C, 100% CO_2/690°C, 100% Ar	30/30	11	0.22	Radfarnia and Iliuta (2011)
	K-Li_2ZrO_3[b]	SS	550°C, 60% CO_2/800°C, 100% N_2	100/30	6	0.2	Pannocchia et al. (2007)
	K-Li_2ZrO_3[c]	SS	500°C, 100% CO_2/750°C, 100% N_2	190/230	5	0.26	Iwan et al. (2009)
	Li_4SiO_4	SS	680°C, 100% CO_2,/800°C, 100% N_2	15/10	15	0.3	Wang et al. (2014b)
	Li_4SiO_4	SS	700°C, 50% CO_2/700°C, 100% N_2	30/30	16	0.28	Shan et al. (2013)
	Li_4SiO_4[d]	SS	550°C, 100% CO_2/550°C, 100% N_2	90/90	10	~0.154	Romero-Ibarra et al. (2013)

Na based	Na_2ZrO_3	SS	600°C, 100% CO_2/780°C, 100% N_2	30/65	2	0.24	Lopez-Ortiz et al. (2004)
	Na_2ZrO_3	LS	550°C, 100% CO_2/800°C, 100% N_2	25/55	2	0.15	Yi et al. (2006)
	Na_2ZrO_3	LS	575°C, 50% CO_2/680°C, 100% Ar	n.a.	8	~0.15	Zhao et al. (2007)
	Na_2ZrO_3	SU	575°C, 100% CO_2/840°C, 100% Ar	10/30	4	~0.13	Radfarnia and Iliuta (2012a)

[a]Synthesis method: CSG (citrate sol–gel), CP (coprecipitation), LS (liquid state), LSSC (liquid-state soft chemistry), SS (solid-state), SU (surfactant template/ultrasound assisted).
[b]Li_2CO_3:ZrO_2:K_2CO_3 (1.15:1.0:0.2).
[c]91.3% Li_2ZrO_3 + 3.4% Y_2O_3 + 0.2% Al_2O_3 + 5.1% K_2O.
[d]Ball milled.

In a working temperature range of 450–600°C, its theoretical uptake capacity is 0.28 g of CO_2/g of sorbent.

Different methods have been applied for the preparation of this kind of material. The most usual is the solid-state reaction between ZrO_2 and alkali salts (carbonate) (Ida and Lin, 2003; Ida et al., 2004; Nakagawa and Ohashi, 1998; Ohashi and Nakagawa, 1999; Pannocchia et al., 2007 Wang et al., 2014a). It has been first used by Nakagawa and Ohashi (1998) in the synthesis of the proposed Li_2ZrO_3 sorbent from ZrO_2 and Li_2CO_3 as precursors.

Using the solid-state reaction method, Ohashi and Nakagawa (1999) developed a potassium(carbonate)-doped Li_2ZrO_3 (K-Li_2ZrO_3) whose sorption rate was significantly improved compared to the undoped material. The formation of a eutectic molten carbonate layer of Li_2CO_3 and K_2CO_3 above 500°C which led to the increase of the CO_2 diffusion rate was considered the foundation of the significant enhancement of sorption kinetics. Unfortunately, the presence of potassium has a negative effect on the capture capacity and cyclic stability of the K-doped sorbent (Ochoa-Fernandez et al., 2008). A double-shell model was proposed to explain the mechanism of CO_2 chemisorption on both Li_2ZrO_3 and K-doped Li_2ZrO_3 sorbents (Ida and Lin, 2003; Fig. 12). The improvement of K-doped Li_2ZrO_3 sorption rate is on account of the formation of a liquid layer surrounding Li_2ZrO_3 (molten carbonate), through which the CO_2 diffusion is much faster compared to that in a solid carbonate layer.

The sorbents prepared by solid-state reaction present however slow kinetic rate and low capture capacity due to the large particle size and low porosity. The research in this area has therefore been directed toward the improvement of sorbents capture properties, especially by controlling the size of precursors to succeed in reducing the particle size of these materials. Xiong et al. (2003) reported that the particle size can decrease if ZrO_2 with smaller particle size is used. By applying the solid-state reaction method to prepare K-doped Li_2ZrO_3 using Li_2CO_3, K_2CO_3, and ZrO_2 with different sizes of 1 and 45 μm, they noticed that the sorbent with smaller particle size led to higher sorption rates. Despite this improvement, mixing ZrO_2 and Li_2CO_3 powders at high temperature inevitably results in material sintering. This led to the conclusion that the reduction of sorbent particle size by decreasing the particle sizes of precursors is a problematic issue and the research was focused on improving synthesis methods (Iwan et al., 2009; Nair et al., 2004; Ochoa-Fernandez et al., 2006; Radfarnia and Iliuta, 2011; Xiao et al., 2011a,b; Yi and Eriksen, 2006). The performance of Li_2ZrO_3 produced by applying liquid-state precipitation and

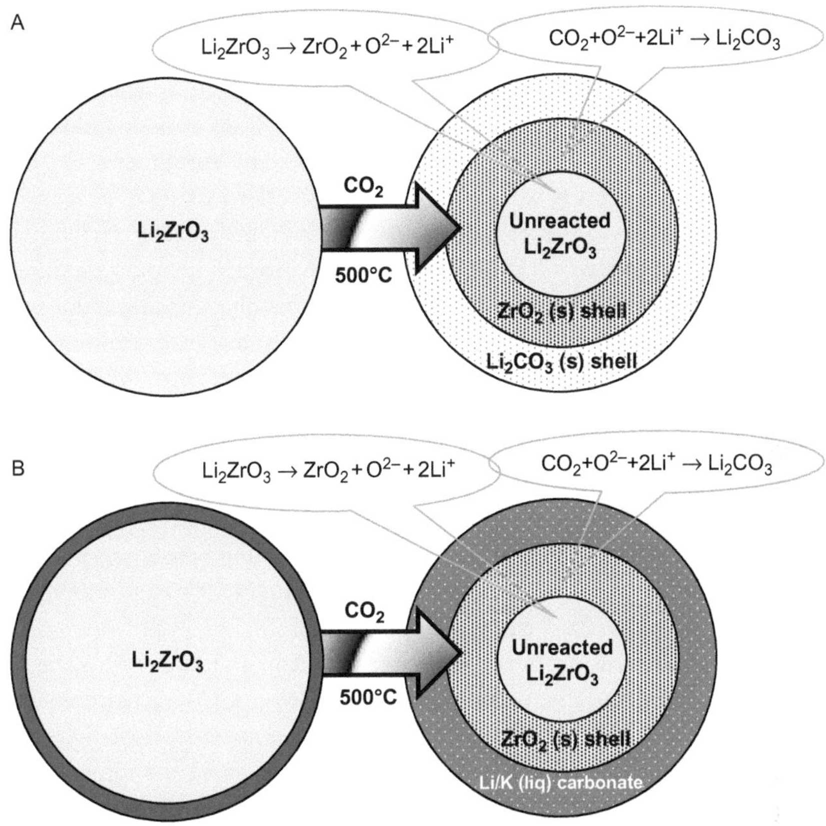

Fig. 12 Double-shell model of the mechanism of CO_2 chemisorption on (A) Li_2ZrO_3 and (B) K-doped Li_2ZrO_3 sorbents. *Adapted from Ida J, Lin YS: Mechanism of high-temperature CO_2 sorption on lithium zirconate,* Environ Sci Technol *37:1999–2004, 2003.*

solid-state reaction methods was investigated by Yi and Eriksen (2006). A comparison between the two techniques revealed that the sorbent obtained by precipitation led to a sorption rate 10 times higher. Higher CO_2 diffusion rate on account of the presence of smaller particle size in the precipitated Li_2ZrO_3 seemed to explain this significant improvement. Steam addition (>20%) could increase even more the sorption kinetics. Cyclic sorption/desorption experiments revealed that the capture rate and capacity were kept constant for three cycles.

A novel liquid-state soft-chemistry route employing zirconoxy nitrate and lithium acetate for the preparation of nanocrystalline tetragonal and monoclinic Li_2ZrO_3 was proposed by Ochoa-Fernandez et al. (2006). Compared to the solid-state reaction, higher sorption and regeneration rates were obtained for the nanocrystalline tetragonal Li_2ZrO_3 due to presence of small crystallite size (about 13 nm) in the material synthesized by the liquid-state soft-chemistry method. The stability in cyclic operation was improved as well, the sorbent being able to maintain above 90% of its initial intake capacity after 100 sorption/regeneration cycles. The use of Li_2CO_3 and $Zr(OH)_2$ instead of ZrO_2 was shown to offer a better mixing of precursors in the aqueous slurry (Iwan et al., 2009). Moreover, the calcination temperature could be reduced (700–750°C) in comparison with the conditions requited in the solid-state method ($\geq$900°C). This modification helps prevent the product agglomeration, leading therefore to materials (promoted or not by potassium carbonate) with higher surface area (~11 m^2/g) and reaction rate. Compared to the tetragonal Li_2ZrO_3 sorbent prepared via the liquid-state soft-chemistry route (Ochoa-Fernandez et al., 2006), the sorption rate was however lower.

A citrate sol–gel method was developed by Xiao et al. (2011a) to produce Li_2ZrO_3 nanocrystals with a tetragonal phase, revealing better performance (faster kinetics, improved capture capacity (~0.26 g of CO_2/g of sorbent vs stoichiometric value 28.7 wt.%), and good stability in cyclic operation) compared to materials produced by coprecipitation and liquid-phase techniques. However, the uptake rate declines at lower CO_2 partial pressures (below 0.25 bar), and the capacity becomes insignificant under 0.15 bar. Optimized K-doped Li_2ZrO_3 sorbents synthesized using the same citrate sol–gel method were however shown to present better capture performance especially at low CO_2 partial pressures, as well as good stability in cyclic operation (Xiao et al., 2011b).

Another method based on surfactant template/sonication was applied by Radfarnia and Iliuta (2011) to produce porous nanoparticle Li_2ZrO_3 by using zirconoxy nitrate, lithium acetate, and cetyl trimethylammonium bromide above critical micelle concentration (CMC). The performance of the sorption efficiency in cyclic operation was compared with that of sorbents prepared by the simple surfactant template method (porous-without sonication) or conventional soft-chemistry route. By ultrasound application, the acoustic cavitation of bubbles (formation, growth, and implosive collapse of bubbles) creates local hot spots that can produce very high pressures and temperatures (about 1000 atm. and 5000°C), as well as heating and cooling rates above 1010 K/s, which is extremely effective in breaking aggregated particles and forming nanocrystals (Suslick et al., 1991). Whether assisted by ultrasound or not, a better sorption capacity and rate were obtained for porous nanopowders in comparison with samples prepared by the conventional method, mainly due to a less agglomerated structure and porous framework facilitating the gas and ion diffusions to/from particle layers. However, the porous material prepared without sonication presented instability during cyclic operation. Integrating the surfactant template method with sonication was found to significantly reduce the particle aggregation on the material surface. Both surfactant concentration and sonication time were shown to strongly control the material properties (crystallite size and BET surface area) and stability. The porous sorbent prepared by the ultrasound-assisted surfactant-template method using less surfactant and irradiation time led to significantly higher uptake capacity compared to other samples. However, in spite of the good sorption capacity, the porous Li_2ZrO_3 was found to suffer from slow sorption kinetics at low CO_2 partial pressure (below 0.75 bar).

2.4.2 Lithium Orthosilicate (Li_4SiO_4)

Lithium orthosilicate was suggested by Kato and Nakagawa (2001) as a new high-temperature CO_2 sorbent:

$$Li_4SiO_4(s) + CO_2(g) \leftrightarrow Li_2CO_3(s) + Li_2SiO_3(s) \quad \Delta H^{\circ}_{298} = -143 \text{kJ/mol} \tag{4}$$

Prepared from Li_2CO_3 and SiO_2 as precursors and using the solid-state method, the sorbent demonstrated adequate sorption capacity in the range of 450–700°C (up to 0.36 g of CO_2/g of sorbent). With few exceptions, the solid-state reaction method was mainly used in different works to synthesize Li_4SiO_4.

At low CO_2 partial pressure, the reports on the efficiency of Li_4SiO_4 are quite inconsistent. While Kato et al. (2005b) reported that both sorption capacity and rate were improved for Li_4SiO_4 compared to Li_2ZrO_3 (around 50% more capacity and 30 times higher kinetics), Rodriguez-Mosqueda and Pfeiffer (2010) noticed a low kinetic rate compared to other alkaline ceramics like lithium and sodium zirconates. The study performed on Li_4SiO_4 prepared by solid-state reaction (Rodriguez-Mosqueda and Pfeiffer, 2010) suggested that the sorption process depends on both temperature and CO_2 flow. In addition, the kinetic analysis revealed that CO_2 sorption kinetics is controlled by lithium diffusion, independent of the CO_2 flow.

The particle size of the sorbent was also found to exert a significant effect on the uptake efficiency. By applying different synthesis methods (solid-state reaction, precipitation, and sol–gel), Venegas et al. (2007) reported that the material obtained by precipitation showed the highest reactivity toward CO_2. This improvement was considered to be on account of the small particle size (~ 3 mm) which led to more active lithium atoms on the particles' surface. The use of ball milling is equally another possibility to improve the sorbent efficiency by decreasing its particle size without further sintering effects and increasing the surface area (Romero-Ibarra et al., 2013). Li_4SiO_4 prepared from LiOH and SiO_2 by solid-state reaction and further mechanically modified by dry ball milling showed a significant decrease in the crystal size from >500 to 175 Å and an increase of specific surface area from 0.4 to 4.9 m^2/g, thus leading to better behavior in cyclic operation by reducing the sintering phenomenon.

The effect of the type of silica precursor on the performance of developed materials is also a noteworthy subject. Wang et al. (2014b) investigated the sorption behavior of Li_4SiO_4-based sorbents prepared via a solid-state reaction method by using Li_2CO_3 and different sources of silica (citric acid pretreatment rice husk ash (CRHA), nanostructured aerosil, and crystalline quartz powders). CRHA-Li_4SiO_4 exhibited the best performance from the point of view of capture capacity (30.5 wt.%), kinetics, and regenerability in cyclic operation (2.1 wt.% decrease of intake capacity after 15 cycles). This performance was attributed to nanoparticle microstructure (lower crystallite size of pure Li_4SiO_4, smaller particle size, and larger specific surface area) and CRHA strong sintering-resistant character. The possibility to use diatomite, a low-cost source of silicon, in the synthesis of Li_4SiO_4-based sorbents via the solid-state reaction was investigated by Shan et al. (2013). The best results were obtained for a Li_2CO_3/SiO_2 molar ratio of 2.6 (sorption capacity and stability during 16 cycles).

Several attempts were directed on the improvement of Li_4SiO_4 capture performance by doping with various elements like Na, K, Ca, Mg, Al, Fe, Cr, Ce, and V (Chen et al., 2016; Mejía-Trejo et al., 2008; Ortiz-Landeros et al., 2012; Seggiani et al., 2011, 2013; Wang et al., 2017; Yang et al., 2016; Zhang et al., 2014). An interesting study was performed by Seggiani et al. (2013) on synthesis of Li_4SiO_4-based sorbents with 10, 20, and 30 wt.% of alkali carbonates (K_2CO_3, Na_2CO_3) as well as binary (K_2CO_3/Li_2CO_3, Na_2CO_3/Li_2CO_3) and ternary ($K_2CO_3/Na_2CO_3/Li_2CO_3$) eutectic carbonate mixtures. All promoted sorbents showed improved performance in comparison with Li_4SiO_4 due to the faster diffusion of CO_2 through the molten carbonate shell compared to the solid Li_2CO_3 shell present in no-promoted Li_4SiO_4. Among the promoted sorbents, those containing alkali carbonates presented higher sorption rate and capacity compared to those containing binary or ternary eutectic carbonate mixtures. While the eutectic melts $(K/Li)CO_3$ and $(Na/Li)CO_3$ are formed during the sorption process on Li_4SiO_4 surface of alkali carbonate-promoted sorbents, a eutectic liquid phase is already present at the beginning of CO_2 sorption on Li_4SiO_4 surface of promoted sorbents containing binary and ternary eutectic carbonate mixtures which implies an additional diffusional resistance. Na_2CO_3-promoted sorbent showed however a significant degeneration of CO_2 capture capacity in cyclic operation due to sintering. On the contrary, K_2CO_3-promoted material exhibited very good cyclic stability. The existence of eutectic carbonates $(K/Li)CO_3$ and $(Na/Li)CO_3$ was confirmed by Yang et al. (2016) by combining differential scanning calorimetry and in situ powder XRD.

The performance of a Ca-doped Li_4SiO_4 synthesized by solid-state reaction was evaluated by Chen et al. (2016). Doping by Ca resulted in an important increase of the specific surface area (up to five times), CO_2 sorption capacity (up to 35.1 wt.% for 6 mol.% Ca in the sorbent) and cyclic performance (26% after 15 cycles for the same material composition). A modified double-shell model was proposed for the mechanism of CO_2 sorption/desorption over Ca–Li_4SiO_4. The transformation between the Ca species involved in the system ($CaSiO_4$–$CaCO_3$–Li_2CaSiO_4) during the sorption/desorption cyclic process was found to increase the sorption efficiency by facilitating the transfer of CO_2 to Li_4SiO_4 ($CaSiO_4 \rightarrow Li_2CaSiO_4$) or accelerating the desorption by the reaction between Li_2CaSiO_4 and Li_2CO_3 ($Li_2CaSiO_4 \rightarrow CaSiO_4$). Among different metallic (K, Mg, Cr, or Ce)-doped Li_4SiO_4 prepared by the sol–gel method, doping by Ce was found to inhibit the growth of crystal aggregation efficiently, thus resulting in a foam-like morphology with small particle size (Wang et al., 2017). Ce_2–Li_4SiO_4 (0.02 mol of Ce ion doped into 1 mol of Li_4SiO_4) exhibited the best performance compared to Li_4SiO_4: higher CO_2 capacity (34.57 wt.%), faster sorption kinetics, and improved cyclic performance. More recently, the impregnation of TiO_2 nanotubes into Li_4SiO_4 structure by the solid-state method resulted in a significant improvement of stability (10 cycles) and kinetics of CO_2 sorption at 700°C in comparison with conventional Li_4SiO_4, due to the strong channeling effects of TiO_2 nanotubes that improve the lithium ions diffusion through the sorbent structure (Lee and Yavuz, 2017). Li_4SiO_4–TiO_2 material was prepared using TiO_2 nanotubes (from P25 TiO_2), LiOH, and fumed silica, followed by thermal treatment.

2.4.3 Sodium Zirconate (Na_2ZrO_3)

Na_2ZrO_3 was proposed as an interesting sorbent material due to its higher sorption rate at high temperatures and lower cost of precursors:

$$Na_2ZrO_3(s) + CO_2(g) \leftrightarrow Na_2CO_3(s) + ZrO_2(s) \quad \Delta H^{\circ}_{298} = -149\ \text{kJ/mol} \quad (5)$$

Among different sorbents developed by solid-state reaction (Li_2ZrO_3, Na_2ZrO_3, Li_4SiO_4, Na_2TiO_3, and Na_3SbO_4), Na_2ZrO_3 was found to exhibit the best sorption kinetics at high temperature, but a poor regeneration performance in comparison with the most investigated sorbents, Li_2ZrO_3 and Li_4SiO_4 (Lopez-Ortiz et al., 2004). Its maximum theoretical sorption capacity is 0.2376 g CO_2/g sorbent.

The most important drawbacks associated with Na_2ZrO_3 are the high regeneration temperature and the inherent weakness in the regeneration process. Most studies were directed toward the improvement of capture performance by incorporating various active metals into its structure and/or the producing materials with smaller particle size. Except for the solid-state reaction method which has been mainly used but with little success, different synthesis techniques have been developed to improve the sorbent performance.

A novel soft-chemical route employing zirconoxy nitrate and sodium citrate for the preparation of nanocrystalline Na_2ZrO_3 was proposed by Zhao et al. (2007). The amorphous zirconium complex formed in the reaction is calcined in a controlled atmosphere. The reaction between nitrate and citrate results in the in situ carbon formation, which is further burned to form an open pore structure. The calcination atmosphere and temperature control the Na_2ZrO_3 crystal phase and favor the formation of pure monoclinic phase, which shows faster kinetics and better stability compared to the hexagonal one.

A kinetic analysis of sorption and diffusion of CO_2 on Na_2ZrO_3 (Alcérreca-Corte et al., 2008) revealed that while a fast kinetics occurs between 550°C and 700°C, the sorption behavior below 300°C was atypical: 17 wt.% at 150°C, 10 wt.% at 200°C, and only 4 wt% at 250°C (5-h process in all cases). This was associated with powder sintering produced by a thermal shock during the heating process, accompanied by a decrease of the surface area, thus inhibiting the CO_2 sorption. The process was found to be controlled by sodium diffusion.

Based on the improvement of capture performances of Li_2ZrO_3 by applying a surfactant template/sonication method and the fact that Na_2ZrO_3 is supposed to offer better CO_2 sorption kinetics compared to Li_2ZrO_3 (Lopez-Ortiz et al., 2004; Zhao et al., 2007), Radfarnia and Iliuta (2012a) used the same method to develop porous Na_2ZrO_3. Interestingly, the behavior of the developed Na_2ZrO_3 was unexpected. The results were inconsistent with those obtained for porous Li_2ZrO_3, which displayed a remarkably improved sorption behavior. The sorbents synthesized by applying the surfactant template/sonication method were less active during cyclic operation compared to the Na_2ZrO_3 prepared by liquid-state mixing. The instability of sorbents prepared by a surfactant template technique was interpreted by particle agglomeration due to sintering during the high temperature required in the regeneration step and, consequently, the loss of material porosity.

The CO_2 chemisorption on Na_2ZrO_3 synthesized by solid-state reaction was studied in depth as a function of temperature by Martinez-dlCruz and Pfeiffer (2012) to evaluate the structural and/or microstructural modifications occurring during the sorption process. It was demonstrated that during the chemisorption Na_2CO_3–ZrO_2 was produced on the external shell of Na_2ZrO_3. Moreover, the microstructural properties were influenced by temperature significantly. Between 300°C and 550°C, Na_2CO_3–ZrO_2 external shell is mesoporous, allowing the CO_2 molecules to diffuse and continue the sorption process. At higher temperatures, the chemisorption process is kinetically controlled by sodium ionic diffusion through the sodium carbonate crystals due to (i) sintering of Na_2CO_3–ZrO_2 external shell which does not facilitate the CO_2 diffusion and consequently, and (ii) the loss of mesoporosity.

2.4.4 Other Ceramic Sorbents

Other than Li_2ZrO_3, Li_4SiO_4, and Na_2ZrO_3, several other ceramic materials have been studied, but to a much lesser extent. They include lithium oxosilicate (Li_8SiO_6) (Duan et al., 2013; Durán-Muñoz et al., 2013; Romero-Ibarra et al., 2014), penta-lithium aluminate (Li_5AlO_4) (Ávalos-Rendón et al., 2009, 2012), lithium cuprate (Li_2CuO_2) (Matsukura et al., 2010; Palacios-Romero and Pfeiffer, 2008), lithium orthotitanate (Li_4TiO_4) (Togashi et al., 2007; Ueda et al., 2011), lithium ferrite ($LiFeO_2$) (Kato et al., 2005a; Yanase et al., 2010), barium ferrite ($Ba_2Fe_2O_5$) (Fujishiro et al., 2011, 2012), and sodium metatitanate (Na_2TiO_3) (Lopez-Ortiz et al., 2004; Sanchez-Camacho et al.,

2014). The CO_2 sorption results in the formation of the corresponding alkaline carbonate and the residual oxide or secondary alkaline phase, in the same manner as with Li_2ZrO_3, Na_2ZrO_3, and Li_4SiO_4.

The interest of Li_5AlO_4 lies in its high Li/Al molar ratio and the lower atomic mass of Al in comparison with other elements included in lithium ceramics. Despite its outstanding theoretical uptake capacity of 0.87 g of CO_2/g of sorbent (the highest among all ceramics) and the capability to react with CO_2 in a very wide temperature range (200–700°C), it has the drawback of poor stability in cyclic operation (Ávalos-Rendón et al., 2009, 2012). Under the same operation conditions (sorption at 700°C under pure CO_2/desorption at 750°C under pure N_2), β-Li_5AlO_4 phase exhibits higher uptake capacity but more important capacity reduction after 20 cycles (from 62.3 wt.% to 8.1 wt.%) in comparison with α-Li_5AlO_4 (47.7 wt.% to 22.1 wt.%). For both lithium aluminates polymorphs, Li_2O sublimation during regeneration seems to be on account of the capacity reduction during cyclic operation. Li_2CuO_2 has a quite high theoretical uptake capacity of 0.402 g of CO_2/g of sorbent and is also able to react with CO_2 in a very wide temperature range of 120–690°C. For an economic point of view, it might be an appealing alternative to Li_2ZrO_3. When the coprecipitation method is used for this sorbent preparation, the real uptake capacity is much lower (0.136 g of CO_2/g of sorbent at 650°C) due to the impossibility to synthesize the single-phase Li_2CuO_2 (Palacios-Romero and Pfeiffer, 2008). However, the application of the solid-state reaction at 680–685°C by using Li_2CO_3 and CuO leads to a single-phase Li_2CuO_2, whose sorption capacity is identical to the theoretical value (Matsukura et al., 2010). Li_8SiO_6 is another ceramic material able to react with CO_2 between 300°C and 700°C. It was first synthesized by Durán-Muñoz et al. (2013) by solid-state reaction from silicon dioxide (SiO_2) and lithium oxide (Li_2O). The sorbent was shown to present faster kinetics in comparison with Li_4SiO_4, as well as high uptake capacity of 0.521 g of CO_2/g of sorbent at 650°C. This capacity can become even higher by adding sodium and/or potassium carbonate, which leads to the formation of eutectic phases (Romero-Ibarra et al., 2014). The working temperature for CO_2 sorption on Li_4TiO_4 is around 300–850°C. The kinetics was found to be faster than that of Li_4SiO_4 (Togashi et al., 2007). Compared to the above-mentioned ceramic materials, Na_2TiO_3, $Ba_2Fe_2O_5$, and $LiFeO_2$ are characterized by very low capture capacity. In addition to a low capture capacity (0.08 g of CO_2/g of sorbent at 600°C), Na_2TiO_3 has a sorption/desorption rate lower than Na_2ZrO_3, which is the most investigated sodium-based sorbent, due to the presence of high molecular titanates (produced in the system at temperatures higher than 700°C) with low reactivity toward CO_2 (Lopez-Ortiz et al., 2004; Sanchez-Camacho et al., 2014). The low capacity of $LiFeO_2$ (0.01 g of CO_2/g of sorbent at 500°C (Kato et al., 2005a)) is mainly due to phase transitions occurring above 425°C, which limit the CO_2 capture (Yanase et al., 2010). The CO_2 capture on $Ba_2Fe_2O_5$ takes place between 500°C and 1000°C, with regeneration above 1000°C. Despite the low capacity of 0.094 g of CO_2/g of sorbent at 1000°C, this material shows better kinetics than Li_4SiO_4 (Fujishiro et al., 2011).

2.4.5 Kinetic Models Applied to Alkaline Ceramic Materials

Double-shell, multiple-exponential, shrinking core, Avrami–Erofeev, and rate law models have been proposed to analyze CO_2 sorption mechanism on alkaline ceramic materials.

Ida and Lin (2003) used a *double-shell model* to describe the mechanism of CO_2 sorption, a chemical sorption process followed by a CO_2 diffusion process, on pure and K-doped Li_2ZrO_3 (Fig. 12). CO_2 sorption mechanism assumes that CO_2 moves to the surface of the lithium zirconate and reacts with Li_2ZrO_3 to form Li_2CO_3 and ZrO_2. In the initial fast period of the CO_2 sorption process, an external dense solid Li_2CO_3 shell is formed followed by the generation of a dense ZrO_2 shell in the middle around the unreacted Li_2ZrO_3 core. CO_2 diffuses through the solid Li_2CO_3 shell and reacts with Li^+ and O^{2-} on the outer surface of the middle ZrO_2 shell. Li^+ and O^{2-} are generated by another reaction on the surface of Li_2ZrO_3. Similarly, Kato et al. (2005b) and Essaki et al. (2006) applied the same model to explain CO_2 sorption mechanism on Li_4SiO_4.

The *double-exponential model* is used to explain the direct CO_2 capture by superficial reaction and the CO_2 capture kinetically controlled by the diffusion process, which follows the external shell generation. This kind of models has been considered in the description of CO_2 sorption mechanism on different alkaline ceramics including Li_4SiO_4, Na_2ZrO_3, Li_8SiO_6, and Li_5AlO_4. According to the model, two different processes take place during CO_2 capture: (1) CO_2 chemisorption at the particle surface, which generates an external shell composed of lithium/sodium carbonate and a metal oxide or a lithium/sodium secondary phase; and (2) lithium/sodium diffusion through the external carbonate layer (which begins once the first stage is complete) followed by the reaction with CO_2 at the surface (Alcérreca-Corte et al., 2008; Ávalos-Rendón et al., 2009; Duan et al., 2013; Mejía-Trejo et al., 2008; Rodriguez-Mosqueda and Pfeiffer, 2010; Venegas et al., 2007; Zhang et al., 2013b).

Zhang et al. (2013b) analyzed the reaction mechanism of CO_2 sorption/desorption process on Li_4SiO_4 via *shrinking core*, *double-exponential*, and *Avrami–Erofeev models*. The shrinking core model, based on the assumption that the reaction rate is controlled by the chemical reaction rate, does not fit the experimental data well in the entire temperature range (550–700°C), indicating that the sorption process is not limited only by the chemical reaction rate. The comparison between the experimental data and the double-exponential model, characterized by CO_2 chemisorption process followed by a lithium diffusion process which occurs as soon as the carbonate-oxide external shell is totally formed, showed that the lithium diffusion process is the limiting step of the CO_2 sorption process. However, this remark is not consistent with the model assumption that the lithium diffusion process takes place once the external shell is entirely formed. The Avrami–Erofeev model was found to be most appropriate to describe CO_2 sorption/desorption process and when linked with the double-shell model undoubtedly explains the mechanism of the CO_2 sorption/desorption on Li_4SiO_4. The sorption process is controlled by the rate of formation and growth of the crystals with a Li_2CO_3/Li_2SiO_3 double-shell structure. The desorption process is limited by the rate of formation and growth of the Li_4SiO_4 crystals.

Pfeiffer and coworkers have published several papers about the application of the *rate law model* for analyzing the mechanism of CO_2 sorption on different ceramics such as Li_8SiO_6, Na_2TiO_3, and Na_2ZrO_3 (Alcérreca-Corte et al., 2008; Durán-Muñoz et al., 2013; Sanchez-Camacho et al., 2014). The rate law model is used when several processes are involved in the mechanism of CO_2 sorption.

3. APPLICATION OF HIGH-TEMPERATURE SORBENTS IN THE SORPTION-ENHANCED PROCESS FOR HYDROGEN PRODUCTION BY STEAM REFORMING: SORBENT-CATALYST SYSTEMS FOR DIFFERENT PROCESSES

The sorption-enhanced steam reforming is an innovative intensified technology, which integrates the reforming reaction (H_2 production) with a selective in situ gas separation (CO_2 sorption) in a single step, in order to shift thermodynamically the reforming reaction and increase the production of hydrogen. There are numerous advantages to this emerging process compared to the traditional steam reforming, mainly:

(i) a highly pure H_2 production (very low CO and CO_2 concentrations) without the need for an individual WGS reactor;
(ii) higher conversions to hydrogen;
(iii) lower risk of coke formation and deposition in the reformer;
(iv) limitation of CO_2 release to the atmosphere along with the production of relatively pure CO_2 for sequestration or further use in different processes;
(v) lower steam/carbon ratios required.

The research interest related to SERP development has been mostly directed toward hydrogen production by different processes like steam reforming of (i) hydrocarbons (methane, ethane, propane, etc.), (ii) ethanol, and (ii) biomass-derived chemicals (in particular, glycerol). Among them, most attention was focussed on methane steam reforming. Table 6 highlights some investigations in this area.

The key element in the SESR processes is the selection of appropriate sorbent materials. The choice of efficient sorbents for selective in situ CO_2 removal from the reaction medium is still a big challenge due to the severe operating conditions, especially the high temperature and the presence of steam during the sorption process. In addition, these sorbent materials need to provide adequate CO_2 sorption capacity and kinetics at relatively low CO_2 partial pressures. Finally, the ease of regeneration, the long-term cyclic stability, and a reasonable production cost are also essential parameters to be considered in the sorbent selection. The high-temperature solid sorbents available in the literature mainly include CaO-based, hydrotalcite (HTlc)-based, and alkaline-based (e.g., Li_2ZrO_3, Na_2ZrO_3, and Li_4SiO_4) materials. Among them, CaO-based sorbents that have found to offer the best performance are the object of most research works.

3.1 Steam Reforming of Methane

As hydrogen is mostly produced by methane (natural gas) steam reforming, most attention has been paid to the application of high-temperature CO_2 sorbents in the sorption-enhanced methane steam reforming (SEMSR). Desulfurized natural gas leads first to a mixture of H_2, CO_2, CO, water, and unreacted CH_4, which is then enriched in H_2 and CO_2 by combining the reforming process with the WGS reaction. Through implementation of the in situ carbon capture, the overall reaction (9) is a combination of the endothermic methane reforming, the exothermic WGS, and the CO_2 reaction with a metal oxide (MeO):

Table 6 Brief Summary of Different Investigations on SESR

Feedstock	Sorbent and Catalyst (Pattern)[a]	NiO Loading[b] (wt.%)	Reaction Conditions[c] Reforming/Regeneration	Cycle Number	H_2 Conc.[d–f] (%)	References
Methane	Ni com[g]. cat./com. CaO (M)	22.0	450–750°C, 15 atm., S/C 3–5/n.a.	1	>95	Balasubramanian et al. (1999)
	Ni com. cat./dolomite (M)	18	650°C; 15 atm., S/C 4/800–950°C, 100% N_2, 4% O_2/N_2, 100% CO_2	25	>95	Ortiz and Harrison (2001)
	Ni com. cat./CaO: $Ca_{12}Al_{14}O_{33}$ (M)	20	630°C; 1 atm., S/C 5/850°C, 100% Ar	12	>90	Li et al. (2006b)
	Ni com. cat./dolomite (M)	(a)	600°C, 1 atm., S/C 3/850°C, 100% N_2	4	>98	Johnsen et al. (2006)
	Ni com. cat./CaO: $Ca_{12}Al_{14}O_{33}$ (M)	18	650°C, 1 atm., S/C 3.4/850°C, 100% He	13	>92	Martavaltzi et al. (2010)
	NiO/CaO/$Ca_{12}Al_{14}O_{33}$ (H)	20	650°C, 1 atm., S/C 3.4/850°C, 100% He	1	90	Martavaltzi and Lemonidou (2010)
	NiO/CaO (H)	(b)	600°C, 1 atm., S/C 3/n.a.	1	80	Chanburanasiri et al. (2011)
	NiO/CaO–Zr (H)	13–20.5	650°C, 1 atm, S/C 4/800°C, 11.1% H_2 in Ar	10	>95	Radfarnia and Iliuta (2014b)
	NiO/CaO–$Ca_9Al_6O_{18}$ (H)	12–25	650°C, 1 atm., S/C 4/800°C, 11.1% H_2 in Ar	10/30	>95	Radfarnia and Iliuta (2014a)
	La_2O_3/NiO/CaO/Al_2O_3 (H)	n.a.	600°C, 1 atm., S/C 4/800°C, 100% N_2	30	>92	Feng et al. (2012)

	$ZrO_2/NiO/CaO/Al_2O_3$ (H)	(c)	600°C, 1 atm., S/C 4/800°C, 100% N_2	20	>90	Wu and Wang (2010)
	NiO/CaO/HTlc (Al–Mg) (H)	(d)	550°C, 1 atm., S/C 4/750°C, 100% N_2	10	99	Broda et al. (2012)
	Ni–CaO–$Ca_{12}Al_{14}O_{33}$ (H)	7	630°C, S/C 3/780°C, 100% N_2	4	95	Kim et al. (2013)
	NiO–HTlc (Al–Mg)/Al-stabilized CaO (M)	(e)	550°C, 1 atm., S/C 4/750°C, 100% N_2	10	99	Broda et al. (2013b)
	NiO–HTlc (Al–Mg)/ CaO:$Ca_9Al_6O_{18}$ (M)	(f)	550°C, 1 atm., S/C 4.2/800°C, 20% H_2/80% N_2	1	97	Xie et al. (2012)
	Ni com. cat./Li_2ZrO_3 (M)	n.a.	505°C, 1 atm., S/C 4/n.a.	1	85	Yi et al. (2006)
	Ni com. cat./Na_2ZrO_3 (M)	n.a.	600°C, 1 atm., S/C 4/n.a.	1	97	Yi et al. (2006)
	NiO–HTlc (Al–Mg)/ Na_2ZrO_3 (M)	(g)	575°C, 5 atm., S/C 5/750°C, 100% Ar	1	97	Ochoa-Fernandez et al. (2007b)
Ethanol	Ni–Al_2O_3 com. cat./ Li_4SiO_4 (M)	58	577°C, 1 atm., S/C 3/n.a.	1	99	Essaki et al. (2008)
	Co–Ni–HTlc/dolomite (M)	(h)	550°C, 1 atm., S/C 3/770°C, 100% Ar	8	>99(Y)	He et al. (2010a)
	Ni/Al_2O_3/com. HTlc (H)	19	400°C, 1–5 atm., S/C 1.5–5/420°C, inert gas	2	~99(Y)	Cunha et al. (2012)
	Ni–MCM-41/CaO (M)	(i)	600°C, 1 atm., S/C 1.6/n.a.	1	~94	Gunduz and Dogu (2012)

Continued

Table 6 Brief Summary of Different Investigations on SESR—cont'd

Feedstock	Sorbent and Catalyst (Pattern)	NiO Loading (wt.%)	Reaction Conditions Reforming/Regeneration	Cycle Number	H_2 Conc.[d–f] (%)	References
	Cu–HTlc (H)	5(Cu)	550°C, 1 atm., S/C 3/770°C, 100% Ar	1	>90	Cunha et al. (2013)
	Ni–HTlc (H)	5.9	500°C, 1 atm., S/C 3/500°C, 100% Ar	5	>99	Wu et al. (2013)
	K–Ni–Cu–HTlc (H)	16	500°C, 1 atm., S/C 5/n.a.	1	99.8	Cunha et al. (2014)
Glycerol	Ni/Al_2O_3/dolomite (M)	18	500°C, 1 atm., S/C 3/n.a.	1	97	Dou et al. (2009)
	Co–Ni/dolomite (M)	(h)	575°C, 1 atm., S/C 3/770°C, 100% Ar	1	~99	He et al. (2010b)
	Ni/(Ca/Al_2O_3) (H)	41.21	550°C, 1 atm., S/C 3/900°C, 100% N_2	5	90	Wang et al. (2015b)
	NiO–$NiAl_2O_4$/K–Li_2ZrO_3 (M)	15.2	550°C, 1 atm., S/C 3/900°C, 100% N_2	5	Up to 99.59	Wang et al. (2015a)
	Co–CaO $Ca_{12}Al_{14}O_{33}$ (H)	15(Co) (j)	525°C, 1 atm., S/C 4/770°C, 100% Ar	50	96.4	Dang et al. (2016)
	La–Ni/Al_2O_3/CaO (M)	16	600°C, 1 atm., S/C 4/770°C, 100% Ar	5	70(wet)[h]	Sanchez et al. (2016)
	Pt–NiCuHTlc[i] (H)	12.7	550°C, 1 atm., S/C 4.5/550°C, steam	20	98.7	Dewoolkar and Vaidya (2016a)
	NiO–-$NiAl_2O_4$/limestone (M)	12.5	550°C, 1 atm., S/C n.a./900°C	6	98	Dou et al. (2016a)

	Ni-, Zn-, or Cu-modified Co–CaO $Ca_{12}Al_{14}O_{33}$ (H)	(k)	525°C, 1 atm., S/C 4/700°C, 100% Ar	10	Up to 99.27	Dang et al. (2017)
	Ni–CaO–MMT (H)	20	600°C, 1 atm., S/C 1/700°C, 100% N_2	10	95	Jiang et al. (2017)
	$Ca_9Al_6O_{18}$–CaO/xNiO (x: 15, 20, 25 wt.%) (H)	15–25	550°C, 1 atm., S/C 3/800°C, 100% Ar	5	~98	Shokrollahi Yancheshmeh et al. (2017)
	$Ca_9Al_6O_{18}$–CaO/ 20NiO–$y$$CeO_2$ (y: 5, 10, 15 wt.%) (H)	20	550°C, 1 atm., S/C 3/800°C, 100% Ar	20	~98	Shokrollahi Yancheshmeh et al. (2017)
Butanol	K–Ni–CaO/HTlc (H)	12.7	550°C, 1 atm., S/C 3/800°C, 100% Ar	21	98.7	Dewoolkar and Vaidya (2017)
Methanol	Ni com. cat./CaO (M)	11	453°C, 3 atm., S/C 4/800°C, 100% Ar	600	99.6	Lysikov et al. (2008)

[a]Mixture (M) or hybrid (H).
[b]Or loading of other metal, when specified.
[d]Conditions other than those specified in the table are presented in some publications; the references should be consulted for more data and details.
[e]Dry basis (if not otherwise specified).
[f]Or H_2 yield (Y) when specified; as the mentioned values can correspond to the first cycle and/or a maximum, the corresponding references should be consulted for more data and details.
[g]Commercial product.
[h]In the total exit flow (wet basis).
[i]Cyclic performance (20 cycles) evaluated for several materials: NiMgHTlc, NiCuHTlc, Pt–NiMgHTlc, and Pt–NiCuHTlc.
(a) Ni loading: >12 wt.%; (b) Ni loading: 12.5 wt.%; (c) Ni loading: 15 wt.%; (d) Ni loading: 45 wt.%; (e) Ni loading: 47 wt.%; (f) Ni:Mg:Al (atomic ratio in HTlc catalyst) = 0.5:2.5:1; (g) Ni loading: 40 wt.%; (h) Co + Ni loading 40%; (i) Ni/Si = 0.12; (j) 15 wt.% Co; (k) 4.05 wt.% Ni, 4.41 wt.% Zn, 5.05–15.34 wt.% Cu, depending on material.

Reforming:

$$CH_4(g) + H_2O(g) \leftrightarrow 3H_2(g) + CO(g) \quad \Delta H^\circ_{298} = 205.9\,\text{kJ/mol} \tag{6}$$

WGS:

$$CO(g) + H_2O(g) \leftrightarrow H_2(g) + CO_2(g) \quad \Delta H^\circ_{298} = -41.1\,\text{kJ/mol} \tag{7}$$

CO_2 sorption:

$$MeO(s) + CO_2(g) \leftrightarrow MeCO_3(s) \quad \Delta H^\circ < 0 \tag{8}$$

Overall:

$$CH_4(g) + 2H_2O(g) + MeO(s) \leftrightarrow 4H_2(g) + MeCO_3(s) \quad \Delta H^\circ \approx 0 \tag{9}$$

In the conventional SMR process, the exit gas flow usually contains CO_2, CO, and nonreacted methane. As mentioned earlier, the in situ removal of CO_2 by application of the SE process results in a highly pure H_2 production with very low concentrations of CO and CO_2. For most applications involving hydrogen, further very costly purifications of the gas can thus be avoided.

CaO-based sorbents are the most used sorbents in the SEMSR process so far. In a previous study, Balasubramanian et al. (1999) investigated the H_2 production by using CaO sorbent (from high-purity $CaCO_3$) and the influence of several operating parameters (temperature, steam to carbon ratio, and feedstock composition) to identify the optimum operation conditions. H_2 with a purity >95% was produced at 650°C and S/C of 4. However, the uptake capacity was found to decrease in cyclic operation, particularly under severe regeneration conditions (presence of CO_2) due to sintering. Many attempts have therefore been made to develop efficient (thermal stable) CaO-based sorbents.

In the application of CaO (75%)/$Ca_{12}Al_{14}O_{33}$ sorbent in a continuous cyclic process for 400 min in two parallel fixed-bed reactors at 630°C, 1 atm., and S/C of 5, Li et al. (2006b) concluded that the prebreakthrough period between the two reactors was a key parameter for an efficient hydrogen production and reported the production of hydrogen with 95% purity. Using Al_2O_3-stabilized Ca-based sorbent and Ni–HTlc-derived catalyst (47 wt.% Ni) for 10 successive cycles at 550°C and S/C of 4, Broda et al. (2013a) reported the production of hydrogen with a purity of 99%. They noticed that the rate of decrease in hydrogen production in cyclic operation was 275% lower in the case of Al-stabilized CaO-based sorbents

(Ca:Al = 80:20) compared to Ni–HTlc/limestone mixture due to the good dispersion of high Tammann temperature support ($Ca_{12}Al_{14}O_{33}$), which hinders the structural sintering and stabilizes the nanostructured morphology. Xu et al. (2013) tested $Ca_9Al_6O_{18}$–CaO sorbent (90 wt.% CaO) derived from calcium lactate by the sol–gel method and Ni-based catalyst for 10 cycles at 600°C and S/C of 4 and reported the production of 98% purity hydrogen during the prebreakthrough period. Broda et al. (2013b) synthesized sorbet pellets from limestone and commercial calcium aluminate (71% Al_2O_3/28% CaO) and mixed them with a Ni–HTlc-derived catalyst (47 wt.% Ni) and obtained a hydrogen purity of 99% (dry and N_2-free basis) at 550°C and S/C = 4. A good thermal stability of the prepared sorbent by means of CaO stabilization through a uniform dispersion of $Ca_{12}Al_{14}O_{33}$ among CaO particles led to a significant improvement in the cyclic sorption capacity (less capacity loss) in respect to limestone (0.41 vs 0.22 g of CO_2/g of sorbent after 10 cycles).

Besides using simple catalyst and sorbent mixtures, hybrid materials sorbent-catalyst which integrate catalytic reaction and CO_2 sorption in a single pellet have the advantages of eliminating mass diffusional limitations, enhancing their activity, and reducing the reactor volume. High sorption efficiency (capacity and kinetics), appropriate Ni dispersion, and adequate surface area are crucial issues in the development of these bifunctional materials.

Satrio et al. (2005) combined catalyst and sorbent into a single pellet with the aim of using particles with larger size and facilitate the reforming process scale-up. The core-in-shell pellets consisting of a CaO core (prepared from limestone (Ames, Iowa) and dolomite (Graymont, Ohio)) and an alumina-based shell (porous shell of Ni/Al_2O_3) exhibited 96% H_2 purity at 650°C compared to 75% in the conventional steam reforming, with low amounts of CO and CO_2. Martavaltzi and Lemonidou (2010) proposed a hybrid material to overcome difficulties in handling mixtures of two solid materials, catalyst and sorbent, and reduce the production cost in industrial scale. Ni–$Ca_{12}Al_{14}O_{33}$–CaO sorbent-catalyst prepared from calcium acetate (Ca precursor), aluminum nitrate, and nickel nitrate demonstrated excellent stability compared to pure CaO during 45 cycles of sorption/desorption mainly due to the synergetic effect of NiO and $Ca_{12}Al_{14}O_{33}$. The steam reforming tests performed in a single reaction cycle at 650°C and S/C of 3.4 led to a H_2 yield of 90% (compared to 84% in conventional steam reforming) with low CO_2 and CO concentrations (2.8% and 2%, respectively), for an optimum NiO loading of 20 wt.%. A similar hybrid material (Ni–$Ca_{12}Al_{14}O_{33}$–CaO) but from another Ca precursor (calcium nitrate) was obtained by applying a combination of precipitation and hydration methods (Kim et al., 2013). While a loading of 7 wt.% Ni led to the best results (H_2 concentration of 94%–95% for 70 min in the prebreakthrough period), the higher Ni loading of 10 wt.% resulted in Ni self-agglomeration.

To improve the material efficiency even more (stability in reforming cyclic operation, higher yield, and lower CO_2 and CO concentrations), especially for suitability in industrial applications, several works have been carried out to develop various sorbent-catalyst hybrid materials with stabilized CaO-based sorbents in order to provide appropriate stability and activity in cyclic operation.

Different Al-stabilized CaO–Ni hybrid sorbent-catalyst materials integrated in a single particle (nickel loadings of 12, 18, and 25 wt.% NiO) were developed by Radfarnia and Iliuta (2014a) by simple wet mixing (limestone acidification coupled with two-step calcination (Radfarnia and Iliuta, 2013a) and tested during 10 SEMSR cycles. Al stabilizer incorporated into the CaO structure improved the sorbent stability. CO_2 sorption and sorption-enhanced reforming results showed that the hybrid sorbent-catalyst with the topmost level of nickel (25 wt.% NiO) offered the best performance: in the CO_2 sorption process CaO molar conversion attained 41.2% at the end of the 25th sorption cycle and in the steam methane reforming CH_4 conversion and H_2 production efficiency were 99.1% and 96.1%, respectively, over 10 cycles. No important carbon deposition was observed; all developed hybrid materials demonstrated very good stability, as well as a good dispersion of nickel on the substrate.

Typical profiles of the reaction products (H_2, CO_2, and CO) obtained in the first cycle of the reforming process operated with the most efficient hybrid sorbent-catalyst (25 wt.% NiO) are illustrated in Fig. 13. The very high H_2 yield achieved in the prebreakthrough period (99%) confirms the large effectiveness of the developed hybrid materials in both reforming and CO_2 sorption processes. The H_2 yield is significantly larger than that of conventional reforming reaction (~80%) attained in the post-breakthrough period (in the absence of CO_2 sorption). For the hybrid sorbent-catalyst with the topmost level of nickel (25 wt.% NiO), the effect of CH_4 flow rate and steam to carbon ratio (S/C) was also studied, as well as its comportment in the long-term cyclic operation of sorption-enhanced steam methane reforming process (30 cycles). The larger S/C ratios favor the production of more H_2, while a larger CH_4 flow rate reduces the prebreakthrough interval and decreases the H_2 yield. The high potential

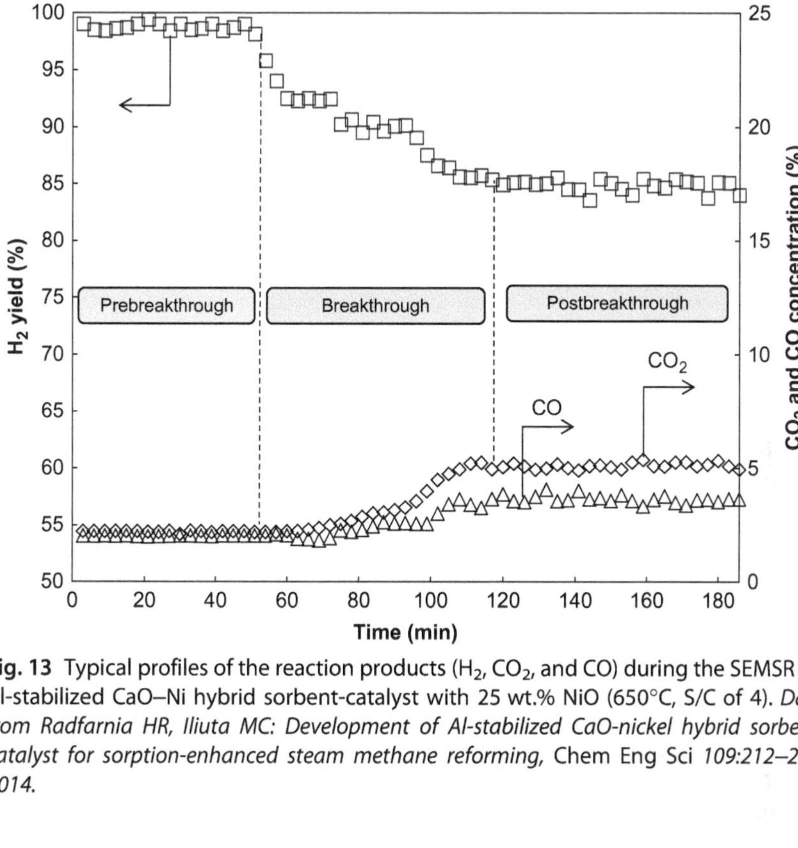

Fig. 13 Typical profiles of the reaction products (H_2, CO_2, and CO) during the SEMSR for Al-stabilized CaO–Ni hybrid sorbent-catalyst with 25 wt.% NiO (650°C, S/C of 4). *Data from Radfarnia HR, Iliuta MC: Development of Al-stabilized CaO-nickel hybrid sorbent-catalyst for sorption-enhanced steam methane reforming,* Chem Eng Sci *109:212–219, 2014.*

of the developed hybrid sorbent-catalyst in the hydrogen production by SEMSR process is demonstrated in the long-term operation (30 cycles) where the reaction time was limited to the prebreakthrough period. An average 97.3% H_2 yield was obtained for all cycles, indicating a very good activity of the proposed catalyst. However, the extent of prebreakthrough period starts to diminish after the 20th cycle of reaction. Another bifunctional CaO–Zr/Ni (13, 18, and 20.5 wt.% NiO) sorbent-catalyst was synthesized via the wet-mixing/sonication technique (Radfarnia and Iliuta, 2011) and then tested for hydrogen production during 10 reaction cycles (Radfarnia and Iliuta, 2014b). As in the case of Al-stabilized CaO–Ni (Radfarnia and Iliuta, 2014a), the results showed that the bifunctional material with the topmost level of nickel (20.5 wt.% NiO) gave the most appropriate activity. A maximum H_2 yield of ~92% was obtained using the bifunctional sorbent-catalyst with 20.5 wt.% NiO (average of ~89% over cyclic operation), which is significantly larger than the equilibrium H_2 yield of conventional steam methane reforming obtained under the same operational conditions (~70%). The presence of enough nickel in the bifunctional material was found essential to accelerate the gasification of deposited coke. For the sorbent-catalyst with the lowest level of nickel, TPO analysis confirmed the generation of a large amount of carbon due to the lack of favorable carbon gasification during the cracking of CH_4. In the case of sorbent-catalysts with higher nickel loading, they were sufficiently active to promote the conversion of CH_4 to syngas, leading to much less generation of inactive carbon and accelerating the coke gasification. Additionally, $CaZrO_3$ in the catalyst support helped the gasification of deposited carbon by facilitating the water adsorption on the catalyst support.

ZrO_2-stabilized NiO–CaO/Al_2O_3 (Wu and Wang, 2010) and La_2O_3-stabilized NiO–CaO/Al_2O_3 (Feng et al., 2012) also exhibited good efficiency in producing hydrogen with high yields (90% for 20 cycles and 92% for 30 cycles, respectively). ZrO_2 incorporation offered favorable stability and activity by avoiding the formation of $NiAl_2O_4$ phase and La_2O_3 improved both stability and nickel grain dispersion over the substrate. More recently, several Ni/CaO–$CaZrO_3$ hybrid materials with Ni, CaO, and $CaZrO_3$ acting as catalyst, sorbent, and stabilizer, respectively, were prepared by the sol–gel method using citric acid as a metal-complex agent (Zhao et al., 2016). The stabilizing effect of $CaZrO_3$ was reflected in an improved uptake capacity of the developed material in cyclic operation (50 carbonation–calcination cycles), as well as in higher stability during the cyclic reforming process. Despite its slow deactivation during the long-term stability test (90 reaction cycles), the catalyst with 15 wt.% Ni, 60 wt.% CaO, and 25 wt.% $CaZrO_3$ presented the best performance in H_2 production (prebreakthrough duration and catalytic activity and stability). Another bifunctional material containing both Ni-reforming catalyst and Ca-based CO_2 sorbent synthesized from a hydrotalcite-based precursor by coprecipitation was found to perform better than limestone mixed with Ni–SiO_2 or Ca-free Ni/HTlc-derived catalyst (Broda et al., 2012). The high H_2 purity (99 vol.% (dry and inert free)) and adequate thermal stability in SEMSR cyclic operation (10 cycles) were on account of the appropriate specific surface area (54 m^2/g) and high Ni and Ca dispersion into the HTlc structure. However, compared to other developed materials, it contains very high amount of Ni (45 wt.%) and very low content of Ca (21 wt.%). The corresponding low capture capacity of 0.074 g CO_2/g sorbent (over 10 cycles) would require an important hybrid material loading in the reactor in order to attain acceptable sorption capacity.

Ceramic sorbents have also been proposed as potential candidates for in situ CO_2 removal in the SEMSR, but only few works are available compared to CaO-based materials and only some sorbents made the object of investigations. In particular, a research group from the Norwegian University of Science and Technology (NTNU) showed special interest in synthesis and application of this ceramic sorbents in SEMSR. Ochoa-Fernandez et al. (2007a) evaluated theoretically the H_2 production by SEMSR by comparing the performance of several sorbents: CaO from calcination of $CaCO_3$ prepared using calcium nitrate and sodium carbonate and ceramics (commercial Li_4SiO_4 and Li_2ZrO_3, K-Li_2ZrO_3, Na_2ZrO_3 obtained by applying a novel soft-chemistry route). As also reported in several investigations, while CaO showed the highest H_2 yield (above 98%) at 575°C, 10 bar, and S/C of 5, there was a significant decrease of its capture capacity in cyclic operation. For comparison, the simulation results obtained for the other considered sorbents were K-Li_2ZrO_3 (93%), Na_2ZrO_3 (90%), Li_2ZrO_3 (89%), and Li_4SiO_4 (82%). By considering the uptake behavior of the sorbents only, Na_2ZrO_3 was considered a good alternative to CaO because of its appropriate kinetics and stability in cyclic sorption/desorption process. The effect of steam on stability, uptake capacity, and regeneration properties of several ceramic materials like Li_2ZrO_3, Na_2ZrO_3, Li_4SiO_4, and K-Li_2ZrO_3 was studied by Ochoa-Fernandez et al. (2009) in the reforming conditions. While a positive effect was found on both capture and regeneration rates, an important loss in the uptake capacity was recorded possibly on account of sintering under the high steam pressure, alkali metal vaporization, and/or phase segregation. Thorough investigations on the effect of the steam on ceramic sorbents during cyclic operation would be helpful in clarifying this behavior. The application of Li_2ZrO_3 and Na_2ZrO_3 sorbents obtained by liquid–liquid synthesis was also investigated by Ochoa-Fernandez et al. (2007a,b). The use of Na_2ZrO_3 led to a H_2 yield above 97% at 575°C, 5 bar, and S/C of 5, with sorbent regeneration at 750°C under Ar. Concerning Li_2ZrO_3, even though H_2 was produced with a yield that was very close to the thermodynamic equilibrium of conventional steam reforming (67.9%), a significant reduction of H_2 yield was noticed in the second cycle of reaction, possibly due to catalyst poisoning in the presence of some sorbent impurities.

3.2 Other Feedstocks for Steam Reforming

Alternatively, the use of other different feedstocks of interest has been investigated, like propane, butane, methanol, ethanol, butanol, acetic acid, and glycerol, but to a lesser extent. Along with methane, Table 6 also presents some works involving hydrogen production by steam reforming of different other feedstocks, mainly ethanol and glycerol.

Compared to MSR, the global steam reforming process of other compounds consists of more complex reactions, which lead to the formation of several by-products.

3.2.1 Steam Reforming of Ethanol

Ethanol has received more attention, most probably for its interest as a renewable feedstock (e.g., bioethanol from biowaste and cellulose). Based on the overall reaction, 6 mol of H_2 is theoretically produced per mole of ethanol:

$$C_2H_6O(g) + 3H_2O(g) \leftrightarrow 6H_2(g) + 2CO_2(g) \quad \Delta H^\circ_{298} = 173.3\,\text{kJ/mol} \tag{10}$$

Depending on operating conditions and catalyst type, along with the main reactions of reforming and WGS, several side reactions can also occur, negatively influencing the hydrogen purity (Cunha et al., 2014; Dewoolkar and Vaidya, 2016b; Gunduz and Dogu, 2012; Wu et al., 2013, 2014b):

Ethanol dehydrogenation:

$$C_2H_6O(g) \leftrightarrow C_2H_4O(g) + H_2(g) \quad \Delta H^\circ_{298} = 68.4\,\text{kJ/mol} \tag{11}$$

Ethanol dehydration:

$$C_2H_6O(g) \leftrightarrow C_2H_4(g) + H_2O(g) \quad \Delta H^\circ_{298} = 45.4\,\text{kJ/mol} \tag{12}$$

Acetaldehyde decomposition:

$$C_2H_4O(g) \leftrightarrow CO(g) + CH_4(g) \quad \Delta H^\circ_{298} = -19\,\text{kJ/mol} \tag{13}$$

Direct ethanol decomposition:

$$C_2H_6O(g) \leftrightarrow CO(g) + CH_4(g) + H_2(g) \quad \Delta H^\circ_{298} = 49.4\,\text{kJ/mol} \tag{14}$$

Methane steam reforming:

$$CH_4(g) + H_2O(g) \leftrightarrow 3H_2(g) + CO(g) \quad \Delta H^\circ_{298} = 205.9\,\text{kJ/mol} \tag{15}$$

WGS:

$$CO(g) + H_2O(g) \leftrightarrow H_2(g) + CO_2(g) \quad \Delta H^\circ_{298} = -41.1\,\text{kJ/mol} \tag{7}$$

Acetic acid formation:

$$C_2H_6O(g) + H_2O(g) \leftrightarrow C_2H_4O_2(g) + H_2(g) \quad \Delta H^o_{298} = 41.7\,kJ/mol \tag{16}$$

Acetic acid decomposition:

$$C_2H_6O(g) \leftrightarrow CH_4(g) + CO_2(g) \quad \Delta H^o_{298} = -233.3\,kJ/mol \tag{17}$$

Coke formation:

$$2CO \leftrightarrow CO_2 + C(s) \quad \Delta H^o_{298} = -86\,kJ/mol\,CO \tag{18}$$

Methane reforming with CO_2 (dry reforming):

$$CH_4 + CO_2 \leftrightarrow 2CO + 2H_2 \quad \Delta H^o_{298} = 247\,kJ/mol \tag{19}$$

Hydrotalcite-like compounds are predominantly used in the preparation of catalysts and hybrid sorbent-catalyst materials for ESR given that this process occurs at intermediate temperatures (less than 500°C), which are required for preservation of the structure of the materials. Hybrid materials are more attractive for their bifunctional role, as support for impregnating the actives phases and CO_2 sorbent at moderate and high temperatures.

Based on different studies on the use of HTlc-based material as CO_2 sorbents at moderately high temperatures (Oliveira et al., 2008), Cunha et al. (2012) investigated the sorption-enhanced ethanol steam reforming (SEESR) using a mixture of Ni/Al_2O_3 catalyst and commercial HTlc. At 400°C the effect of the in situ CO_2 removal was clearly observed, but the high H_2 concentration was kept for a very short prebreakthrough period. An S/C ratio of 3 (instead of the optimum of 5; Wu et al., 2013) results in the increase of CO concentration in the prebreakthrough time and considerable carbon deposition. In spite of the positive effect of a high S/C ratio (significant reduction of carbon deposition and improvement of cyclic sorption capacity of HTlc-based sorbents), an important excess of steam (i) leads to a decrease of CO_2 concentration along with its partial pressure, with a negative effect on the sorption process and (ii) makes the reforming process more expensive and energy intensive. Excess of water vapors can cause partial pore mouth closure, which increases the diffusion resistance in CO_2 transport to sorbent active sites, with a negative effect on the uptake capacity (Ficicilar and Dogu, 2006).

Several HTlc-based hybrid materials with improved features have been developed (Cunha et al., 2013, 2014; Dewoolkar and Vaidya, 2016b; Wu et al., 2012b, 2013, 2014a,b) and investigated in the SEESR. CuMgAl materials prepared by Cu (active phase) impregnation on HTlc showed adequate stability in the reforming process during 4 h, by producing a high H_2 purity over 90 mol.% in the prebreakthrough period (Cunha et al., 2013). However, compared to other possible active phases (like Ni, Co, and Cu), Cu is prone to deactivation because of thermal sintering as Cu presents a low Hüttig temperature (Spencer, 1986), which is in agreement with the proposed order of stability for different metals: Ag < Cu < Au < Pd < Fe < Ni < Co < Pt < Rh < Ru < Ir < Os < Re (Bartholomew, 2001). Cationic modified (Mg^{2+}, Ca^{2+}, Cu^{2+}, and Zn^{2+}) Ni/HTlc hybrid materials were synthesized by coprecipitation and used for studying the effect of different parameters (reaction temperature, S/C ratio, sorbent mass fraction) on hydrogen production in cyclic operation (Dewoolkar and Vaidya, 2016b). Cu^{2+}- and Mg^{2+}-based Ni/HTlc showed better performance (respectively, better cyclic stability for 21 and 18 cycles, and higher hydrogen purity of 95.1 mol.% and 89.3 mol.% in the prebreakthrough period) compared to Ca^{2+}- and Zn^{2+}-based materials. At 300°C, Cu^{2+}-based Ni/HTlc exhibit the highest sorption capacity of 0.053 g CO_2/g sorbent. A uniform distribution of Ni, Ca, and Al in the Ni–CaO–Al_2O_3 HTlc-based hybrid catalyst improves to a high extent the sorption capacity and reforming activity compared to CaO + Ni/Al_2O_3 mixture (Wu et al., 2012a). Using a developed NiMgAl hybrid material by Ni impregnation on HTlc, Wu et al. (2013) investigated numerically the effect of different parameters (hybrid composition and operation conditions) and compared them to experimental data. The results revealed the improvement of H_2 yield and thermal efficiency compared to conventional steam reforming. At 500°C, initial H_2 purity of more than 99 mol.% could be obtained for an optimum S/C of 5 (around 95 mol.% was kept for the first 500 s). K-promoted HTlc was later used by the same research group as CO_2 sorbent and support for Ni and Co incorporation to prepare K–Ni–Cu–hydrotalcite hybrid material and investigate its efficiency (catalytic and sorption properties) up to 500°C (Cunha et al., 2014). At higher temperatures an irreversible destruction of the structure of HTlc occurs. While Cu can preferentially catalyze WGS and ethanol dehydrogenation, Ni is more appropriate for acetaldehyde decomposition and methane steam reforming. The alloy formed between Cu and Ni ($Ni_{0.5}Cu_{0.5}$) offers the advantage of preventing the interactions between the adsorbed species and the catalytic decomposition of CO and methane. At 500°C and for S/C of 5, a high-purity hydrogen (99.8 mol.%) was obtained during the prebreakthrough period after which it decreased to 67.1 mol.%.

In addition to HTlc-based sorbent-catalysts, other materials can also be used in this reforming process. For example, Gunduz and Dogu (2012) incorporated Ni and Co into mesoporous MCM-41 with Ni/Si and Co/Si molar ratios of 0.12 and evaluated the effectiveness of the synthesized material in both SEESR and conventional ESR, for comparison purposes. Ni–MCM-41 leads to a better performance compared to Co–MCM-41 in the ESR. By using Ni–MCM-41 in the reaction tests, no liquid product was detected in the condensed stream except for water. In the ESR at 500°C, ~60% H_2, 20% CO_2, 20% CH_4, and 2.0% CO were found in the gaseous stream. The presence of methane proves that the reaction (14) occurred in the reaction conditions, but the very low CO concentration seems to indicate that most CO produced in reaction (14) was converted through WGS to H_2 and CO_2. In the presence of CaO as CO_2 sorbent, the H_2 concentration is significantly increased to around 90% in the first 40 min of the process, after which a continuous decrease occurs until reaching about 60% (CaO saturation) over 100 min. Interestingly, the presence of CaO results in an increase of CH_4 concentration, especially at higher temperatures, possibly on account of a limitation of reaction (19) when CO_2 is removed in situ from the reaction medium. The use of Co–MCM-41 results in a much reduced hydrogen yield and an average ethanol conversion of 80% (ESR) and 94% (SEESR) at 500°C, in comparison with complete conversion of ethanol over Ni–MCM-41. Moreover, in the SER, besides methane, CO, and CO_2, ethylene, formaldehyde, and trace of acetone in some initial experiments were detected. First, the presence of ethylene (~25%) indicates that along with the reforming reaction, ethanol dehydration could occur in the reaction medium. Second, the presence of formaldehyde shows that another side reaction (20) could take place in the reaction system:

$$C_2H_6O + H_2O \leftrightarrow 2CH_2O + 2H_2 \tag{20}$$

Interestingly, the presence of several by-products was much reduced by the in situ CO_2 removal, demonstrating the great advantage of SEESR: formaldehyde, CO, and CO_2 were not detected in the first 80 min of reaction. Besides CO_2, the formaldehyde also reacts with CaO. Finally, the results revealed that coke formation was also reduced in the SEESR in the presence of both catalysts.

3.2.2 Steam Reforming of Glycerol

Conversion of biomass (abundant, clean, and renewable) is one of the promising H_2 production approaches. However, renewable vegetable oils and animal fats conversion into biofuels through catalytic transesterification (process that is undergoing a boom through new biorefineries worldwide as part of global efforts to combat climate change and reduce dependency on fossil fuels) produces a significant amount of glycerol as by-product. On the other side, glycerol is seen as a very promising source of hydrogen. In the context of a sustainable development, the application of high-temperature CO_2 sorbents in the sorption-enhanced glycerol steam reforming (SEGSR) attracted special attention in the last years. The high surplus of glycerol on the market made it become a waste problem and opened significant areas of research for the development of technologies for its valorization into value-added compounds. Using glycerol as a feedstock for hydrogen production has two main advantages: valorizing a large amount of redundant glycerol produced as an inevitable by-product of biodiesel production and achieving the full environmental benefit of using hydrogen as an energy carrier by using a carbon-neutral source for its generation. Moreover, the sorption-enhanced steam reforming process is applied to produce a hydrogen-rich gas stream in one single step and increase the efficiency of steam reforming process by driving the equilibrium limited WGS reaction toward hydrogen production. This integrated process intents to minimize the abundant glycerol by-product streams via an energy efficient alternative for producing hydrogen. It offers an opportunity to improve the economics of green hydrogen production via an enhanced thermal efficiency process when the exothermic CO_2 chemisorption provides the heat to the endothermic steam glycerol reforming, while reducing the capital cost (the processing steps for subsequently CO_2 separation are generally not required).

The overall SEGSR reaction (23) is a combination of the endothermic glycerol pyrolysis, the exothermic WGS, and the CO_2 reaction with a metal oxide (MeO):

Reforming:

$$C_3H_8O_3 \leftrightarrow 4H_2 + 3CO \quad \Delta H^{\circ}_{298} = 251\,\text{kJ/mol} \tag{21}$$

WGS:

$$CO + H_2O \leftrightarrow H_2 + CO_2 \quad \Delta H^{\circ}_{298} = -41.1\,\text{kJ/mol} \tag{7}$$

CO_2 sorption:

$$MeO(s) + CO_2(g) \leftrightarrow MeCO_3(s) \quad \Delta H^{\circ} < 0 \tag{22}$$

$$C_3H_8O_3 + 3H_2O + MeO \leftrightarrow 7H_2 + 3CO_2 + MeCO_3 \tag{23}$$

However, while glycerol pyrolysis and WGS are the main reactions, several side reactions can occur (24–33). The complexity of steam glycerol reforming is reflected in the formation of CH_4 and carbon deposition, along with the presence of high CO and CO_2 concentrations in the exit flow which affects the purity of hydrogen produced. Hydrogen purification is also noteworthy for being a costly and complex process.

Glycerol hydrogenolysis:

$$2C_3H_5(OH)_3 + H_2 \leftrightarrow 3CH_4 + 3CO + 3H_2O \quad \Delta H^{\circ}_{298} = -55.8\,\text{kJ/mol glycerol} \tag{24}$$

$$C_3H_5(OH)_3 + 2H_2 \leftrightarrow 2CH_4 + CO + 2H_2O \quad \Delta H^{\circ}_{298} = -158.8\,\text{kJ/mol} \tag{25}$$

Methanation:

$$CO + 3H_2 \leftrightarrow CH_4 + H_2O \quad \Delta H^{\circ}_{298} = -205.9\,\text{kJ/mol} \tag{26}$$

$$CO_2 + 4H_2 \leftrightarrow CH_4 + 2H_2O \quad \Delta H^{\circ}_{298} = -165\,\text{kJ/mol} \tag{27}$$

Methane reforming:

$$CO_2 + CH_4 \leftrightarrow 2CO + 2H_2 \quad \Delta H^{\circ}_{298} = 247\,\text{kJ/mol} \tag{28}$$

$$CH_4 + H_2O \leftrightarrow CO + 3H_2 \quad \Delta H^{\circ}_{298} = 205.9\,\text{kJ/mol} \tag{29}$$

Coke formation:

$$CH_4 \leftrightarrow 2H_2 + C(s) \quad \Delta H^{\circ}_{298} = 74.8\,\text{kJ/mol} \tag{30}$$

$$CO + H_2 \leftrightarrow C(s) + H_2O \quad \Delta H^{\circ}_{298} = -131\,\text{kJ/mol} \tag{31}$$

$$2CO \leftrightarrow CO_2 + C(s) \quad \Delta H^{\circ}_{298} = -86\,\text{kJ/mol CO} \tag{32}$$

$$CO_2 + 2H_2 \leftrightarrow C(s) + 2H_2O \quad \Delta H^{\circ}_{298} = -90\,\text{kJ/mol} \tag{33}$$

The SEGSR has the advantage of producing highly pure hydrogen in a single step (single reactor), increasing glycerol conversion, decreasing the coke formation, and requiring a relatively low steam/carbon ratios (Dou et al., 2014; Fermoso et al., 2012b; He et al., 2010b; Lin, 2013; Wang et al., 2010). Dou et al. (2014) provided an overview of some issues and challenges to achieve a highly efficient hydrogen production from the SEGSR process. Much fewer studies on SEGSR are available in comparison with SEMSR. An important issue associated with this integrated processes is the proper selection of operating conditions. Several works have therefore been performed to find the optimum conditions for SEGSR (Chen et al., 2009a; He et al., 2010b; Iliuta et al., 2013; Li et al., 2010; Wess et al., 2015).

On the basis of a thermodynamic investigation of the effects of different parameters (temperature, pressure, steam to carbon (S/C) ratio, percentage of CO_2 sorption, and carrier gas to feed reactants molar ratio) on the in situ CO_2 removal by sorption and product distribution in the SEGSR, Chen et al. (2009a) found that in the most favorable conditions (527–577°C at 1 atm. and $S/C \approx 3$), the maximum hydrogen yield could be increased from 6 to 7 mol per mole of glycerol when CO_2 is completely removed from the reaction media. The most favorable temperature is approximately 100°C lower compared to traditional steam glycerol reforming (without in situ CO_2 removal). An important advantage of this intensified process is the possibility to limit the carbon formation at lower S/C ratio.

To investigate the effect of operating parameters on the performance of SEGSR process with CaO-based sorbent derived from natural limestone, Iliuta et al. (2013) developed a two-scale, nonisothermal, unsteady-state model taking into account the gas dynamics which tied steam glycerol reforming/CO_2 capture kinetics, thermodynamics, thermal effects, and variable gas flow rate due to chemical and physical contractions. The results of simulations performed between 873K and 923K, ratio sorbent/catalyst of 1–4, inlet gas velocity of 0.1 and 0.2 m/s, and glycerol partial pressure of 10 and 15 kPa for 50 cycles (corresponding to 6000 min) showed that coupling the SGR with in situ CO_2 separation using solid sorbents (SEGSR) leads to hybrid configuration that substantially improves the process efficiency. This work offers a very useful general overview of the effect of in situ CO_2 removal on glycerol steam reforming process.

In practical applications, the *cyclic behavior* of the SEGSR process (CO_2 sorption/desorption) is a key element in the evaluation of the efficiency of this integrated technology. As illustrated in Fig. 14, in the prebreakthrough period associated with the first cycle the dry basis hydrogen concentration (mole fraction) is higher than 99% during 80 min. Similar data were obtained experimentally by He et al. (2010b) using Co–Ni-reforming catalysts derived from hydrotalcite-like material (HTls) and dolomite as CO_2 sorbent. This period, characterized by a very high concentration of

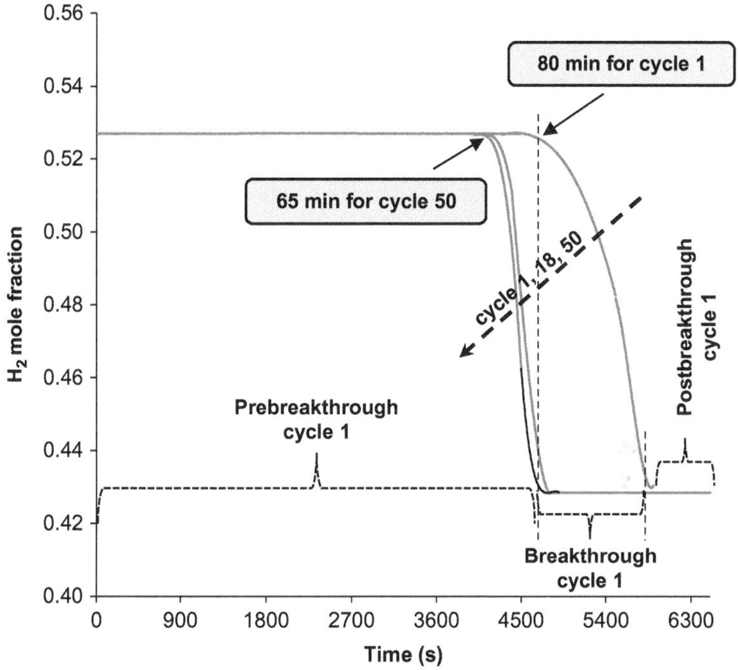

Fig. 14 Typical outlet H_2 mole fraction during cyclic SEGSR. *Data from Iliuta I, Radfarnia HR, Iliuta MC: Hydrogen production by sorption-enhanced steam glycerol reforming: sorption kinetics and reactor simulation,* AIChE J *59:2105–2118, 2013.*

hydrogen in the exit flow (near the equilibrium value) due to the in situ removal by chemical sorption of CO_2 produced in the reforming process, is a key parameter in the design and operation of SEGSR. The prebreakthrough time depends on the sorbent characteristics (material type and thermal stability). After this period of maximum CO_2 removal from the reaction media, the sorbent removal efficiency starts decreasing in the breakthrough period characterized by an increase of CO_2 concentration in the reaction media and a corresponding decrease of H_2. The completion of sorbent capacity to react with CO_2 (postbreakthrough period) corresponds to the traditional process when only the glycerol reforming reactions occur (hydrogen, CO_2, and CO concentrations match the usual values) and the sorbent regeneration is required to restart the cyclic process. Interestingly, data simulation revealed that the operation time to produce a highly pure hydrogen decreased with the increase of the number of cycles, following the tendency of activity loss of the sorbent. Moreover, after a certain number of cycles (e.g., 50 in the case of the sorbent developed in this modeling study), the prebreakthrough time seems to keep a constant value, following the usual tendency of a sorbent to reach its final residual capture capacity (Li and Cai, 2007).

As an integrated process, there is a strong relationship between reforming reaction and CO_2 chemical sorption through the *ratio sorbent/catalyst* selected to assure that the reforming reaction is not the limiting step (production rate of CO_2 larger than its sorption rate). Indeed, the prebreakthrough time when the SEGSR reactor operates at the highest hydrogen purity decreases with the decrease of the sorbent/catalyst ratio (lower sorbent volume fraction in the reactor), which corresponds to a lower capacity of CO_2 sorbent (dry basis H_2 mole fraction is higher than 99% during 80 min for a sorbent/catalyst volume ratio of 4 and 54 min for a sorbent/catalyst volume ratio of 1) (Fig. 15).

The length of the prebreakthrough time is also affected by the inlet superficial *gas velocity*. The decrease of the inlet superficial gas velocity has a positive effect on the prebreakthrough time. Indeed, at high gas amounts in the feed, the reaction rate (CO_2 formation) exceeds the CO_2 sorption rate

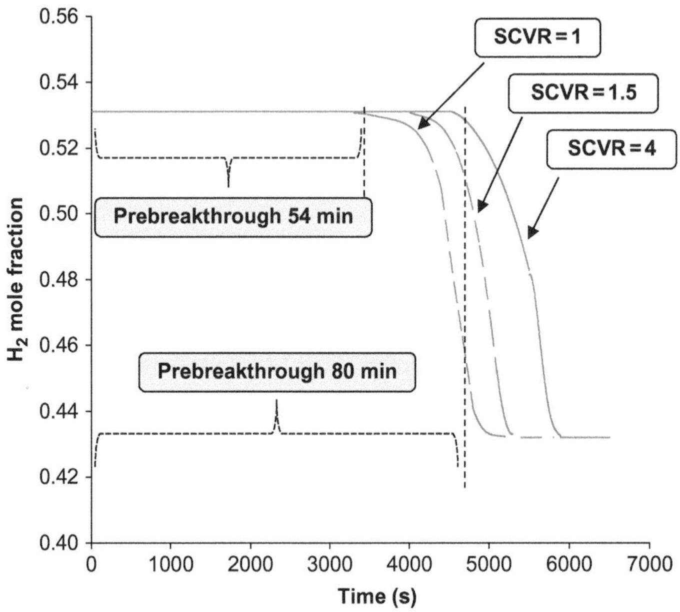

Fig. 15 Effect of sorbent/catalyst volume ratio (SCVR) on outlet H_2 mole fraction during SEGSR (first reaction cycle). *Data from Iliuta I, Radfarnia HR, Iliuta MC: Hydrogen production by sorption-enhanced steam glycerol reforming: sorption kinetics and reactor simulation,* AIChE J *59:2105–2118, 2013.*

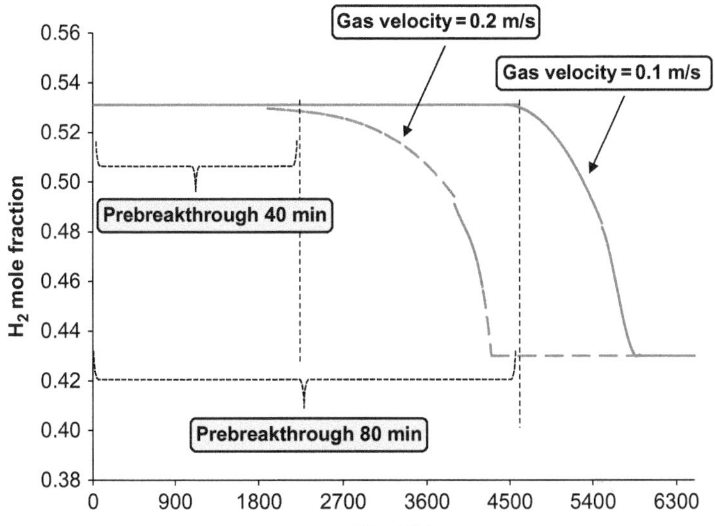

Fig. 16 Effect of inlet gas velocity on outlet H_2 mole fraction during SEGSR (first reaction cycle). *Data from Iliuta I, Radfarnia HR, Iliuta MC: Hydrogen production by sorption-enhanced steam glycerol reforming: sorption kinetics and reactor simulation,* AIChE J *59:2105–2118, 2013.*

and the in situ CO_2 removal efficiency starts decreasing after a shorter prebreakthrough time due to the high amount of CO_2 in the reaction media. Consequently, the period corresponding to a maximum hydrogen concentration in the exit flow is shorter. As can be seen in Fig. 16, when the inlet superficial gas velocity changes from 0.1 to 0.2 m/s, the CO_2 removal efficiency decreases and the prebreakthrough time corresponding to mole fraction of hydrogen higher than 99% diminishes from 80 to 40 min.

The *glycerol partial pressure* in the feed is another process parameter affecting the prebreakthrough time. Indeed, the decrease of the glycerol partial pressure has a positive effect on the prebreakthrough time. According to the simulation results (Fig. 17), when the glycerol partial pressure increases from 10 to 15 kPa, the prebreakthrough time corresponding to a mole fraction of hydrogen higher than 99% diminishes from 80 to 70 min, following the experimental results by He et al. (2010b). An increase of glycerol partial pressure leads to an increase of CO_2 produced in the reforming reaction and, as mentioned earlier, decreases the CO_2 removal efficiency and, consequently, the prebreakthrough time.

An interesting aspect is the *thermal behavior* of the SEGSR process that could be investigated by modeling of the reforming reactor. The thermal

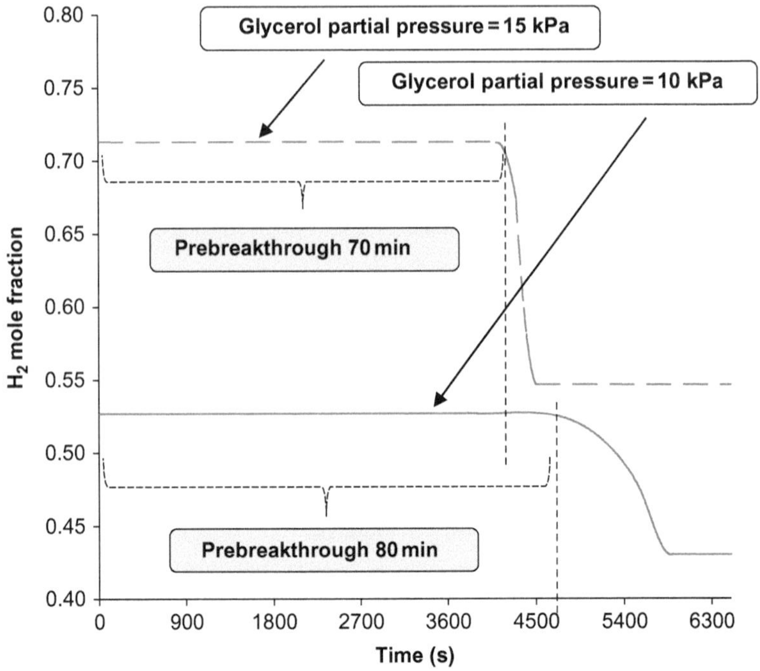

Fig. 17 Effect of glycerol partial pressure on outlet H_2 mole fraction during SEGSR (first reaction cycle). *Data from Iliuta I, Radfarnia HR, Iliuta MC: Hydrogen production by sorption-enhanced steam glycerol reforming: sorption kinetics and reactor simulation,* AIChE J *59:2105–2118, 2013.*

behavior of the sorption-enhanced steam glycerol reforming process, at a wall reactor temperature of 873K, is given in Fig. 18. In the entrance fixed-bed reactor zone the endothermic steam glycerol reforming reaction is faster than the exothermic CO_2 sorption reaction, and as result a low-temperature region is developed. In the ensuing fixed-bed reactor zone, CO_2 sorption reaction becomes dominant and the heat generated is used to heat up this region which moves toward the reactor exit with the increase of reaction time. With the amplification of the sorption/desorption (SESGR/regeneration) cyclic number, the low-temperature fixed-bed zones expand and the high-temperature fixed-bed zones move quickly and this explain the decrease of the operation time of producing high-purity hydrogen. Sorbent conversion is not significant in the low-temperature entrance region of the fixed-bed reactor because of the low CO_2 sorption rate (Fig. 19). In the succeeding high-temperature region, the fractional sorbent conversion rises due to the heat generated in CO_2 sorption process and,

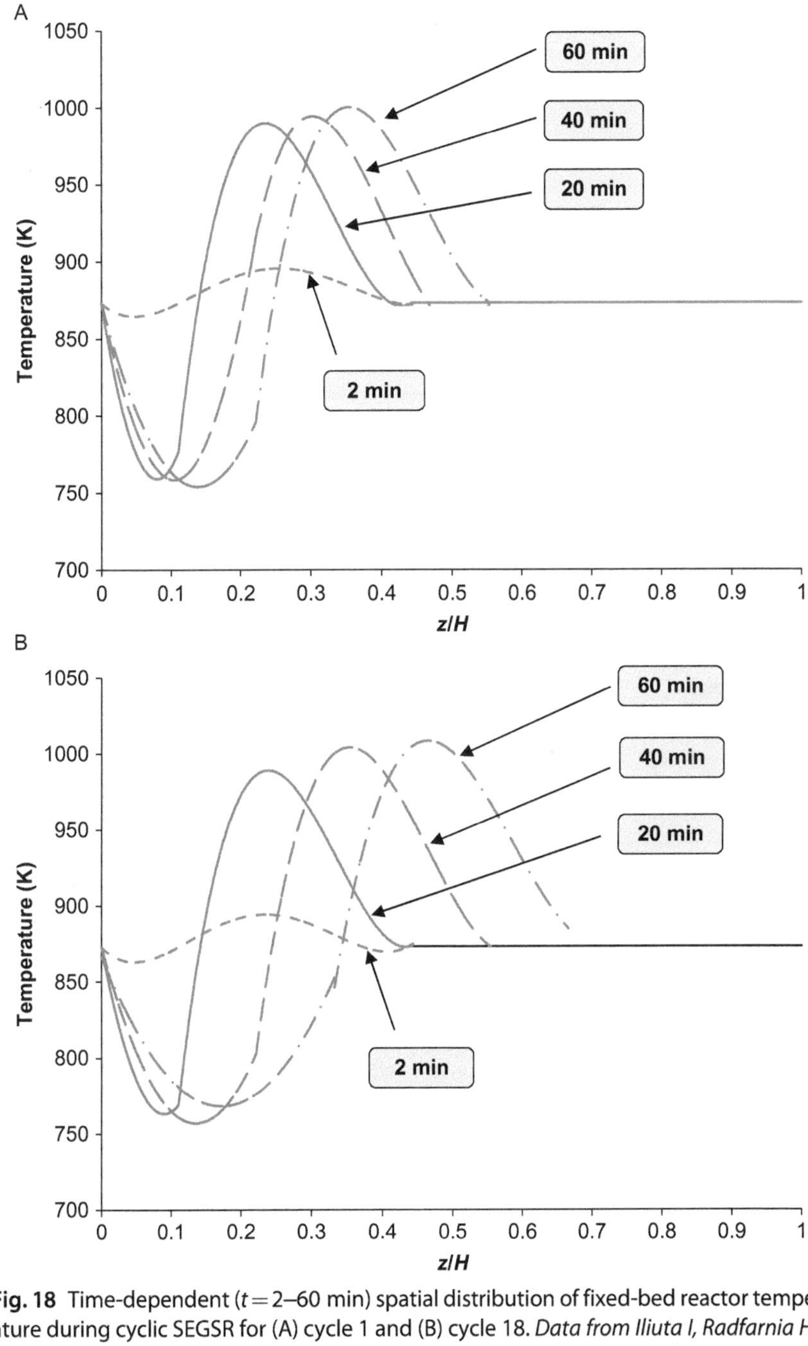

Fig. 18 Time-dependent ($t = 2$–60 min) spatial distribution of fixed-bed reactor temperature during cyclic SEGSR for (A) cycle 1 and (B) cycle 18. *Data from Iliuta I, Radfarnia HR, Iliuta MC: Hydrogen production by sorption-enhanced steam glycerol reforming: sorption kinetics and reactor simulation,* AIChE J *59:2105–2118, 2013.*

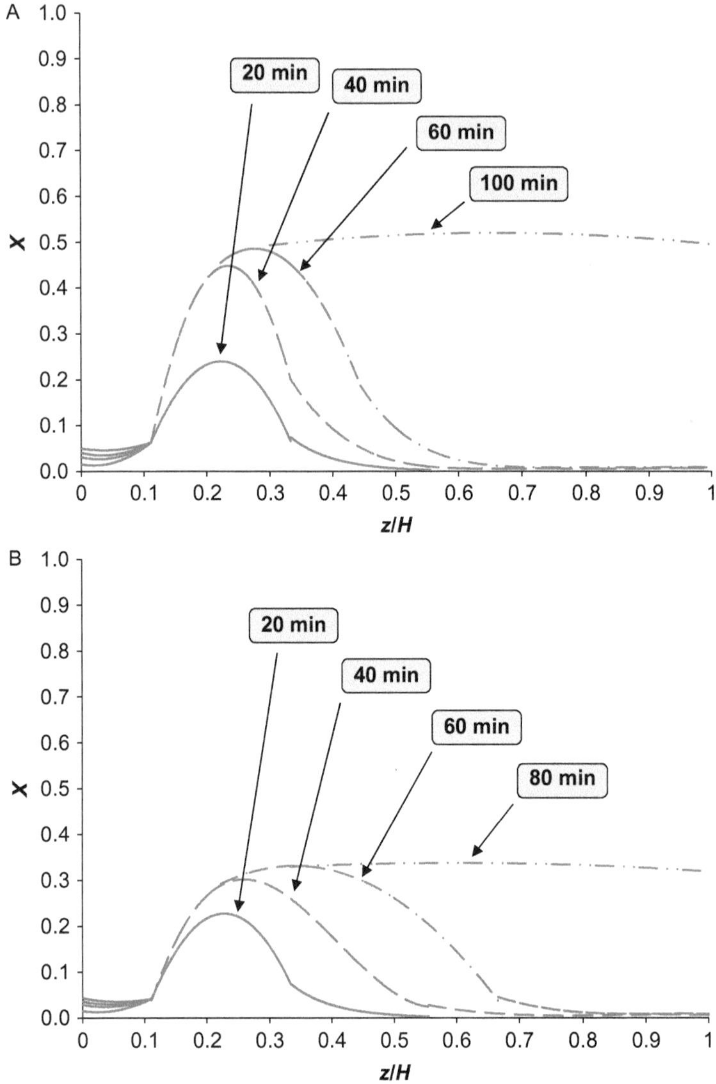

Fig. 19 Time-dependent axial distribution of sorbent conversion for (A) cycle 1 and (B) cycle 18. *Data from Iliuta I, Radfarnia HR, Iliuta MC: Hydrogen production by sorption-enhanced steam glycerol reforming: sorption kinetics and reactor simulation,* AIChE J *59:2105–2118, 2013.*

following the temperature profile, the bed zone with maximum in conversion moves toward the reactor exit with the increase of reaction time.

The thermal behavior of the sorption-enhanced steam glycerol reforming process is altered by the sorbent/catalyst ratio (Fig. 20). CO_2

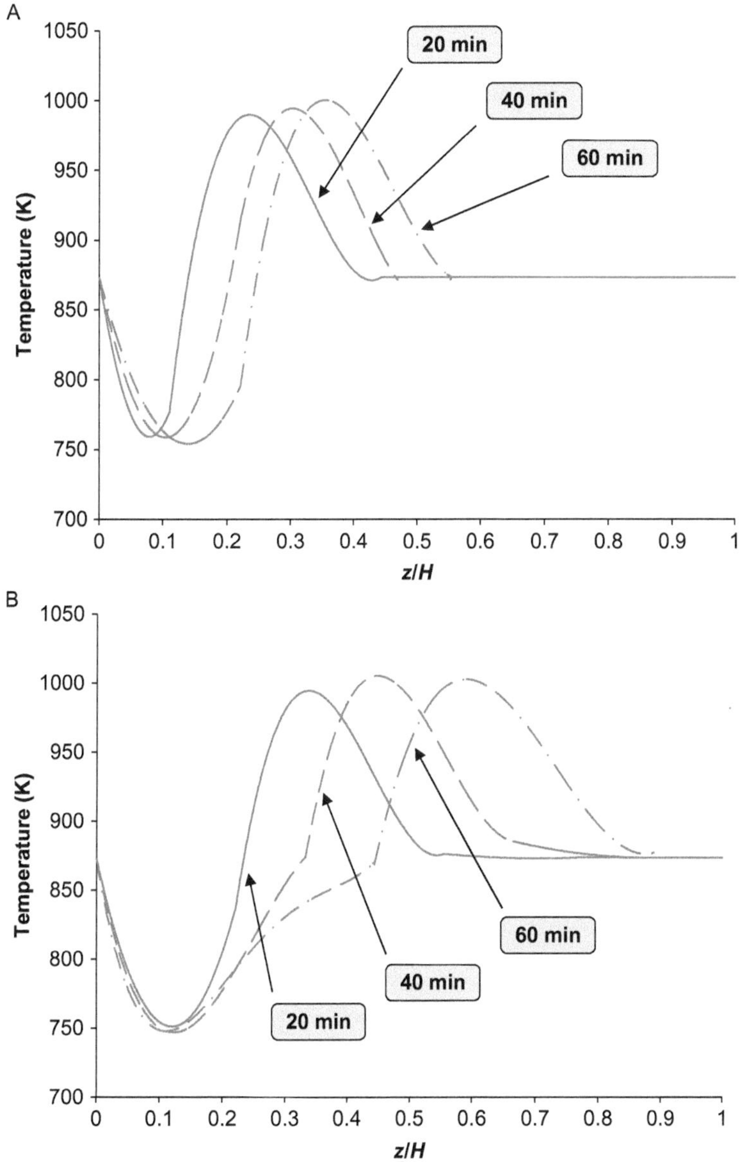

Fig. 20 Reactor temperature as a function of time and space (reactor axial direction) at the first reaction cycle for sorbent/catalyst volume ratio of (A) 4 and (B) 1. *Data from Iliuta I, Radfarnia HR, Iliuta MC: Hydrogen production by sorption-enhanced steam glycerol reforming: sorption kinetics and reactor simulation,* AIChE J *59:2105–2118, 2013.*

sorption capacity diminishes with the reduction of the sorbent/catalyst volume ratio, and due to the relatively insufficient quantity of sorbent in the entrance fixed-bed reactor zone, the low-temperature bed zone enlarges and the bed zones with a maximum in temperature move quickly toward the reactor exit with the increase of reaction time. Following the thermal regime, the entry region of the fixed bed with low sorbent conversion expands with the reduction of sorbent/catalyst volume ratio (Fig. 21). Also, at lower sorbent/catalyst volume ratio, the reactor zones with a maximum in sorbent conversion move quickly toward the reactor exit with the increase of reaction time and the sorbent attains its maximum conversion after a reduced period of time.

The increase of the inlet reactor temperature improves the thermal comportment of the sorption-enhanced steam glycerol reforming process (Fig. 22) and increases the length of the hydrogen high-purity time period (Fig. 23). CO_2 sorption capacity increases with the amplification of the inlet temperature because fixed-bed zones with a maximum in temperature move slowly toward the reactor exit with the increase of reaction time and low-temperature fixed-bed reactor zones contract.

The integrated system is limited by the sorbent sorption capacity. At large inlet gas velocity CO_2 removal efficiency starts to decline after a small period of time (Fig. 24) because of the relatively insufficient quantity of sorbent to support the large CO_2 production rate in glycerol reforming process, and as a result, the low-temperature fixed-bed zone expands and the bed zones with a maximum in temperature move quickly toward the reactor outlet with the increase of reaction time (Fig. 25).

The positive effects of in situ CO_2 removal from the reaction media in different conditions resulting from the simulation data agree well with several experimental investigations. Most works have considered the physical mixing of sorbent and catalyst and only few studies have focussed on development and application of bifunctional sorbent-catalyst materials. Among available works, only very few have studied the SESRG process for more than 10 cycles (Dang et al., 2016; Dewoolkar and Vaidya, 2016a; Shokrollahi Yancheshmeh et al., 2017).

As mentioned earlier, CaO is the most investigated CO_2 sorbent. In spite of its drawbacks, the numerous advantages of this material pushed the researchers to invent and develop different approaches to improve its performance in long-term cyclic operation. Several studies revealed that the use of CaO as CO_2 sorbent can offer significant improvement of H_2 production along with an important reduction of CO concentration in the exit flow

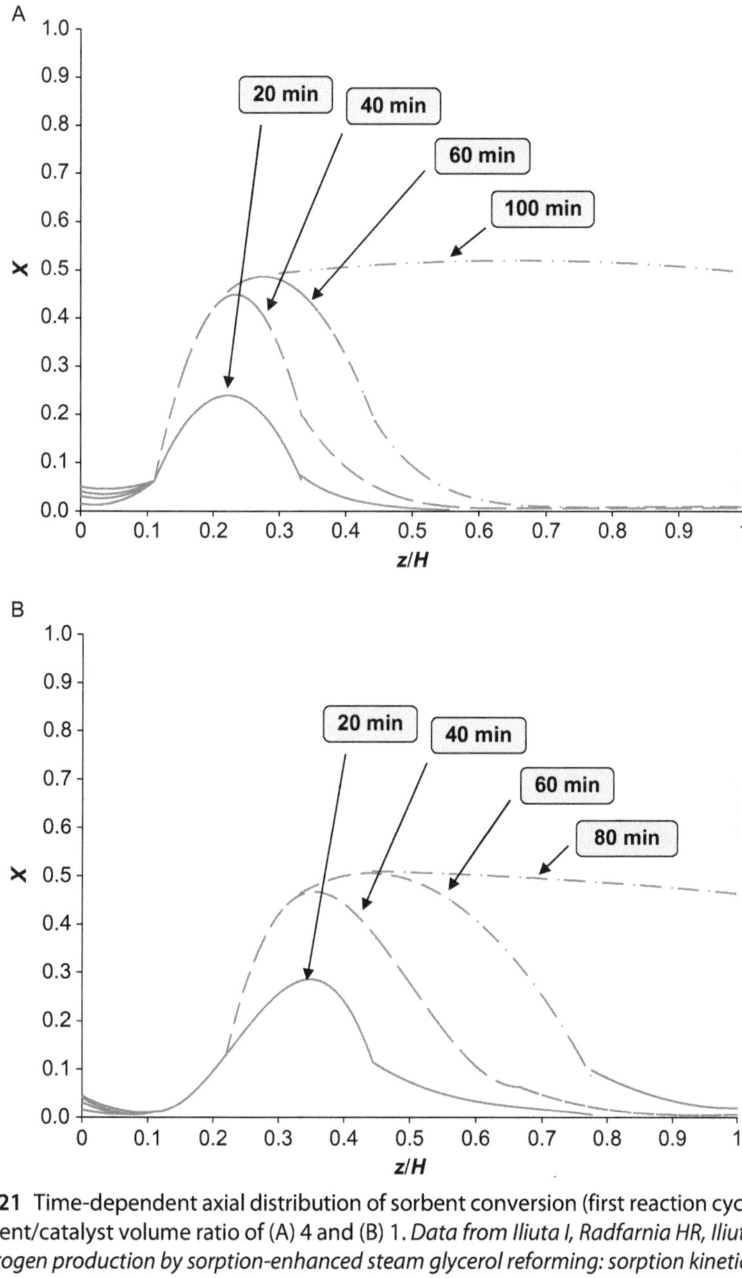

Fig. 21 Time-dependent axial distribution of sorbent conversion (first reaction cycle) for sorbent/catalyst volume ratio of (A) 4 and (B) 1. *Data from Iliuta I, Radfarnia HR, Iliuta MC: Hydrogen production by sorption-enhanced steam glycerol reforming: sorption kinetics and reactor simulation,* AIChE J *59:2105–2118, 2013.*

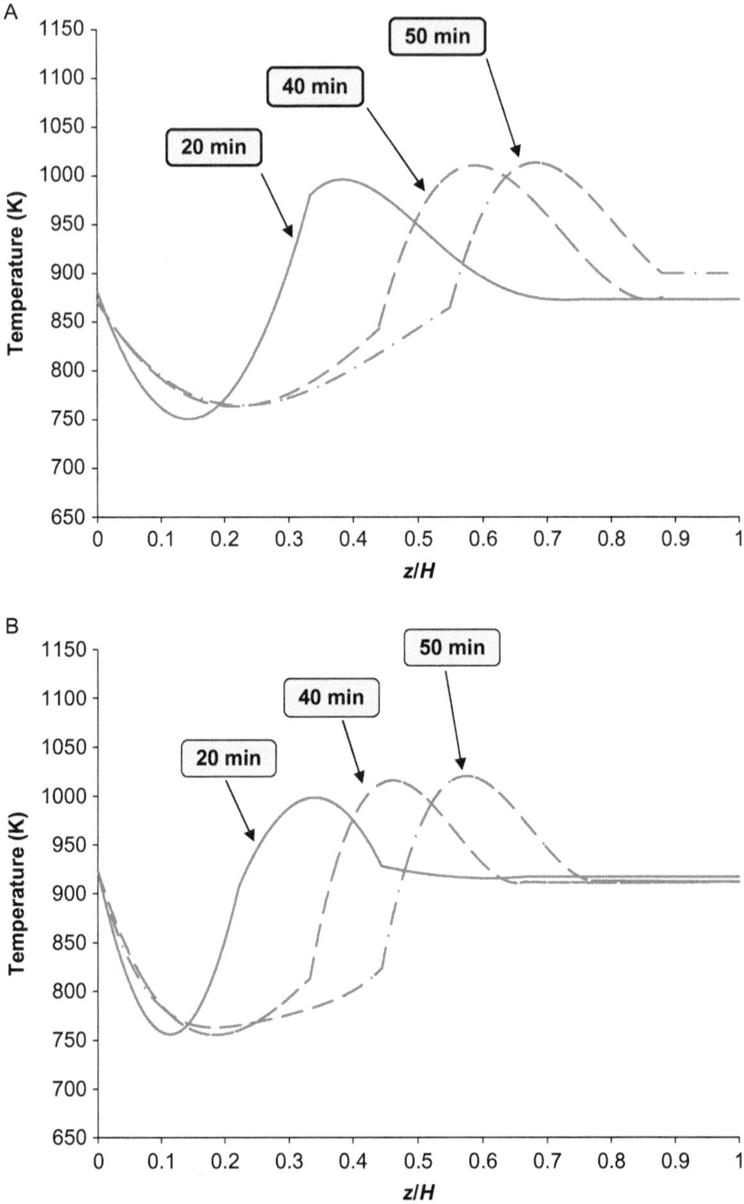

Fig. 22 Variation of reactor temperature function of time and space (reactor axial direction) for inlet values of (A) 873K and (B) 923K (first reaction cycle). *Data from Iliuta I, Radfarnia HR, Iliuta MC: Hydrogen production by sorption-enhanced steam glycerol reforming: sorption kinetics and reactor simulation,* AIChE J *59:2105–2118, 2013.*

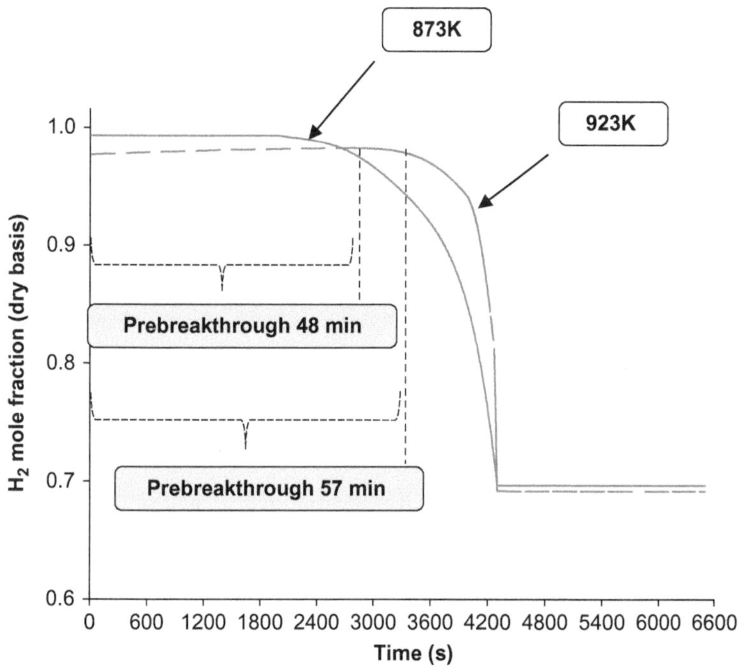

Fig. 23 Effect of inlet reactor temperature on outlet H_2 mole fraction (dry basis) during SEGSR (first reaction cycle). *Data from Iliuta I, Radfarnia HR, Iliuta MC: Hydrogen production by sorption-enhanced steam glycerol reforming: sorption kinetics and reactor simulation,* AIChE J *59:2105–2118, 2013.*

(Díaz Alvarado and Gracia, 2012; He et al., 2015; Wang et al., 2010). Dou et al. (2009) performed the reaction over commercial catalyst (18 wt.% NiO/82 wt.% Al_2O_3) mixed with CaO-based sorbent derived from dolomite at 400–700°C, 1 atm., and S/C of 3. The application of the sorption-increased process led to an increase of H_2 to more than 90%. Performing the process at 500°C (considered the optimum reaction temperature) provided the longest prebreakthrough time corresponding to 97% purity of hydrogen. According to the experimental work by He et al. (2010b) performed on Co–Ni/HTls (25%–30% Co–10%–15% Ni/HTls) as reforming catalysts and calcined dolomite as CO_2 sorbent, a hydrogen purity close to equilibrium could be obtained beyond 575°C and S/G > 4, due to the enhancement of methane steam reforming and WGS, as well as the reduction of noncatalytic reactions. As example, while hydrogen purity and yield reached around 99% for S/G of 9, the SEGSR efficiency dropped significantly below 550°C and for S/G of 3 due to the tendency in increasing pyrolysis reactions and decreasing the steam reforming. Wang et al. (2015b) developed a

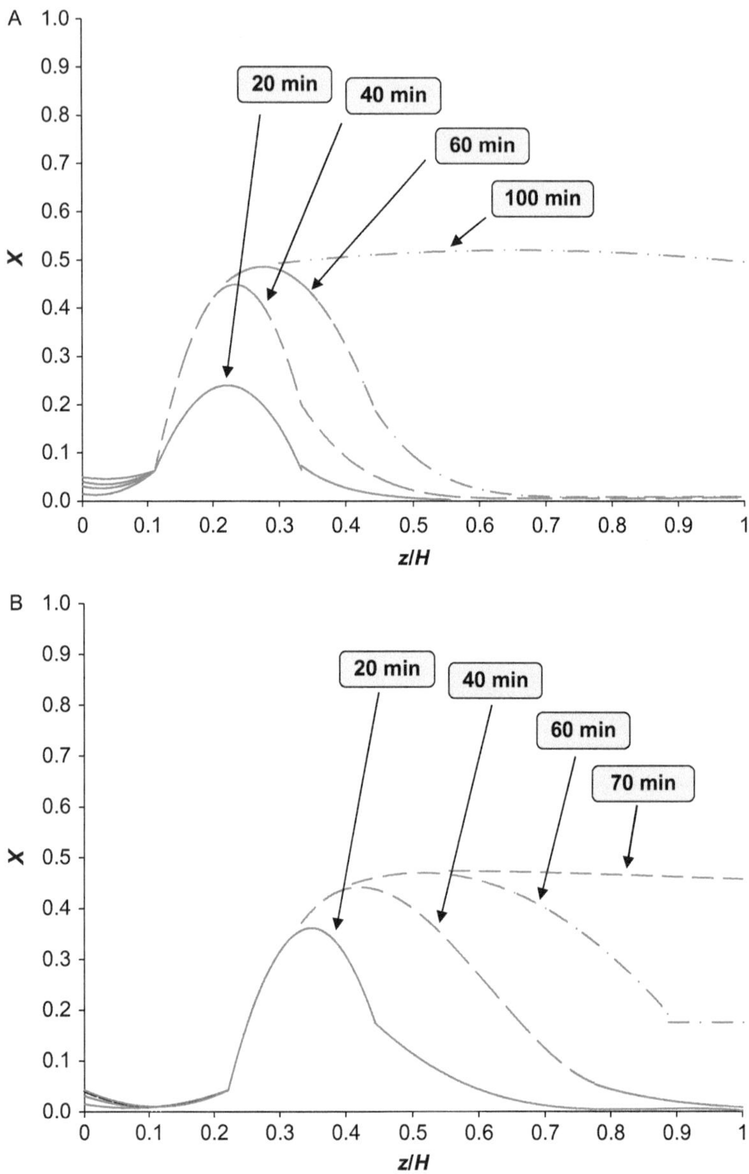

Fig. 24 Time-dependent axial distribution of sorbent conversion for two inlet superficial gas velocities (v_{sg}), at the first cycle: (A) 0.1 m/s; (B) 0.2 m/s. *Data from Iliuta I, Radfarnia HR, Iliuta MC: Hydrogen production by sorption-enhanced steam glycerol reforming: sorption kinetics and reactor simulation,* AIChE J *59:2105–2118, 2013.*

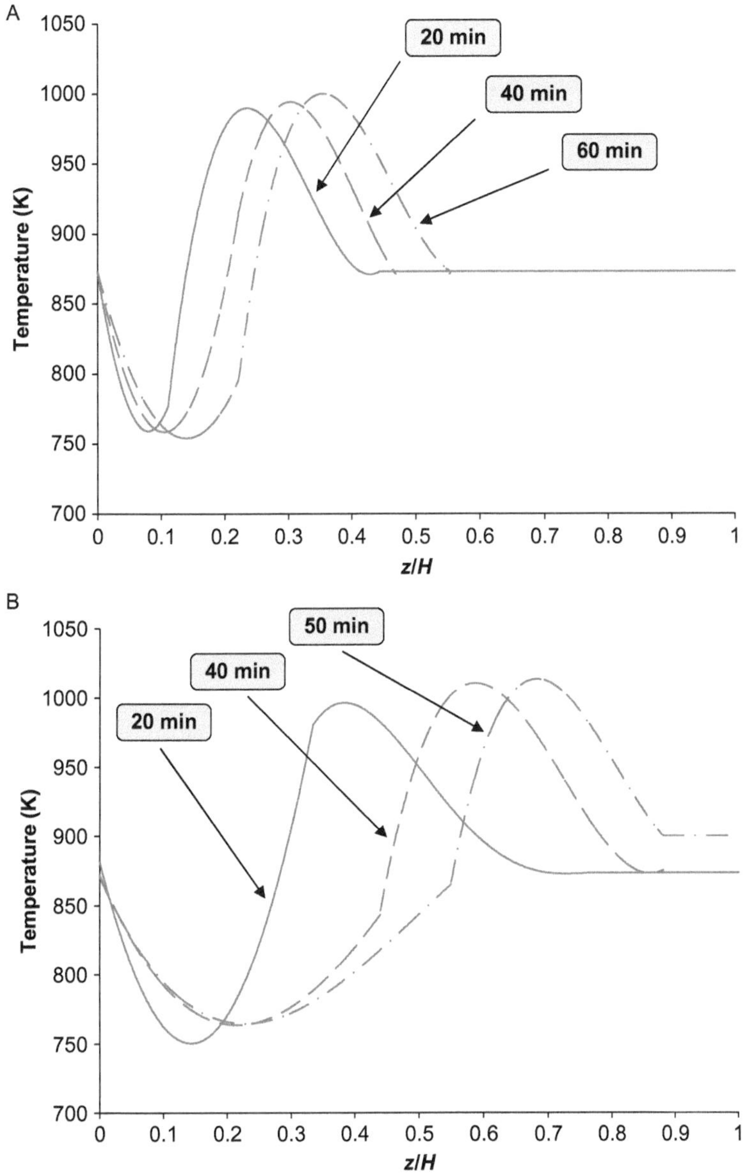

Fig. 25 Reactor temperature as a function of time and space (reactor axial direction) for two inlet superficial gas velocities (v_{sg}), at the first reaction cycle: (A) 0.1 m/s; (B) 0.2 m/s. *Data from Iliuta I, Radfarnia HR, Iliuta MC: Hydrogen production by sorption-enhanced steam glycerol reforming: sorption kinetics and reactor simulation,* AIChE J *59:2105–2118, 2013.*

bifunctional catalyst containing 41.21 wt.% NiO, 28.02 wt.% Al_2O_3, and 30.77 wt.% CaO. The tests performed at 550°C (S/C of 3) revealed that even if the H_2 yield did not significantly decrease during five cycles (reaction/sorbent regeneration), the yield and the prebreakthrough time (10 min) were not satisfactory. Among two Pt-based bifunctional materials (Pt–NiMgHTlc and Pt–NiCuHTlc) investigated by Dewoolkar and Vaidya at 550°C (S/C of 4.5), Pt–NiCuHTlc showed a better performance with H_2 concentration of 98.7% and uptake capacity of 5.6 wt.% for the first cycle (Dewoolkar and Vaidya, 2016a). In cyclic operation (20 cycles) Pt–NiCuHTlc kept almost stable performance during the first 17 cycles, decreasing thereafter over the last 3 cycles to ~78%. Dang et al. (2016) developed an efficient Co–CaO–$Ca_{12}Al_{14}O_{33}$ bifunctional material derived from HTl to produce a hydrogen purity of 96.4% which remained stable over 50 reaction–calcination cycles at 525°C and S/C of 4. However, the prebreakthrough time was shortened from 12 to 6 min during the first 20 cycles. The addition of Cu to bimetallic Co-HTlc-based catalysts was found to result in the production of high-purity hydrogen by helping to suppress the methanation reaction (Dang et al., 2017). A ultrasound-assisted cation-exchange method was employed by Jiang et al. (2017) to develop a series of bifunctional catalysts from lamellar montmorillonite (MMT) substance (Ni–MMT–CaO with different Ca/MMT ratio). For Ca/MMT molar ratio of 5, the maximum H_2 concentration of 16% during the breakthrough period at the first cycle was reduced to 15% after 10 consecutive cycles.

More recently, Shokrollahi Yancheshmeh et al. (2017) developed $Ca_9Al_6O_{18}$–CaO/xNiO (x=15, 20, and 25 wt.%) and $Ca_9Al_6O_{18}$–CaO/20NiO–$y$$CeO_2$ (y=5, 10, and 15 wt.%) hybrid sorbent–catalyst materials by applying a wet-mixing method (limestone acidification coupled with two-step calcination (Radfarnia and Iliuta, 2013a)). Addition of CeO_2 as promoter into the most efficient bifunctional $Ca_9Al_6O_{18}$–CaO/20NiO material significantly improved its performance owing to the thin shell-connected structure formed by the addition of CeO_2, the controlled number of adjacent Ni active sites, and the oxygen mobility characteristics of CeO_2. Fig. 26 illustrates the significant effect of CeO_2 loading. For Ni20Ce5, the prebreakthrough time decreased by ~34% over five cycles possibly due to CaO sintering and coke deposition. For Ni20Ce15, a decrease by ~34% in the prebreakthrough period was mainly due to CaO sintering resulted from coke deposition; the presence of CeO_2 could not contribute successfully to the removal of carbonaceous species because of low interaction

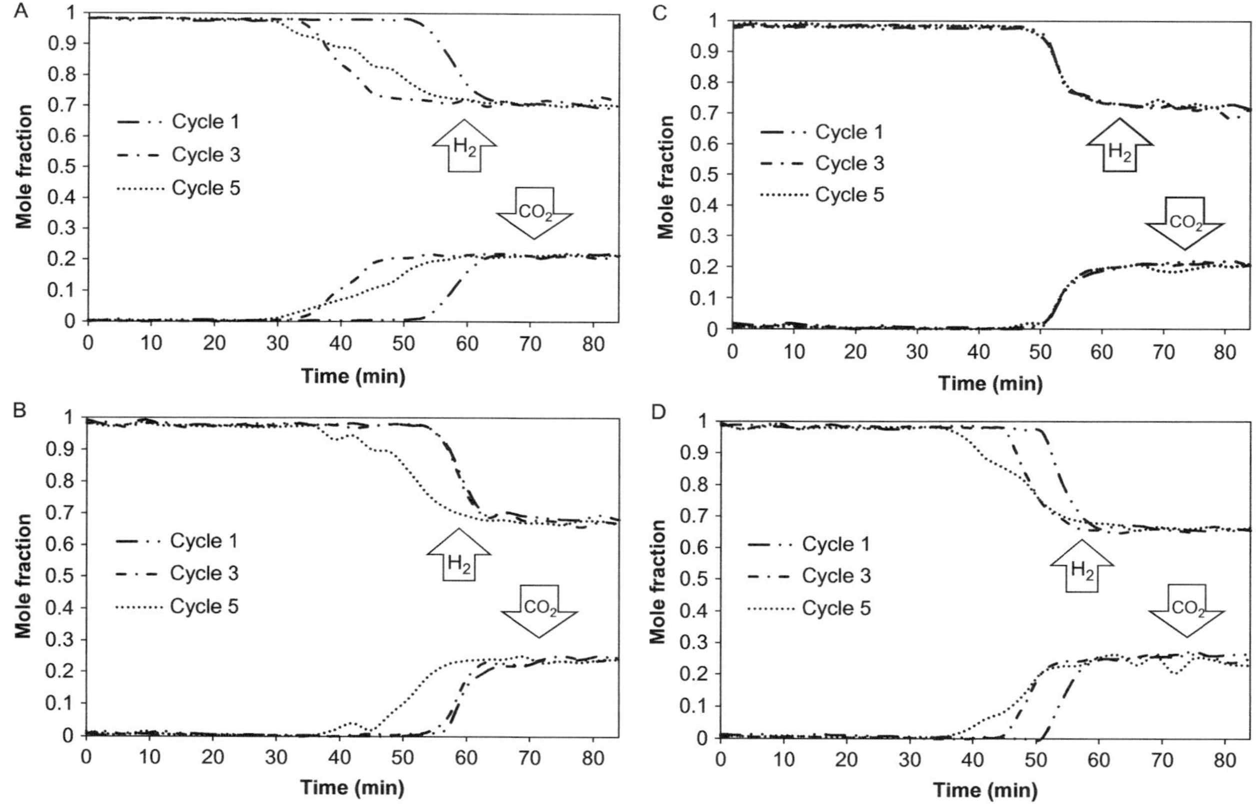

Fig. 26 Breakthrough behavior of $Ca_9Al_6O_{18}$–CaO/20NiO–yCeO$_2$ ($y=5$, 10, and 15 wt.%) hybrid sorbent-catalyst in SEGSR over five cycles (550°C, S/C of 3): (A) Ni20, (B) Ni20Ce5, (C) Ni20Ce10, (D) Ni20Ce15. *Data from Shokrollahi Yancheshmeh M, Radfarnia HR, Iliuta MC: Sustainable production of high-purity hydrogen by sorption enhanced steam reforming of glycerol over CeO_2-promoted $Ca_9Al_6O_{18}$-CaO/NiO bifunctional material,* ACS Sustainable Chem Eng, *2017, https://doi.org/10.1021/acssuschemeng.7b01627.*

of NiO with CeO_2. 10 wt.% of CeO_2 offered the best performance over 20 SESRG/regeneration cycles, with a stable H_2 purity of ~98%, H_2 yield of ~91%, and prebreakthrough time of 30 min (Figs. 26 and 27). The appropriate contact between CeO_2 and Ni helped gasifying the carbonaceous species and promoted the reforming reaction by decreasing the coke formation. Hydrogen concentration of more than 98% and negligible CO concentration in the prebreakthrough stage undoubtedly open the opportunities for the application of the produced gas stream in fuel cell systems without any need for additional purification.

Compared to the market price of purified glycerol (55–78 $/ton), the one for crude glycerol varies generally between 25 and 46 $/ton (Fermoso et al., 2012b). From an economic point of view, the valorization of crude glycerol should therefore be more attractive for producing hydrogen of high purity by SEGSR. However, due to the diversity of crude glycerol composition (Table 7) and the high complexity of multiple possible reactions that might occur in glycerol steam reforming, only very few studies concern the SESR of crude glycerol (Dou et al., 2010a; Fermoso et al., 2012a). The available studies show that, compared to the sorption-enhanced reforming of pure glycerol, hydrogen purity and yield obtained using crude

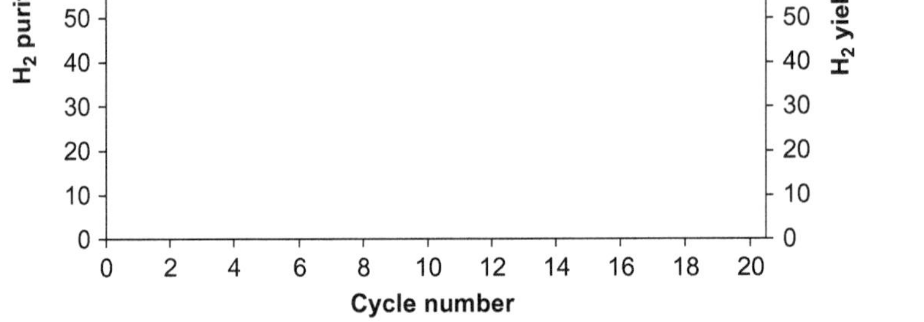

Fig. 27 Average H_2 purity and yield as a function of cycle number for $Ca_9Al_6O_{18}$–CaO/20NiO–10CeO_2 hybrid sorbent-catalyst in SEGSR (550°C, S/C of 3). *Data from Shokrollahi Yancheshmeh M, Radfarnia HR, Iliuta MC: Sustainable production of high-purity hydrogen by sorption enhanced steam reforming of glycerol over CeO_2-promoted $Ca_9Al_6O_{18}$-CaO/NiO bifunctional material,* ACS Sustainable Chem Eng, *2017, https://doi.org/10.1021/acssuschemeng.7b01627.*

Table 7 Main Liquid Components of Crude Glycerol From Different Sources (wt.%)

Component	Source #1	Source #2	Source #3
Glycerol	63.0	22.9	57.1
Methanol	6.2	10.9	11.3
Water	28.7	18.2	1.0
Soap	n.d.	26.2	31.4
FATTY acid methyl esters	n.d.	.21.3	0.5

Data from Xiao Y, Xiao GM, Varma A: A universal procedure for crude glycerol purification from different feedstocks in biodiesel production: experimental and simulation study, *Ind Eng Chem Res* 52:1-4291–14296, 2013.

glycerol are lower especially due to coke formation from heavy components like fatty acid methyl esters (Table 7).

For example, Dou et al. (2010a) investigated the crude glycerol steam reforming in a fixed-bed reactor over commercial Ni/Al_2O_3 catalyst (18 wt.% NiO) at 400–700°C, 1 atm., for a ratio S/C of 3, with and without in situ CO_2 removal. At 600°C, H_2 purity could increase considerably from 68% (conventional steam reforming without CO_2 removal) to 88% (sorption-enhanced steam reforming of crude glycerol with in situ CO_2 removal using a CaO-based sorbent derived from dolomite). Similarly, Fermoso et al. (2012a) evaluated the steam reforming of crude glycerol in a fixed-bed reactor (atmospheric pressure, 550–600°C, S/C of 3) using a mixture of catalyst (Ni–Co/HTlc) and sorbent (calcined dolomite) with a sorbent/catalyst mass ratio of 5. The in situ CO_2 removal resulted in a significant increase of H_2 yield and purity of 88% and 99.7 vol.%, respectively, compared to conventional steam reforming. Undoubtedly, the production of almost pure H_2 from crude glycerol is a great challenge mainly because of the complexity of crude glycerol composition.

4. CONCLUSIONS AND PERSPECTIVES

As an integrated process combining reforming, water gas shift, and in situ CO_2 capture in the same reactor, the sorption-enhanced steam reforming is a very appealing alternative to conventional steam reforming for the production of highly pure H_2 in a single step. Compared to the traditional steam reforming (maximum 80% H_2 purity on a dry basis), it was demonstrated that the in situ removal of CO_2 could usually result in very high purity reaching 95%–98%, with very low content of CO and CO_2.

In most applications, subsequent purifications of hydrogen should therefore not be generally required.

Numerous promising sorbents have been developed, as potential candidates for sorption-enhanced steam reforming processes, including Ca-based, HTlc-based, and alkaline ceramic sorbents. CaO-based sorbents derived from natural precursors like limestone and dolomite are especially advantageous due to their suitable uptake capacity and kinetics, low cost and availability. However, in spite of the high capacity and kinetics of fresh (unused) materials, they are affected by a high instability in long-term cyclic operation due to sintering during continuous calcination/regeneration processes at high temperatures. Although several strategies have been investigated to improve their stability and performance, it continues to represent a significant challenge. Their cyclic stability in long-time operation still needs to be much improved to make them appropriate candidates for different SESR processes. The alkaline ceramics like Li_2ZrO_3, Li_4SiO_4, and Na_2ZrO_3 and corresponding doped materials have in general acceptable capacities and better stability in cyclic operation, but most of them are affected by slow kinetics especially at low CO_2 partial pressures, an essential requirement in the SESR process. Additional work on the improvement of their properties as well as applications in cyclic reforming reactions in realistic conditions, to highlight the effect of different operation parameters, would be of high interest.

Although the development and application of bifunctional (hybrid sorbent-catalyst) materials appear to be an interesting option over simple physical mixtures of catalyst and sorbent, with essential practical and economic benefits, much work still needs to be done for all kinds of combinations sorbent–catalyst–reforming process, including semipilot and pilot plant tests (especially under severe operation conditions and long-term cyclic operation) before an eventual commercialization.

The SEMSR has undoubtedly received most attention. Concerning the application of high-temperature sorbents in steam reforming process of feedstocks other than methane, like ethanol, butanol, and glycerol, few works have been completed until now compared to methane. More attention was given to ethanol and glycerol, but the number of investigations is still scarce especially on glycerol. This is well understood from a traditional point of view (main reforming process for hydrogen production all over the world), but most probably also because of the difficulties to manage the reforming of other feedstocks (mainly, side reactions with complexity depending on the compound structure, more complex analysis of the reaction media composed by gaseous and liquid multicomponent systems, more complex and costly installations and process design, more difficulties in performing long-time cyclic tests, and increased probability for coke formation). However, with the upcoming unavoidable emptying of natural gas resources, the interest will focus inevitably on other possible feedstocks for hydrogen generation, mainly issued from renewable sources. For example, taking into account that ethanol can be produced from renewable sources and the huge amounts of glycerol (also produced from valorization of renewable sources) which is believed to become a significant waste problem in the near future, their valorization into value-added compounds (like hydrogen in the context of the steam reforming process) appears as a very appealing subject of high interest. In particular, the use of crude feedstocks, like ethanol and glycerol, although much more challenging in comparison with the purified compounds, should be more often included in future investigations. Finally, by considering that in the sorption-enhanced process the production of highly pure hydrogen is accompanied by simultaneous production of very pure CO_2 (which can be sequestrated or used in numerous industrial applications), new developments of both novel sorbents and improved process designs would also be very appealing research areas of interest, especially in the context of a sustainable development.

REFERENCES

Abanades, J. C. (2002). The maximum capture efficiency of CO_2 using a carbonation/calcination cycle of $CaO/CaCO_3$. *Chem Eng J, 90*, 303–306.

Abanades, J. C., & Alvarez, D. (2003). Conversion limits in the reaction of CO_2 with lime. *Energy Fuel, 17*, 308–315.

Albrecht, K. O., Wagenbach, K. S., Satrio, J. A., Shanks, B. H., & Wheelock, T. D. (2008). Development of a CaO-based CO_2 sorbent with improved cyclic stability. *Ind Eng Chem Res, 47*, 7841–7848.

Alcérreca-Corte, I., Fregoso-Israel, E., & Pfeiffer, H. (2008). CO_2 absorption on Na_2ZrO_3: a kinetic analysis of the chemisorption and diffusion processes. *J Phys Chem C, 112*, 6520–6525.

Alvarez, D., & Abanades, J. C. (2005a). Determination of the critical product layer thickness in the reaction of CaO with CO_2. *Ind Eng Chem Res, 44*, 5608–5615.

Alvarez, D., & Abanades, J. C. (2005b). Pore-size and shape effects on the recarbonation performance of calcium oxide submitted to repeated calcination/recarbonation cycles. *Energy Fuel, 19*, 270–278.

Angeli, S. D., Martavaltzi, C. S., & Lemonidou, A. A. (2014). Development of a novel-synthesized Ca-based CO_2 sorbent for multicycle operation: parametric study of sorption. *Fuel, 127*, 62–69.

Arias, B., Abanades, J. C., & Anthony, E. J. (2011). Model for self-reactivation of highly sintered CaO particles during CO_2 capture looping cycles. *Energy Fuel, 25*, 1926–1930.

Arias, B., Grasa, G., Abanades, J. C., Manovic, V., & Anthony, E. J. (2012a). The effect of steam on the fast carbonation reaction rates of CaO. *Ind Eng Chem Res, 51*, 2478–2482.

Arias, B., Grasa, G. S., Alonso, M., & Abanades, J. C. (2012b). Post-combustion calcium looping process with a highly stable sorbent activity by recarbonation. *Energ Environ Sci, 5*, 7353–7359.

Ávalos-Rendón, T., Casa-Madrid, J., & Pfeiffer, H. (2009). Thermochemical capture of carbon dioxide on lithium aluminates ($LiAlO_2$ and Li_5AlO_4): a new option for the CO_2 absorption. *J Phys Chem A, 113*, 6919–6923.

Ávalos-Rendón, T., Lara, V. H., & Pfeiffer, H. (2012). CO_2 chemisorption and cyclability analyses of lithium aluminate polymorphs (α- and β-Li_5AlO_4). *Ind Eng Chem Res, 51*, 2622–2630.

Balasubramanian, B., Ortiz, A. L., Kaytakoglu, S., & Harrison, D. P. (1999). Hydrogen from methane in a single-step process. *Chem Eng Sci, 54*, 3543–3552.

Bandi A, Specht M, Sichler P, Nicoloso N, 2002. In situ gas conditioning in fuel reforming for hydrogen generation, 5th international symposium on gas cleaning at high temperature. US DOE National Energy Technology Laboratory, Morgantown, USA.

Barelli, L., Bidini, G., Gallorini, F., & Servili, S. (2008). Hydrogen production through sorption-enhanced steam methane reforming and membrane technology: a review. *Energy, 33*, 554–570.

Barker, R. (1974). The reactivity of calcium oxide towards carbon dioxide and its use for energy storage. *J Appl Chem Biotechnol, 24*, 221–227.

Bartholomew, C. H. (2001). Mechanisms of catalyst deactivation. *Appl Catal A, 212*, 17–60.

Bhatia, S. K., & Perlmutter, D. D. (1983). Effect of the product layer on the the kinetics of the CO_2-lime reaction. *AIChE J, 29*, 79–86.

Blamey, J., Al-Jeboori, M. J., Manovic, V., Fennell, P. S., & Anthony, E. J. (2016). CO_2 capture by calcium aluminate pellets in a small fluidized bed. *Fuel Process Technol, 142*, 100–106.

Borgwardt, R. H. (1989a). Calcium oxide sintering in atmospheres containing water and carbon dioxide. *Ind Eng Chem Res, 28*, 493–500.

Borgwardt, R. H. (1989b). Sintering of nascent calcium-oxide. *Chem Eng Sci, 44*, 53–60.

Broda, M., & Müller, C. R. (2012). Synthesis of highly efficient, Ca-based, Al_2O_3-stabilized, carbon gel-templated CO_2 sorbents. *Adv Mater, 24*, 3059–3064.

Broda, M., & Müller, C. R. (2014). Sol-gel-derived, CaO-based, ZrO_2-stabilized CO_2 sorbents. *Fuel, 127*, 94–100.

Broda, M., Kierzkowska, A. M., Baudouin, D., Imtiaz, Q., Coperet, C., & Muller, C. R. (2012). Sorbent-enhanced methane reforming over a Ni-Ca-based, bifunctional catalyst sorbent. *ACS Catal, 2*, 1635–1646.

Broda, M., Kierzkowska, A. M., & Müller, C. R. (2013a). Sorbent-enhanced steam methane reforming reaction studied over a Ca-based CO_2 sorbent and Ni catalyst. *Chem Eng Technol, 36*, 1496–1502.

Broda, M., Manovic, V., Imtiaz, Q., Kierzkowska, A. M., Anthony, E. J., & Müller, C. R. (2013b). High-purity hydrogen via the sorption-enhanced steam methane reforming reaction over a synthetic CaO-based sorbent and a Ni catalyst. *Environ Sci Technol, 47*, 6007–6014.

Broda, M., Manovic, V., Anthony, E. J., & Müller, C. R. (2014). Effect of pelletization and addition of steam on the cyclic performance of carbon-templated, CaO-based CO_2 sorbents. *Environ Sci Technol, 48*, 5322–5328.

Centi, G., & Perathoner, S. (2009). Opportunities and prospects in the chemical recycling of carbon dioxide to fuels. *Catal Today, 148*, 191–205.

Champagne, S., DY, L., Macchi, A., Symonds, R. T., & Anthony, E. J. (2013). Influence of steam injection during calcination on the reactivity of CaO-based sorbent for carbon capture. *Ind Eng Chem Res, 52*, 2241–2246.

Chanburanasiri, N., Ribeiro, A. M., Rodrigues, A. E., et al. (2011). hydrogen production via sorption enhanced steam methane reforming process using Ni/CaO multifunctional catalyst. *Ind Eng Chem Res, 50*, 13662–13671.

Chang, P.-H., Chang, Y.-P., Chen, S.-Y., C-T, Y., & Chyou, Y.-P. (2011). Ca-rich Ca–Al-Oxide, high-temperature-stable sorbents prepared from hydrotalcite precursors: synthesis, characterization, and CO_2 capture capacity. *ChemSusChem, 4*, 1844–1851.

Chang, P. H., Lee, T. J., Chang, Y. P., & Chen, S. Y. (2013). CO_2 sorbents with scaffold-like CaAl layered double hydroxides as precursors for CO_2 capture at high temperatures. *ChemSusChem, 6*, 1076–1083.

Chang, P. H., Chang, Y. P., Lai, Y. H., Chen, S. Y., CT, Y., & Chyou, Y. P. (2014). Synthesis, characterization and high temperature CO_2 capture capacity of nanoscale Ca-based layered double hydroxides via reverse microemulsion. *J Alloys Compd, 586*, S498–S505.

Chen, H., Zhang, T., Dou, B., et al. (2009a). Thermodynamic analyses of adsorption-enhanced steam reforming of glycerol for hydrogen production. *Int J Hydrogen Energy, 34*, 7208–7222.

Chen, Z., Song, H. S., Portillo, M., Lim, C. J., Grace, J. R., & Anthony, E. J. (2009b). Long-term calcination/carbonation cycling and thermal pretreatment for CO_2 capture by limestone and dolomite. *Energy Fuel, 23*, 1437–1444.

Chen, H., Zhao, C., & Yang, Y. (2013). Enhancement of attrition resistance and cyclic CO_2 capture of calcium-based sorbent pellets. *Fuel Process Technol, 116*, 116–122.

Chen, X. X., Xiong, Z., Qin, Y. D., et al. (2016). High-temperature CO_2 sorption by Ca-doped Li_4SiO_4 sorbents. *Int J Hydrogen Energy, 41*, 13077–13085.

Coppola, A., Salatino, P., Montagnaro, F., & Scala, F. (2014). Reactivation by water hydration of the CO_2 capture capacity of a calcium looping sorbent. *Fuel, 127*, 109–115.

Cunha, A. F., YJ, W., Diaz Alvarado, F. A., Santos, J. C., Vaidya, P. D., & Rodrigues, A. E. (2012). Steam reforming of ethanol on a Ni/Al_2O_3 catalyst coupled with a hydrotalcite-like sorbent in a multilayer pattern for CO_2 uptake. *Can J Chem Eng, 90*, 1514–1526.

Cunha, A. F., YJ, W., Santos, J. C., & Rodrigues, A. E. (2013). Sorption enhanced steam reforming of ethanol on hydrotalcite-like compounds impregnated with active copper. *Chem Eng Res Des, 91*, 581–592.

Cunha, A. F., YJ, W., Li, P., JG, Y., & Rodrigues, A. E. (2014). Sorption-enhanced steam reforming of ethanol on a novel K-Ni-Cu-hydrotalcite hybrid material. *Can J Chem Eng, 53*, 3842–3853.

Dang, C., Yu, H., Wang, H., Peng, F., & Yang, Y. (2016). A bi-functional Co–CaO–$Ca_{12}Al_{14}O_{33}$ catalyst for sorption-enhanced steam reforming of glycerol to high-purity hydrogen. *Chem Eng J, 286*, 329–338.

Dang, C. X., Wang, H. J., Yu, H., & Peng, F. (2017). Co-Cu-CaO catalysts for high-purity hydrogen from sorption-enhanced steam reforming of glycerol. *Appl Catal A, 533*, 9–16.

Dennis, J. S., & Pacciani, R. (2009). The rate and extent of uptake of CO_2 by a synthetic, CaO-containing sorbent. *Chem Eng Sci, 64*, 2147–2157.

Dewoolkar, K. D., & Vaidya, P. D. (2016a). Sorption-enhanced steam reforming of glycerol over Ni–hydrotalcite: effect of promotion with Pt. *ChemCatChem, 8*, 3499–3509.

Dewoolkar, K. D., & Vaidya, P. D. (2016b). Tailored hydrotalcite-based hybrid materials for hydrogen production via sorption-enhanced steam reforming of ethanol. *Int J Hydrogen Energy, 41*, 6094–6106.

Dewoolkar, K. D., & Vaidya, P. D. (2017). New hybrid materials for improved hydrogen production by the sorption-enhanced steam reforming of butanol. *Energ Technol, 5*, 1–12.

Díaz Alvarado, F., & Gracia, F. (2012). Oxidative steam reforming of glycerol for hydrogen production: thermodynamic analysis including different carbon deposits representation and CO_2 adsorption. *Int J Hydrogen Energy, 37*, 14820–14830.

Ding, Y., & Alpay, E. (2000). Adsorption-enhanced steam-methane reforming. *Chem Eng Sci, 55*, 3929–3940.

Donat, F., Florin, N. H., Anthony, E. J., & Fennell, P. S. (2012). Influence of high-temperature steam on the reactivity of CaO sorbent for CO_2 capture. *Environ Sci Tech, 46*, 1262–1269.
Dou, B., Dupont, V., Rickett, G., et al. (2009). Hydrogen production by sorption-enhanced steam reforming of glycerol. *Bioresour Technol, 100*, 3540–3547.
Dou, B., Rickett, G. L., Dupont, V., et al. (2010a). Steam reforming of crude glycerol with in situ CO_2 sorption. *Bioresour Technol, 101*, 2436–2442.
Dou, B., Song, Y., Liu, Y., & Feng, C. (2010b). High temperature CO_2 capture using calcium oxide sorbent in a fixed-bed reactor. *J Hazard Mater, 183*, 759–765.
Dou, B., Song, Y., Wang, C., Chen, H., & Xu, Y. (2014). Hydrogen production from catalytic steam reforming of biodiesel byproduct glycerol: issues and challenges. *Renew Sustain Energy Rev, 30*, 950–960.
Dou, B., Jiang, B., Song, Y., et al. (2016a). Enhanced hydrogen production by sorption-enhanced steam reforming from glycerol with in-situ CO_2 removal in a fixed-bed reactor. *Fuel, 166*, 340–346.
Dou, B., Wang, C., Song, Y., et al. (2016b). Solid sorbents for in-situ CO_2 removal during sorption-enhanced steam reforming process: a review. *Renew Sustain Energy Rev, 53*, 536–546.
Duan, Y. H., Pfeiffer, H., Li, B. Y., et al. (2013). CO_2 capture properties of lithium silicates with different ratios of Li_2O/SiO_2: an ab initio thermodynamic and experimental approach. *Phys Chem Chem Phys, 15*, 13538–13558.
Durán-Muñoz, F., Romero-Ibarra, I. C., & Pfeiffer, H. (2013). Analysis of the CO_2 chemisorption reaction mechanism in lithium oxosilicate (Li_8SiO_6): a new option for high-temperature CO_2 capture. *J Mater Chem A, 1*, 3919–3925.
Essaki, K., Kato, M., & Nakagawa, K. (2006). CO_2 removal at high temperature using packed bed of lithium silicate pellets. *J Cerma Soc Jpn, 114*, 739–742.
Essaki, K., Muramatsu, T., & Kato, M. (2008). Effect of equilibrium-shift in the case of using lithium silicate pellets in ethanol steam reforming. *Int J Hydrogen Energy, 33*, 6612–6618.
Feng, H. Z., Lan, P. Q., & SF, W. (2012). A study on the stability of a NiO-CaO/Al_2O_3 complex catalyst by La_2O_3 modification for hydrogen production. *Int J Hydrogen Energy, 37*, 14161–14166.
Fermoso, J., He, L., & Chen, D. (2012a). Production of high purity hydrogen by sorption enhanced steam reforming of crude glycerol. *Int J Hydrogen Energy, 37*, 14047–14054.
Fermoso, J., He, L., & Chen, D. (2012b). Sorption enhanced steam reforming (SESR): a direct route towards efficient hydrogen production from biomass-derived compounds. *J Chem Technol Biotechnol, 87*, 1367–1374.
Ficicilar, B., & Dogu, T. (2006). Breakthrough analysis for CO_2 removal by activated hydrotalcite and soda ash. *Catal Today, 115*, 274–278.
Filitz, R., Kierzkowska, A. M., Broda, M., & Müller, C. R. (2011). Highly efficient CO_2 sorbents: development of synthetic, calcium-rich dolomites. *Environ Sci Technol, 46*, 559–565.
Florin, N. H., & Harris, A. T. (2008). Screening CaO-based sorbents for CO_2 capture in biomass gasifiers. *Energy Fuel, 22*, 2734–2742.
Florin, N. H., & Harris, A. T. (2009). Reactivity of CaO derived from nano-sized $CaCO_3$ particles through multiple CO_2 capture-and-release cycles. *Chem Eng Sci, 64*, 187–191.
Florin, N. H., Blamey, J., & Fennell, P. S. (2010). Synthetic CaO-based sorbent for CO_2 capture from large-point sources. *Energy Fuel, 24*, 4598–4604.
Fujishiro, F., Fukasawa, K., & Hashimoto, T. (2011). CO_2 absorption and desorption properties of single phase $Ba_2Fe_2O_5$ and analysis of their mechanism using thermodynamic calculation. *J Am Ceram Soc, 94*, 3675–3678.
Fujishiro, F., Kojima, Y., & Hashimoto, T. (2012). Kinetics and mechanism of chemical reaction of CO_2 and $Ba_2Fe_2O_5$ under various CO_2 partial pressures. *J Am Ceram Soc, 95*, 3634–3637.

German, R. M., & Munir, Z. A. (1976). Surface-area reduction during isothermal sintering. *J Am Ceram Soc*, *59*, 379–383.
Gonzalez, B., Liu, W., Sultan, D. S., & Dennis, J. S. (2016). The effect of steam on a synthetic Ca-based sorbent for carbon capture. *Chem Eng J*, *285*, 378–383.
Grasa, G. S., & Abanades, J. C. (2006). CO_2 capture capacity of CaO in long series of carbonation/calcination cycles. *Ind Eng Chem Res*, *45*, 8846–8851.
Grasa, G., GonzAlez, B., Alonso, M., & Abanades, J. C. (2007). Comparison of CaO-based synthetic CO_2 sorbents under realistic calcination conditions. *Energy Fuel*, *21*, 3560–3562.
Grasa, G., Martinez, I., Diego, M. E., & Abanades, J. C. (2014). Determination of CaO carbonation kinetics under recarbonation conditions. *Energy Fuel*, *28*, 4033–4042.
Gunduz, S., & Dogu, T. (2012). Sorption-enhanced reforming of ethanol over Ni- and Co-incorporated MCM-41 type catalysts. *Ind Eng Chem Res*, *51*, 8796–8805.
Gupta, H., & Fan, L.-S. (2002). Carbonation-calcination cycle using high reactivity calcium oxide for carbon dioxide separation from flue gas. *Ind Eng Chem Res*, *41*, 4035–4042.
Harrison, D. P. (2008). Sorption-enhanced hydrogen production: a review. *Ind Eng Chem Res*, *47*, 6486–6501.
He, L., Berntsen, H., & Chen, D. (2010a). Approaching sustainable H_2 production: sorption enhanced steam reforming of ethanol. *J Phys Chem A*, *114*, 3834–3844.
He, L., Parra, J. M. S., Blekkan, E. A., & Chen, D. (2010b). Towards efficient hydrogen production from glycerol by sorption enhanced steam reforming. *Energ Environ Sci*, *3*, 1046–1056.
He, C., Zheng, J., Wang, K., Lin, H., Wang, J.-Y., & Yang, Y. (2015). Sorption enhanced aqueous phase reforming of glycerol for hydrogen production over Pt-Ni supported on multi-walled carbon nanotubes. *Appl Catal B*, *162*, 401–411.
Hsu, M. J., Lee, K. H., & Chyou, Y. P. (2014). CO_2 capture at high temperature using calcium-based sorbents. *J Chin Inst Eng*, *37*, 152–164.
Ida, J., & Lin, Y. S. (2003). Mechanism of high-temperature CO_2 sorption on lithium zirconate. *Environ Sci Technol*, *37*, 1999–2004.
Ida, J., Xiong, R. T., & Lin, Y. S. (2004). Synthesis and CO_2 sorption properties of pure and modified lithium zirconate. *Sep Purif Technol*, *36*, 41–51.
Iliuta, I., Iliuta, M. C., & Larachi, F. (2011). Sorption-enhanced dimethyl ether synthesis-multiscale reactor modeling. *Chem Eng Sci*, *66*, 2241–2251.
Iliuta, I., Radfarnia, H. R., & Iliuta, M. C. (2013). Hydrogen production by sorption-enhanced steam glycerol reforming: sorption kinetics and reactor simulation. *AIChE J*, *59*, 2105–2118.
Iwan, A., Stephenson, H., Ketchie, W. C., & Lapkin, A. A. (2009). High temperature sequestration of CO_2 using lithium zirconates. *Chem Eng J*, *146*, 249–258.
Jiang, B., Dou, B., Wang, K., et al. (2017). Sorption enhanced steam reforming of biodiesel by-product glycerol on Ni-CaO-MMT multifunctional catalysts. *Chem Eng J*, *313*, 207–216.
Jing, J. Y., Li, T. Y., Zhang, X. W., et al. (2017). Enhanced CO_2 sorption performance of CaO/Ca3Al2O6 sorbents and its sintering-resistance mechanism. *Appl Energy*, *199*, 225–233.
Johnsen, K., Ryu, H. J., Grace, J. R., & Lim, C. J. (2006). Sorption-enhanced steam reforming of methane in a fluidized bed reactor with dolomite as CO_2-acceptor. *Chem Eng Sci*, *61*, 1195–1202.
Kato, M., & Nakagawa, K. (2001). New series of lithium containing complex oxides, lithium silicates, for application as a high temperature CO_2 absorbent. *J Cerma Soc Jpn*, *109*, 911–914.
Kato, M., Essaki, K., Nakagawa, K., Suyama, Y., & Terasaka, K. (2005a). CO_2 absorption properties of lithium ferrite for application as a high-temperature CO_2 absorbent. *J Cerma Soc Jpn*, *113*, 684–686.

Kato, M., Nakagawa, K., Essaki, K., et al. (2005b). Novel CO_2 absorbents using lithium-containing oxide. *Int J Appl Ceram Technol, 2*, 467–475.

Kierzkowska, A. M., Poulikakos, L. V., Broda, M., & Muller, C. R. (2013). Synthesis of calcium-based, Al_2O_3-stabilized sorbents for CO_2 capture using a co-precipitation technique. *Int J Greenhouse Gas Control, 15*, 48–54.

Kim, J. N., Ko, C. H., & Yi, K. B. (2013). Sorption enhanced hydrogen production using one-body CaO-$Ca_{12}Al_{14}O_{33}$-Ni composite as catalytic absorbent. *Int J Hydrogen Energy, 38*, 6072–6078.

Koirala, R., Gunugunuri, K. R., Pratsinis, S. E., & Smirniotis, P. G. (2011). Effect of zirconia doping on the structure and stability of CaO-based sorbents for CO_2 capture during extended operating cycles. *J Phys Chem C, 115*, 24804–24812.

Koirala, R., Reddy, G. K., & Smirniotis, P. G. (2012). Single nozzle flame-made highly durable metal doped Ca-based sorbents for CO_2 capture at high temperature. *Energy Fuel, 26*, 3103–3109.

Lan, P. Q., & Wu, S. F. (2014). Synthesis of a porous Nano-CaO/MgO-based CO_2 adsorbent. *Chem Eng Technol, 37*, 580–586.

Lee, J. S., & Yavuz, C. T. (2017). Enhanced sorption cycle stability and kinetics of CO_2 on lithium silicates using the lithium ion channeling effect of TiO_2 nanotubes. *Ind Eng Chem Res, 56*, 3413–3417.

Lee, W. E., Vieira, W., Zhang, S., Ahari, K. G., Sarpoolaky, H., & Parr, C. (2001). Castable refractory concretes. *Int Mater Rev, 46*, 145–167.

Li, Z. S., & Cai, N. S. (2007). Modeling of multiple cycles for sorption-enhanced steam methane reforming and sorbent regeneration in fixed bed reactor. *Energy Fuel, 21*, 2909–2918.

Li, Z. S., Cai, N. S., Huang, Y. Y., & Han, H. J. (2005). Synthesis, experimental studies, and analysis of a new calcium-based carbon dioxide absorbent. *Energy Fuel, 19*, 1447–1452.

Li, Z. S., Cai, N. S., & Huang, Y. Y. (2006a). Effect of preparation temperature on cyclic CO_2 capture and multiple carbonation-calcination cycles for a new Ca-based CO_2 sorbent. *Ind Eng Chem Res, 45*, 1911–1917.

Li, Z. S., Cai, N. S., & Yang, J. B. (2006b). Continuous production of hydrogen from sorption-enhanced steam methane reforming in two parallel fixed-bed reactors operated in a cyclic manner. *Ind Eng Chem Res, 45*, 8788–8793.

Li, Y., Zhao, C., Qu, C., Duan, L., Li, Q., & Liang, C. (2008). CO_2 capture using CaO modified with ethanol/water solution during cyclic calcination/carbonation. *Chem Eng Technol, 31*, 237–244.

Li, L. Y., King, D. L., Nie, Z. M., & Howard, C. (2009a). Magnesia-stabilized calcium oxide absorbents with improved durability for high temperature CO_2 capture. *Ind Eng Chem Res, 48*, 10604–10613.

Li, Y., Zhao, C., Chen, H., Liang, C., Duan, L., & Zhou, W. (2009b). Modified CaO-based sorbent looping cycle for CO_2 mitigation. *Fuel, 88*, 697–704.

Li, Y., Wang, W., Chen, B., & Cao, Y. (2010). Thermodynamic analysis of hydrogen production via glycerol steam reforming with CO_2 adsorption. *Int J Hydrogen Energy, 35*, 7768–7777.

Li, Y., Sun, R., Liu, H., & Lu, C. (2011). Cyclic CO_2 capture behavior of limestone modified with pyroligneous acid (PA) during calcium looping cycles. *Ind Eng Chem Res, 50*, 10222–10228.

Li, Z. S., Fang, F., Tang, X. Y., & Cai, N. S. (2012). Effect of temperature on the carbonation reaction of CaO with CO_2. *Energy Fuel, 26*, 2473–2482.

Li, Z., Liu, Y., & Cai, N. (2014a). Understanding the enhancement effect of high-temperature steam on the carbonation reaction of CaO with CO_2. *Fuel, 127*, 88–93.

Lide, D. R. (2003). *CRC handbook of chemistry and physics* (ed 84). CRC Press, Taylor & Francis Group, LLC: Boca Raton.

Lin, Y.-C. (2013). Catalytic valorization of glycerol to hydrogen and syngas. *Int J Hydrogen Energy, 38*, 2678–2700.

Linden, I., Backman, P., Brink, A., & Hupa, M. (2011). Influence of water vapor on carbonation of CaO in the temperature range 400-550 °C. *Ind Eng Chem Res, 50*, 14115–14120.

Liu, M., & Evans, B. (1997). Dislocation recovery kinetics in single-crystal calcite. *J Geophys Res Solid Earth, 102*, 24801–24809.

Liu, W. Q., Feng, B., YQ, W., Wang, G. X., Barry, J., & da Costa, J. C. D. (2010a). Synthesis of sintering-resistant sorbents for CO_2 capture. *Environ Sci Technol, 44*, 3093–3097.

Liu, W. Q., Low, N. W. L., Feng, B., Wang, G. X., & da Costa, J. C. D. (2010b). Calcium precursors for the production of CaO sorbents for multicycle CO_2 capture. *Environ Sci Technol, 44*, 841–847.

Liu, W. Q., Yin, J. J., Qin, C. L., Feng, B., & Xu, M. H. (2012). Synthesis of CaO-based sorbents for CO_2 capture by a spray-drying technique. *Environ Sci Technol, 46*, 11267–11272.

Liu, F.-Q., Li, W.-H., Liu, B.-C., & Li, R.-X. (2013). Synthesis, characterization, and high temperature CO_2 capture of new CaO based hollow sphere sorbents. *J Mater Chem A, 1*, 8037–8044.

Lopez-Ortiz, A., Rivera, N. G. P., Rojas, A. R., & Gutierrez, D. L. (2004). Novel carbon dioxide solid acceptors using sodium containing oxides. *Sep Sci Technol, 39*, 3559–3572.

Lourenço, R. R., Angelica, R. S., & Rodrigues, J. D. (2013). Preparation of refractory calcium aluminate cement using the sonochemical process. *Mat Res, 16*, 731–739.

Lu, H., Reddy, E. P., & Smirniotis, P. G. (2006). Calcium oxide based sorbents for capture of carbon dioxide at high temperatures. *Ind Eng Chem Res, 45*, 3944–3949.

Lu, H., Khan, A., & Smirniotis, P. G. (2008). Relationship between structural properties and CO_2 capture performance of CaO-based sorbents obtained from different organometallic precursors. *Ind Eng Chem Res, 47*, 6216–6220.

Lu, H., Khan, A., Pratsinis, S. E., & Smirniotis, P. G. (2009a). Flame-made durable doped-CaO nanosorbents for CO_2 capture. *Energy Fuel, 23*, 1093–1100.

Lu, H., Smirniotis, P. G., Ernst, F. O., & Pratsinis, S. E. (2009b). Nanostructured Ca-based sorbents with high CO_2 uptake efficiency. *Chem Eng Sci, 64*, 1936–1943.

Luo, C., Zheng, Y., Zheng, C., Yin, J., Qin, C., & Feng, B. (2013). Manufacture of calcium-based sorbents for high temperature cyclic CO_2 capture via a sol-gel process. *Int J Greenhouse Gas Control, 12*, 193–199.

Lysikov, A. I., Salanov, A. N., & Okunev, A. G. (2007). Change of CO_2 carrying capacity of CaO in isothermal recarbonation-decomposition cycles. *Ind Eng Chem Res, 46*, 4633–4638.

Lysikov, A. I., Trukhan, S. N., & Okunev, A. G. (2008). Sorption enhanced hydrocarbons reforming for fuel cell powered generators. *Int J Hydrogen Energy, 33*, 3061–3066.

Manovic, V., & Anthony, E. J. (2007). Steam reactivation of spent CaO-based sorbent for multiple CO_2 capture cycles. *Environ Sci Technol, 41*, 1420–1425.

Manovic, V., & Anthony, E. J. (2008). Thermal activation of CaO-based sorbent and self-reactivation during CO_2 capture looping cycles. *Environ Sci Technol, 42*, 4170–4174.

Manovic, V., & Anthony, E. J. (2009). Screening of binders for pelletization of CaO-based sorbents for CO_2 capture. *Energy Fuel, 23*, 4797–4804.

Manovic, V., & Anthony, E. J. (2010). Carbonation of CaO-based sorbents enhanced by steam addition. *Ind Eng Chem Res, 49*, 9105–9110.

Manovic, V., Anthony, E. J., Grasa, G., & Abanades, J. C. (2008). CO_2 looping cycle performance of a high-purity limestone after thermal activation/doping. *Energy Fuel, 22*, 3258–3264.

Manovic, V., Charland, J. P., Blamey, J., Fennell, P. S., DY, L., & Anthony, E. J. (2009). Influence of calcination conditions on carrying capacity of CaO-based sorbent in CO_2 looping cycles. *Fuel, 88*, 1893–1900.

Martavaltzi, C. S., & Lemonidou, A. A. (2008a). Development of new CaO based sorbent materials for CO_2 removal at high temperature. *Microporous Mesoporous Mater, 110*, 119–127.
Martavaltzi, C. S., & Lemonidou, A. A. (2008b). Parametric study of the $CaO - Ca_{12}Al_{14}O_{33}$ synthesis with respect to high CO_2 sorption capacity and stability on multicycle operation. *Ind Eng Chem Res, 47*, 9537–9543.
Martavaltzi, C. S., & Lemonidou, A. A. (2010). Hydrogen production via sorption enhanced reforming of methane: development of a novel hybrid material-reforming catalyst and CO_2 sorbent. *Chem Eng Sci, 65*, 4134–4140.
Martavaltzi, C. S., Pampaka, E. P., Korkakaki, E. S., & Lemonidou, A. A. (2010). Hydrogen production via steam reforming of methane with simultaneous CO_2 capture over CaO-$Ca_{12}Al_{14}O_{33}$. *Energy Fuel, 24*, 2589–2595.
Martinez, I., Grasa, G., Murillo, R., Arias, B., & Abanades, J. C. (2011). Evaluation of CO_2 carrying capacity of reactivated CaO by hydration. *Energy Fuel, 25*, 1294–1301.
Martinez-dlCruz, L., & Pfeiffer, H. (2012). Microstructural thermal evolution of the Na_2CO_3 phase produced during a Na_2ZrO_3-CO_2 chemisorption process. *J Phys Chem C, 116*, 9675–9680.
Materic, V., Hyland, M., Jones, M. I., & Holt, R. (2014a). Investigation of the friability of Ca looping sorbents during and after hydration based reactivation. *Fuel, 127*, 70–77.
Materic, V., Hyland, M., Jones, M. I., & Northover, B. (2014b). High temperature carbonation of $Ca(OH)_2$: the effect of particle surface area and pore volume. *Ind Eng Chem Res, 53*, 2994–3000.
Matsukura, Y., Okumura, T., Kobayashi, R., & Oh-Ishi, K. (2010). Synthesis and CO_2 absorption properties of single-phase Li_2CuO_2 as a CO_2 absorbent. *Chem Lett, 39*, 966–967.
Mejía-Trejo, V. L., Fregoso-Israel, E., & Pfeiffer, H. (2008). Textural, structural, and CO_2 chemisorption effects produced on the lithium orthosilicate by its doping with sodium ($Li_{4-x}Na_xSiO_4$). *Chem Mater, 20*, 7171–7176.
Mohammadi, M., Lahijani, P., & Mohamed, A. R. (2014). Refractory dopant-incorporated CaO from waste eggshell as sustainable sorbent for CO_2 capture: experimental and kinetic studies. *Chem Eng J, 243*, 455–464.
Nair, B. N., Yamaguchi, T., Kawamura, H., Nakao, S. I., & Nakagawa, K. (2004). Processing of lithium zirconate for applications in carbon dioxide separation: structure and properties of the powders. *J Am Ceram Soc, 87*, 68–74.
Nakagawa, K., & Ohashi, T. (1998). A novel method of CO_2 capture from high temperature gases. *J Electrochem Soc, 145*, 1344–1346.
Nieto-Sanchez, A. J., Olivares-Marin, M., Garcia, S., Pevida, C., & Cuerda-Correa, E. M. (2013). Influence of the operation conditions on CO_2 capture by CaO-derived sorbents prepared from synthetic $CaCO_3$. *Chemosphere, 93*, 2148–2158.
Ochoa-Fernandez, E., Ronning, M., Grande, T., & Chen, D. (2006). Synthesis and CO_2 capture properties of nanocrystalline lithium zirconate. *Chem Mater, 18*, 6037–6046.
Ochoa-Fernandez, E., Haugen, G., Zhao, T., et al. (2007a). Process design simulation of H_2 production by sorption enhanced steam methane reforming: evaluation of potential CO_2 acceptors. *Green Chem, 9*, 654–662.
Ochoa-Fernandez, E., Lacalle-Vila, C., Zhao, T., Ronning, M., & Chen, D. (2007b). Experimental demonstration of H_2 production by CO_2 sorption enhanced steam methane reforming using ceramic acceptors. In *Proceedings of the 8th natural gas conversion symposium. Natural gas conversion VIII: vol. 167* (pp. 159–164).
Ochoa-Fernandez, E., Ronning, M., Yu, X., Grande, T., & Chen, D. (2008). Compositional effects of nanocrystalline lithium zirconate on its CO_2 capture properties. *Ind Eng Chem Res, 47*, 434–442.

Ochoa-Fernandez, E., Zhao, T. J., Ronning, M., & Chen, D. (2009). Effects of steam addition on the properties of high temperature ceramic CO_2 acceptors. *J Environ Eng ASCE*, *135*, 397–403.

Ohashi, T., & Nakagawa, K. (1999). Effect of potassium carbonate additive on CO_2 absorption in lithium zirconate powder. In S. M. Kauzlarich, E. M. McCarron, & A. W. Sleight (Eds.), *Solid-state chemistry of inorganic materials II* (pp. 249–254). Cambridge University Press.

Oliveira, E. L. G., Grande, C. A., & Rodrigues, A. E. (2008). CO_2 sorption on hydrotalcite and alkali-modified (K and Cs) hydrotalcites at high temperatures. *Sep Purif Technol*, *62*, 137–147.

Ortiz, A. L., & Harrison, D. P. (2001). Hydrogen production using sorption-enhanced reaction. *Ind Eng Chem Res*, *40*, 5102–5109.

Ortiz-Landeros, J., Gomez-Yanez, C., Palacios-Romero, L. M., Lima, E., & Pfeiffer, H. (2012). Structural and thermochemical chemisorption of CO_2 on Li4+x(Si1-xAlx) O-4 and Li4-x(Si1-xVx)O-4 solid solutions. *J Phys Chem A*, *116*, 3163–3171.

Pacciani, R., Muller, C. R., Davidson, J. F., Dennis, J. S., & Hayhurst, A. N. (2008). Synthetic Ca-based solid sorbents suitable for capturing CO_2 in a fluidized bed. *Can J Chem Eng*, *86*, 356–366.

Palacios-Romero, L. M., & Pfeiffer, H. (2008). Lithium cuprate (Li_2CUO_2): a new possible ceramic material for CO_2 chemisorption. *Chem Lett*, *37*, 862–863.

Pannocchia, G., Puccini, M., Seggiani, M., & Vitolo, S. (2007). Experimental and modeling studies on high-temperature capture of CO_2 using lithium zirconate based sorbents. *Ind Eng Chem Res*, *46*, 6696–6706.

Qin, C., Yin, J., Luo, C., An, H., Liu, W., & Feng, B. (2013). Enhancing the performance of CaO/CuO based composite for CO_2 capture in a combined Ca-Cu chemical looping process. *Chem Eng J*, *228*, 75–86.

Radfarnia, H. R., & Iliuta, M. C. (2011). Surfactant-template/ultrasound-assisted method for the preparation of porous nanoparticle lithium zirconate. *Ind Eng Chem Res*, *50*, 9295–9305.

Radfarnia, H. R., & Iliuta, M. C. (2012a). Application of surfactant-template technique for preparation of sodium zirconate as high temperature CO_2 sorbent. *Sep Purif Technol*, *93*, 98–106.

Radfarnia, H. R., & Iliuta, M. C. (2012b). Development of zirconium-stabilized calcium oxide absorbent for cyclic high-temperature CO_2 capture. *Ind Eng Chem Res*, *51*, 10390–10398.

Radfarnia, H. R., & Iliuta, M. C. (2013a). Limestone acidification using citric acid coupled with two-step calcination for improving the CO_2 sorbent activity. *Ind Eng Chem Res*, *52*, 7002–7013.

Radfarnia, H. R., & Iliuta, M. C. (2013b). Metal oxide-stabilized calcium oxide CO_2 sorbent for multicycle operation. *Chem Eng J*, *232*, 280–289.

Radfarnia, H. R., & Iliuta, M. C. (2014a). Development of Al-stabilized CaO-nickel hybrid sorbent-catalyst for sorption-enhanced steam methane reforming. *Chem Eng Sci*, *109*, 212–219.

Radfarnia, H. R., & Iliuta, M. C. (2014b). Hydrogen production by sorption-enhanced steam methane reforming process using CaO-Zr/Ni bifunctional sorbent-catalyst. *Chem Eng Process*, *86*, 96–103.

Radfarnia, H. R., & Sayari, A. (2015). A highly efficient CaO-based CO_2 sorbent prepared by a citrate-assisted sol–gel technique. *Chem Eng J*, *262*, 913–920.

Reddy, G. K., Quillin, S., & Smirniotis, P. (2014). Influence of the synthesis method on the structure and CO_2 adsorption properties of Ca/Zr sorbents. *Energy Fuel*, *28*, 3292–3299.

Ridha, F. N., Manovic, V., Macchi, A., Anthony, M. A., & Anthony, E. J. (2013a). Assessment of limestone treatment with organic acids for CO_2 capture in Ca-looping cycles. *Fuel Process Technol, 116*, 284–291.

Ridha, F. N., Manovic, V., YH, W., Macchi, A., & Anthony, E. J. (2013b). Pelletized CaO-based sorbents treated with organic acids for enhanced CO_2 capture in Ca-looping cycles. *Int J Greenhouse Gas Control, 17*, 357–365.

Rodriguez-Mosqueda, R., & Pfeiffer, H. (2010). Thermokinetic analysis of the CO_2 chemisorption on Li_4SiO_4 by using different gas flow rates and particle sizes. *J Phys Chem A, 114*, 4535–4541.

Romero-Ibarra, I. C., Ortiz-Landeros, J., & Pfeiffer, H. (2013). Microstructural and CO_2 chemisorption analyses of Li_4SiO_4: effect of surface modification by the ball milling process. *Thermochim Acta, 567*, 118–124.

Romero-Ibarra, I. C., Duran-Munoz, F., & Pfeiffer, H. (2014). Influence of the K-, Na- and K-Na-carbonate additions during the CO_2 chemisorption on lithium oxosilicate (Li_8SiO_6). *Greenhouse Gases Sci Technol, 4*, 145–154.

Rong, N., Wang, Q. H., Fang, M. X., Cheng, L. M., Luo, Z. Y., & Cen, K. F. (2013). Steam hydration reactivation of CaO-based sorbent in cyclic carbonation/calcination for CO_2 capture. *Energy Fuel, 27*, 5332–5340.

Rout, K. R., Fermoso, J., Chen, D., & Jakobsen, H. A. (2014). Kinetic rate of CO_2 uptake of a synthetic Ca-based sorbent: experimental data and numerical simulations. *Fuel, 120*, 53–65.

Sanchez, N., Encinar, J. M., & Gonzalez, J. F. (2016). Sorption enhanced steam reforming of glycerol: use of La-modified Ni/Al_2O_3 as catalyst. *Ind Eng Chem Res, 55*, 3736–3741.

Sanchez-Camacho, P., Romero-Ibarra, I. C., Duan, Y. H., & Pfeiffer, H. (2014). Thermodynamic and kinetic analyses of the CO_2 chemisorption mechanism on Na_2TiO_3: experimental and theoretical evidences. *J Phys Chem C, 118*, 19822–19832.

Sanchez-Jimenez, P. E., Valverde, J. M., & Perez-Maqueda, L. A. (2014). Multicyclic conversion of limestone at Ca-looping conditions: the role of solid-sate diffusion controlled carbonation. *Fuel, 127*, 131–140.

Santos, E. T., Alfonsín, C., Chambel, A. J. S., et al. (2012). Investigation of a stable synthetic sol–gel CaO sorbent for CO_2 capture. *Fuel, 94*, 624–628.

Satrio, J. A., Shanks, B. H., & Wheelock, T. D. (2005). Development of a novel combined catalyst and sorbent for hydrocarbon reforming. *Ind Eng Chem Res, 44*, 3901–3911.

Sayyah, M., Ito, B. R., Rostam-Abadi, M., Lu, Y., & Suslick, K. S. (2013). CaO-based sorbents for CO_2 capture prepared by ultrasonic spray pyrolysis. *RSC Adv, 3*, 19872–19875.

Seggiani, M., Puccini, M., & Vitolo, S. (2011). High-temperature and low concentration CO_2 sorption on Li_4SiO_4 based sorbents: study of the used silica and doping method effects. *Int J Greenhouse Gas Control, 5*, 741–748.

Seggiani, M., Puccini, M., & Vitolo, S. (2013). Alkali promoted lithium orthosilicate for CO_2 capture at high temperature and low concentration. *Int J Greenhouse Gas Control, 17*, 25–31.

Shan, S. Y., Jia, Q. M., Jiang, L. H., Li, Q. C., Wang, Y. M., & Peng, J. H. (2013). Novel Li_4SiO_4-based sorbents from diatomite for high temperature CO_2 capture. *Ceram Int, 39*, 5437–5441.

Shokrollahi Yancheshmeh, M., Radfarnia, H. R., & Iliuta, M. C. (2016a). High temperature CO_2 sorbents and their application for hydrogen production by sorption enhanced steam reforming process. *Chem Eng J, 283*, 420–444.

Shokrollahi Yancheshmeh, M., Radfarnia, H. R., & Iliuta, M. C. (2016b). Influence of steam addition during carbonation or calcination on the CO_2 capture performance of $Ca_9Al_6O_{18}$-CaO sorbent. *J Nat Gas Sci Eng, 36*, 1062–1069.

Shokrollahi Yancheshmeh, M., Radfarnia, H. R., & Iliuta, M. C. (2017). Sustainable production of high-purity hydrogen by sorption enhanced steam reforming of glycerol over CeO_2-promoted $Ca_9Al_6O_{18}$-CaO/NiO bifunctional material. In *ACS Sustainable Chem Eng*. https://doi.org/10.1021/acssuschemeng.7b01627 (in press).

Silaban, A., Narcida, M., & Harrison, D. P. (1992). Calcium acetate as a sorbent precursor for the removal of carbon dioxide from gas streams at high temperature. *Resour Conserv Recycl*, 7, 139–153.

Silaban, A., Narcida, M., & Harrison, D. P. (1996). Characteristics of the reversible reaction between CO_2(g) and calcined dolomite. *Chem Eng Commun*, *146*, 149–162.

Spath, P., & Mann, M. (2000). *Life cycle assessment of hydrogen production via natural gas steam reforming*. Technical report NREL/TP-570-27637, NREL.

Spencer, M. S. (1986). Stable and metastable metal-surfaces in heterogeneous catalysis. *Nature*, *323*, 685–687.

Stendardo, S., Andersen, L. K., & Herce, C. (2013). Self-activation and effect of regeneration conditions in CO_2-carbonate looping with CaO-$Ca_{12}Al_{14}O_{33}$ sorbent. *Chem Eng J*, *220*, 383–394.

Sun, P., Grace, J. R., Lim, C. J., & Anthony, E. J. (2007). The effect of CaO sintering on cyclic CO_2 capture in energy systems. *AIChE J*, *53*, 2432–2442.

Sun, P., Lim, J., & Grace, J. R. (2008). Cyclic CO_2 capture by limestone-derived sorbent during prolonged calcination/carbonation cycling. *AIChE J*, *54*, 1668–1677.

Sun, Z. C., Luo, S. W., Qi, P. P., & Fan, L. S. (2012). Ionic diffusion through calcite ($CaCO_3$) layer during the reaction of CaO and CO_2. *Chem Eng Sci*, *81*, 164–168.

Sun, R., Li, Y., Wu, S., Liu, C., Liu, H., & Lu, C. (2013a). Enhancement of CO_2 capture capacity by modifying limestone with propionic acid. *Powder Technol*, *233*, 8–14.

Sun, R. Y., Li, Y. J., Liu, C. T., Xie, X., & CM, L. (2013b). Utilization of lime mud from paper mill as CO_2 sorbent in calcium looping process. *Chem Eng J*, *221*, 124–132.

Suslick, K. S., Choe, S. B., Cichowlas, A. A., & Grinstaff, M. W. (1991). Sonochemical synthesis of amorphous iron. *Nature*, *353*, 414–416.

Symonds, R. T., DY, L., Hughes, R. W., Anthony, E. J., & Macchi, A. (2009). CO_2 capture from simulated syngas via cyclic carbonation/calcination for a naturally occurring limestone: pilot-plant testing. *Ind Eng Chem Res*, *48*, 8431–8440.

Togashi, N., Okumura, T., & Oh-ishi, K. (2007). Synthesis and CO_2 absorption property of Li_4TiO_4 as a novel CO_2 absorbent. *J Cerma Soc Jpn*, *115*, 324–328.

Ueda, S., Inoue, R., Sasaki, K., Wakuta, K., & Ariyama, T. (2011). CO_2 absorption and desorption abilities of Li_2O-TiO_2 compounds. *ISIJ Int*, *51*, 530–537.

Valverde, J. M., Sanchez-Jimenez, P. E., Perejon, A., & Perez-Maqueda, L. A. (2013). Role of looping-calcination conditions on self-reactivation of thermally pretreated CO_2 sorbents based on CaO. *Energy Fuel*, *27*, 3373–3384.

Valverde, J. M., Sanchez-Jimenez, P. E., & Perez-Maqueda, L. A. (2014a). Effect of heat pretreatment/recarbonation in the Ca-looping process at realistic calcination conditions. *Energy Fuel*, *28*, 4062–4067.

Valverde, J. M., Sanchez-Jimenez, P. E., & Perez-Maqueda, L. A. (2014b). High and stable CO_2 capture capacity of natural limestone at Ca-looping conditions by heat pretreatment and recarbonation synergy. *Fuel*, *123*, 79–85.

Venegas, M. J., Fregoso-Israel, E., Escamilla, R., & Pfeiffer, H. (2007). Kinetic and reaction mechanism of CO_2 sorption on Li_4SiO_4: study of the particle size effect. *Ind Eng Chem Res*, *46*, 2407–2412.

Wang, Y., Lin, S., & Suzuki, Y. (2008). Limestone calcination with CO_2 capture (II): decomposition in CO_2/steam and CO_2/N_2 atmospheres. *Energy Fuel*, *22*, 2326–2331.

Wang, X., Li, M., Li, S., Wang, H., Wang, S., & Ma, X. (2010). Hydrogen production by glycerol steam reforming with/without calcium oxide sorbent: a comparative study of thermodynamic and experimental work. *Fuel Process Technol*, *91*, 1812–1818.

Wang, S. P., Shen, H., Fan, S. S., Zhao, Y. J., Ma, X. B., & Gong, J. L. (2013). Enhanced CO_2 adsorption capacity and stability using CaO-based adsorbents treated by hydration. *AIChE J*, *59*, 3586–3593.

Wang, C., Dou, B. L., Song, Y. C., Chen, H. S., YJ, X., & Xie, B. Z. (2014a). High temperature CO_2 sorption on Li_2ZrO_3 based sorbents. *Ind Eng Chem Res*, *53*, 12744–12752.

Wang, K., Zhao, P. F., Guo, X., Li, Y. M., Han, D. T., & Chao, Y. (2014b). Enhancement of reactivity in Li_4SiO_4-based sorbents from the nano-sized rice husk ash for high-temperature CO_2 capture. *Energy Conv Manag*, *81*, 447–454.

Wang, S., Fan, S., Zhao, Y., Fan, L., Liu, S., & Ma, X. (2014c). Carbonation condition and modeling studies of calcium-based sorbent in the fixed-bed reactor. *Ind Eng Chem Res*, *53*, 10457–10464.

Wang, C., Chen, Y., Cheng, Z., et al. (2015a). Sorption-enhanced steam reforming of glycerol for hydrogen production over a $NiO/NiAl_2O_4$ catalyst and Li_2ZrO_3-based sorbent. *Energy Fuel*, *29*, 7408–7418.

Wang, C., Dou, B., Jiang, B., et al. (2015b). Sorption-enhanced steam reforming of glycerol on Ni-based multifunctional catalysts. *Int J Hydrogen Energy*, *40*, 7037–7044.

Wang, K., Yin, Z. G., Zhao, P. F., Zhou, Z. Y., Su, Z., & Sun, J. (2017). Development of metallic element-stabilized Li_4SiO_4 sorbents for cyclic CO_2 capture. *Int J Hydrogen Energy*, *42*, 4224–4232.

Wess, R., Nores-Pondal, F., Laborde, M., & Giunta, P. (2015). Single stage H_2 production, purification and heat supply by means of sorption-enhanced steam reforming of glycerol. A thermodynamic analysis. *Chem Eng Sci*, *134*, 86–95.

Wu, S. F., & Wang, L. L. (2010). Improvement of the stability of a ZrO_2-modified Ni-nano-CaO sorption complex catalyst for ReSER hydrogen production. *Int J Hydrogen Energy*, *35*, 6518–6524.

Wu, S. F., Beum, T. H., Yang, J. I., & Kim, J. N. (2007). Properties of Ca-Base CO_2 sorbent using $Ca(OH)_2$ as precursor. *Ind Eng Chem Res*, *46*, 7896–7899.

Wu, S. F., Li, Q. H., Kim, J. N., & Yi, K. B. (2008). Properties of a nano CaO/Al_2O_3 CO_2 sorbent. *Ind Eng Chem Res*, *47*, 180–184.

Wu, G. W., Zhang, C. X., Li, S. R., et al. (2012a). Sorption enhanced steam reforming of ethanol on Ni-CaO-Al_2O_3 multifunctional catalysts derived from hydrotalcite-like compounds. *Energ Environ Sci*, *5*, 8942–8949.

Wu, Y. J., Diaz Alvarado, F. A., Santos, J. C., Gracia, F., Cunha, A. F., & Rodrigues, A. E. (2012b). Sorption-enhanced steam reforming of ethanol: thermodynamic comparison of CO_2 sorbents. *Chem Eng Technol*, *35*, 847–858.

Wu, Y. J., Li, P., JG, Y., Cunha, A. F., & Rodrigues, A. E. (2013). Sorption-enhanced steam reforming of ethanol on NiMgAl multifunctional materials: experimental and numerical investigation. *Chem Eng J*, *231*, 36–48.

Wu, Y. J., Li, P., JG, Y., Cunha, A. F., & Rodrigues, A. E. (2014a). Sorption-enhanced steam reforming of ethanol for continuous high-purity hydrogen production: 2D adsorptive reactor dynamics and process design. *Chem Eng Sci*, *118*, 83–93.

Wu, Y. J., Santos, J. C., Li, P., JG, Y., Cunha, A. F., & Rodrigues, A. E. (2014b). Simplified kinetic model for steam reforming of ethanol on a Ni/Al_2O_3 catalyst. *Can J Chem Eng*, *92*, 116–130.

Wu, Y. J., Li, P., JG, Y., Cunha, A. F., & Rodrigues, A. E. (2016). Progress on sorption-enhanced reaction process for hydrogen production. *Rev Chem Eng*, *32*, 271–303.

Xiao, Q., Liu, Y., Zhong, Y., & Zhu, W. (2011a). A citrate sol-gel method to synthesize Li_2ZrO_3 nanocrystals with improved CO_2 capture properties. *J Mater Chem*, *21*, 3838–3842.

Xiao, Q., Tang, X., Liu, Y., Zhong, Y., & Zhu, W. (2011b). Citrate route to prepare K-doped Li_2ZrO_3 sorbents with excellent CO_2 capture properties. *Chem Eng J*, *174*, 231–235.

Xie, M. M., Zhou, Z. M., Qi, Y., Cheng, Z. M., & Yuan, W. K. (2012). Sorption-enhanced steam methane reforming by in situ CO_2 capture on a CaO-$Ca_9Al_6O_{18}$ sorbent. *Chem Eng J, 207*, 142–150.
Xiong, R. T., Ida, J., & Lin, Y. S. (2003). Kinetics of carbon dioxide sorption on potassium-doped lithium zirconate. *Chem Eng Sci, 58*, 4377–4385.
Xu, P., Xie, M. M., Cheng, Z. M., & Zhou, Z. M. (2013). CO_2 capture performance of CaO-based sorbents prepared by a sol-gel method. *Ind Eng Chem Res, 52*, 12161–12169.
Yanase, I., Kameyama, A., & Kobayashi, H. (2010). CO_2 absorption and structural phase transition of α-$LiFeO_2$. *J Cerma Soc Jpn, 118*, 48–51.
Yang, S., & Xiao, Y. (2008). Steam catalysis in CaO carbonation under low steam partial pressure. *Ind Eng Chem Res, 47*, 4043–4048.
Yang, L., Yu, H., Wang, S., Wang, H., & Zhou, Q. (2013a). Carbon dioxide captured from flue gas by modified Ca-based sorbents in fixed-bed reactor at high temperature. *Chin J Chem Eng, 21*, 199–204.
Yang, X., Zhao, L., Yang, S., & Xiao, Y. (2013b). Investigation of natural CaO-MgO sorbent for CO_2 capture. *Asia Pac J Chem Eng, 8*, 906–915.
Yang, X. W., Liu, W. Q., Sun, J., et al. (2016). Alkali-doped lithium orthosilicate sorbents for carbon dioxide capture. *ChemSusChem, 9*, 2480–2487.
Yi, K. B., & Eriksen, D. O. (2006). Low temperature liquid state synthesis of lithium zirconate and its characteristics as a CO_2 sorbent. *Sep Sci Technol, 41*, 283–296.
Yi KB, Meyer J, Eriksen D, 2006. Selection, preparation and study of high temperature novel CO_2 sorbents for sorption enhanced SMR, nordic H_2-seminar, Oslo.
Yong, Z., & Rodrigues, A. E. (2002). Hydrotalcite-like compounds as adsorbents for carbon dioxide. *Energ Conver Manage, 43*, 1865–1876.
Yong, Z., Mata, V., & Rodriguez, A. E. (2001). Adsorption of carbon dioxide onto hydrotalcite-like compounds (HTlcs) at high temperatures. *Ind Eng Chem Res, 40*, 204–209.
Yu, C. T., & Chen, W. C. (2014). Hydrothermal preparation of calcium-aluminum carbonate sorbent for high-temperature CO_2 capture in fixed-bed reactor. *Fuel, 122*, 179–185.
Zeman, F. (2008). Effect of steam hydration on performance of lime sorbent for CO_2 capture. *Int J Greenhouse Gas Control, 2*, 203–209.
Zhang, L., Lu, Y., & Rostam-Abadi, M. (2012). Sintering of calcium oxide (CaO) during CO_2 chemisorption: a reactive molecular dynamics study. *Phys Chem Chem Phys, 14*, 16633–16643.
Zhang, M. M., Peng, Y. X., Sun, Y. Z., Li, P., & JG, Y. (2013a). Preparation of CaO-Al_2O_3 sorbent and CO_2 capture performance at high temperature. *Fuel, 111*, 636–642.
Zhang, Q., Han, D. Y., Liu, Y., Ye, Q., & Zhu, Z. B. (2013b). Analysis of CO_2 sorption/desorption kinetic behaviors and reaction mechanisms on Li_4SiO_4. *AIChE J, 59*, 901–911.
Zhang, S., Zhang, Q., Wang, H. Y., Ni, Y. H., & Zhu, Z. B. (2014). Absorption behaviors study on doped Li_4SiO_4 under a humidified atmosphere with low CO_2 concentration. *Int J Hydrogen Energy, 39*, 17913–17920.
Zhao, T. J., Ochoa-Fernandez, E., Ronning, M., & Chen, D. (2007). Preparation and high-temperature CO_2 capture properties of nanocrystalline Na_2ZrO_3. *Chem Mater, 19*, 3294–3301.
Zhao, M., Bilton, M., Brown, A. P., et al. (2014). Durability of CaO-$CaZrO_3$ sorbents for high-temperature CO_2 capture prepared by a wet chemical method. *Energy Fuel, 28*, 1275–1283.
Zhao, C. J., Zhou, Z. M., Cheng, Z. M., & Fang, X. C. (2016). Sol-gel-derived, $CaZrO_3$-stabilized Ni/CaO-$CaZrO_3$ bifunctional catalyst for sorption-enhanced steam methane reforming. *Appl Catal B, 196*, 16–26.
Zhou, Z. M., Qi, Y., Xie, M. M., Cheng, Z. M., & Yuan, W. K. (2012). Synthesis of CaO-based sorbents through incorporation of alumina/aluminate and their CO_2 capture performance. *Chem Eng Sci, 74*, 172–180.

CHAPTER THREE

Reactor Design for Sorption-Enhanced Reforming Using Ca—Cu Chemical Loops

José R. Fernández, Juan C. Abanades
Spanish Research Council, INCAR-CSIC, C/Francisco Pintado Fe, Oviedo, Asturias, Spain

Contents

Abstract

The growing concern about global warming has stressed the need for novel H_2 production technologies with CO_2 capture. The sorption-enhanced reforming (SER) allows a H_2-rich product gas to be obtained at moderate operating conditions. The Ca—Cu looping process arises as a promising SER method as it provides a viable solution for regenerating the CO_2 sorbent used during the SER with potentially low energy penalties and costs. This chapter compiles the main recent advances in the Ca—Cu process, focusing on reactor, and process design and the experimental work needed to validate the performance of reactors and materials. Significant progress has also been made in simulating this system in power plants and industrial systems.

ABBREVIATIONS

ASU air separation unit
ATR autothermal reforming
BFG blast furnace gas
CCS CO_2 capture and storage
CLC chemical looping combustion

Advances in Chemical Engineering, Volume 51
ISSN 0065-2377
https://doi.org/10.1016/bs.ache.2017.07.003

COG coke oven gas
FTR fired tubular reformer
HF heat exchange front
IEA International Energy Agency
IPCC Intergovernmental Panel on Climate Change
MDEA monoethanolamine
NGCC natural gas combined cycles
POX partial oxidation
PSA pressure swing adsorption
RF reaction front
SCM shrinking core model
SER sorption-enhanced reforming
SEWGS sorption-enhanced water gas shift
SMR steam methane reforming
SPECCA specific primary energy consumption for CO_2 avoided
TGA thermogravimetric analyzer
TRL technology readiness level
WGS water gas shift
XRD X-ray diffraction

NOMENCLATURE

C_i concentration of component *i* in the reactor ($kmol\,m^{-3}$)
$C_{i_{eq}}$ equilibrium concentration of component *i* ($kmol\,m^{-3}$)
C_{pi} specific heat capacity of component *i* ($kJ\,kg^{-1}\,K^{-1}$)
D_{eff} axial mass dispersion coefficient ($m^2\,s^{-1}$)
D_i inner diameter of the reactor (m)
d_p particle diameter (m)
$E_{a,i}$ activation energy ($kJ\,mol^{-1}$)
H_{ri} enthalpy reaction of gas component *i* ($kJ\,mol^{-1}$)
h_w heat transfer coefficient ($W\,m^{-2}\,K^{-1}$)
$k_{0,j}$ preexponential factor of reaction *j* ($mol^{1-n}\,m^{3n-3}\,s^{-1}$)
k_j reaction rate constant of reaction *j* ($mol^{1-n}\,m^{3n-3}\,s^{-1}$)
K_g thermal conductivity of the gas ($W\,K^{-1}\,m^{-1}$)
K_s thermal conductivity of the solids ($W\,K^{-1}\,m^{-1}$)
K_i equilibrium constant
L reactor length (m)
L_i layer thickness of component *i* (m)
M_i molar mass of component *i* ($kg\,kmol^{-1}$)
N reaction order, dimensionless
Nu Nusselt number, dimensionless
P pressure, bar
Pe_{az} Peclet number, dimensionless
Pr Prandt number, dimensionless
R ideal gas constant ($kJ\,kmol^{-1}\,K^{-1}$)
Re Reynolds number, dimensionless

r_i reaction rate of component i (mol m^{-3}s^{-1})
Sc Schmidt number, dimensionless
T time (s)
T temperature (K)
u_g superficial velocity of gas (m s^{-1})
u_e velocity of heat exchange front (m/s^{-1})
u_r velocity of reaction front (RF) (m/s^{-1})
X_i fractional molar conversion of component i, dimensionless
z axial coordinate in bed (m)

GREEK LETTERS

ρ_i apparent density of component i (kg m^{-3})
$\rho_{m,i}$ molar density of component i (mol m^{-3})
ΔT_{max} maximum adiabatic temperature increase (K)
ε bed void fraction, dimensionless
φ stoichiometric coefficient, dimensionless
η effectiveness factor, dimensionless
λ_{eff} effective axial heat dispersion coefficient (W m^{-2}K^{-1})
λ_{0bed} effective axial heat dispersion with no flow conditions (W m^{-2}K^{-1})
τ time for complete solid conversion, dimensionless
υ_i gas-phase mole fraction of CO_2, dimensionless
μ_g gas viscosity (kg s^{-1}m^{-1})

1. INTRODUCTION

Over the last few decades, the demand for hydrogen has increased significantly due to the growth in the production of ammonia and methanol and to the increasing use of H_2 in the oil refining industry (IEA, 2015). Moreover, hydrogen is considered a sustainable energy carrier that offers the advantages of fossil fuels (i.e., a high flexibility and energy density) but with a potentially low carbon footprint (IEA, 2006). Around 7.2 exajoules of H_2 per year are currently produced to satisfy the demand for H_2 in the refining and chemical industry (IEA, 2015). However, the use of H_2 as a feedstock for hydrocracking, steel, and glass industries and also as a clean fuel in modern gas turbines, combustion engines, and fuel cells will inevitably lead to a rise in global H_2 consumption in the coming years (IEA, 2015). Although H_2 can be obtained from a wide variety of sources (natural gas, oil, coal, biomass, electrolysis, solar energy, etc.), around 96% of current H_2 is produced from fossil fuels, resulting in approximately 0.5 Gt of CO_2 emitted into the atmosphere per year (Voldsun et al., 2016). There is also great concern in mitigating the global warming by reducing the CO_2 emissions from large-scale power plants and industrial processes. In a carbon-constrained world, only carbon capture and storage (CCS) has the potential to drastically reduce the CO_2 emissions from large stationary sources (IPCC, 2014), in line with the ambitious climate change mitigation targets toward 1.5°C for the long term, recently agreed at 22nd Conference of the Parties, COP22, held in Marrakesh in 2016.

Steam methane reforming (SMR) is the predominant H_2 production technology at industrial scale (Voldsun et al., 2016). Around 48% of H_2 is currently obtained by this technology (IEA, 2015). SMR is a mature multistep process in which methane and steam (mixed at a steam-to-carbon (S/C) ratio of 2.5–4) are converted to H_2 and CO. The highly endothermic reactions involved in SMR (Eqs. 1 and 2) require high temperatures (around 900°C) to achieve high methane conversions and H_2 yields in the presence of a nickel-based catalyst (Rostrup-Nielsen et al., 2002).

$$CH_4 + H_2O \rightarrow CO + 3H_2 \quad \Delta H_{298K} = 206.2\ \text{kJ/mol} \tag{1}$$

$$CH_4 + 2H_2O \rightarrow CO_2 + 4H_2 \quad \Delta H_{298K} = 165\ \text{kJ/mol} \tag{2}$$

The SMR process takes place typically in externally heated tubular reformers where heat is supplied from the combustion of additional methane (Dybkjaer, 1995). Although low pressure favors the SMR equilibrium, the process is carried out at moderate pressures between 15 and 30 bar, as this allows the operation to be carried out in more compact reactors and facilitates the subsequent use of the H_2 produced (Barrelli et al., 2008). As the product gas obtained from the reformers still contains about 10 vol.% CO (on a dry basis), a subsequent water gas shift (WGS) operation is carried out to convert the CO to CO_2 (and H_2O to H_2) (Eq. 3), thereby enhancing the H_2 purity in the gas stream. The WGS operation is slightly exothermic and can be performed in a single stage or in two stages. The first shift stage is performed at temperatures of around 350°C and is catalyzed by Cr—Fe oxides, while the second stage is carried out at temperatures of around 250°C in the presence of a Cu—Zn catalyst (Rostrup-Nielsen et al., 2002).

$$CO + H_2O \rightarrow CO_2 + H_2 \quad \Delta H_{298K} = -41\ \text{kJ/mol} \tag{3}$$

The composition of the product gas (on a dry basis) leaving the WGS reactors is about 70–75 vol.% H_2 and 15–20 vol.% CO_2 (the rest is unreacted

CH_4 and CO) (Rostrup-Nielsen, 2009). A final purification step is needed for downstream. Several technologies can be used depending on the final use of the hydrogen. Wet scrubbing with monoethanolamine (MDEA) can be used to remove the CO_2 from the product gas. Pressure swing adsorption (PSA) is a reliable and established option when almost totally pure H_2 is needed (Harrison, 2008). The tail gas is usually recirculated to be used as fuel in the reformer burners.

SMR is a highly efficient route compared to other alternatives such as the gasification of coal, partial oxidation (POX), or autothermal reforming (ATR) (Voldsun et al., 2016). However, multiple stages carried out under different conditions with different catalysts are required to enhance the production of H_2. The need for elevated temperatures (around 900°C) to achieve a high methane conversion (imposed by the SMR equilibrium) favors the deposition of carbon over the reforming catalyst (Tavares et al., 1996). This phenomenon may lead to the deactivation of the catalyst and to the blockage of the tubular reformers (resulting in higher pressure drops). Moreover, the severe conditions of high pressure and temperature in the reformer make it necessary to use expensive alloys, which increases the capital cost (Harrison, 2008).

The development of large-scale hydrogen production processes in combination with CO_2 capture at a moderate cost and with a reduced energy penalty constitutes a great step forward in the climate change mitigation route (Boot-Handford et al., 2014). The combination of SMR with the separation in situ of CO_2 appears as an attractive option for the sustainable conversion of methane into high purity H_2 (Abanades et al., 2015). In the sorption-enhanced reforming (SER), a high-temperature sorbent (typically a CaO-based material) is combined with the reforming catalyst to remove the CO_2 from the gaseous phase as soon as it is produced (Harrison, 2008). The steam reforming and WGS reactions are both shifted toward H_2 production in accordance with the Le Chatelier principle (Hufton et al., 1999). As a result, the almost total conversion of methane can be achieved at temperatures between 600 and 700°C. It is also possible to carry out the SMR, WGS, and CO_2 removal in a single reaction stage, obtaining a product gas with more than 90 vol.% of H_2 purity (on a dry basis) and with very low contents in methane and CO (Xiu et al., 2003), which simplifies the downstream purification step. A simplified scheme of the SER process is shown in Fig. 1. The exothermic CO_2 sorption reaction (i.e., the CaO carbonation) (Eq. 4) provides directly the heat necessary in the reformer for the

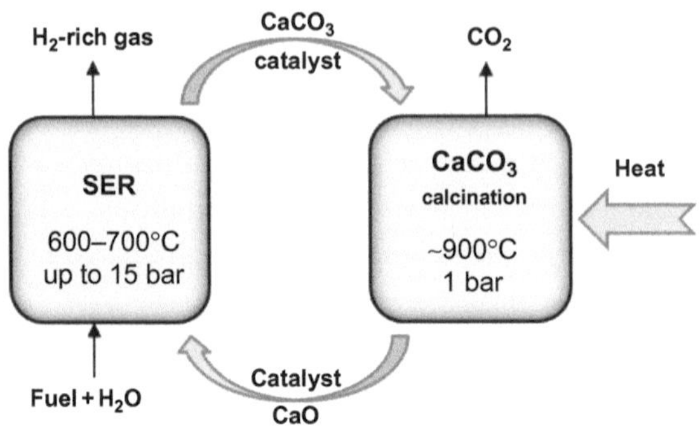

Fig. 1 Simplified scheme of the sorption-enhanced reforming process using a CaO-based solid as CO_2 acceptor.

endothermic reforming reactions to be carried out, so that no additional energy is required to produce the hydrogen (Barrelli et al., 2008). No shift catalysts are required and the operation at moderate temperatures minimizes coke formation, reduces the need for excess steam, and allows the use of cheaper construction materials in the reformer (Harrison, 2008).

$$CaO + CO_2 \rightarrow CaCO_3 \quad \Delta H_{298K} = -178.8\ \text{kJ/mol} \tag{4}$$

The use of calcium oxide to separate CO_2 in gasification processes is not a new concept. Du Montay patented in 1880 a method that uses lime to promote the gasification of coal. In the early 1930s, Williams (1933) patented a process scheme for the production of H_2 from a mixture of methane and steam in the presence of a reforming catalyst and lime. Later, Gorin and Retallick (1963) patented a version of the same process but carried out in a fluidized bed. Brun-Tsekhovoi et al. (1988) reported the experimental results obtained from the SER of methane showing significantly higher H_2 production yields compared to those obtained in the conventional SMR. Many works have been published over the last few years orientated toward finding the best configuration for carrying out the SER process and the optimal operation window for achieving higher H_2 yields and energy efficiencies. Moreover, high-performance materials that fulfill catalytic and CO_2 sorbent properties have been developed (Harrison, 2008).

The use of fluidized beds has been proposed to carry out the SER process (Di Carlo et al., 2010; Johnsen et al., 2006; Martínez et al., 2013a; Meyer et al., 2014; Wang et al., 2011), since this configuration facilitates the control of the temperature in the reformer, the replacement of spent materials by fresh solids and also avoids the need for complex high-temperature valve set-ups, characteristic of competing packed bed systems. However, it is extremely challenging to establish intense circulation of solids between two reactors with incompatible atmospheres at high pressure, which is of particular interest in a large number of H_2 applications. For this reason, multiple fixed bed reactors with periodic switching between the H_2 production and sorbent regeneration stages have been also proposed for the SER operation (Li et al., 2006; Martavaltzi et al., 2011; Rout and Jakobsen, 2012; Solieman et al., 2009). Pressurized fixed beds permit a more compact reactor configuration but they require adequate heat management strategies in order to achieve the complete conversion of the solids and at the same time to avoid the formation of hot spots inside the reactors.

Significant improvements in reactor modeling, sorbent durability, and experimental validation at a laboratory scale have been made, but the SER technology is still under transition to pilot-plant scale. One of the main obstacles for the further scaling-up of SER is the need to develop inexpensive CO_2 sorbents that maintain a reasonably high sorption capacity in the long term and to find a method of supplying the heat required for the calcination of the sorbent under CO_2-rich atmospheres (a condition which is necessary to facilitate the subsequent purification, compression, and storage of CO_2) with the minimum energy penalty and capital cost.

A wide variety of high-temperature sorbents have been studied for the CO_2 capture, including calcium-based materials (Blamey et al., 2010; Erans et al., 2016; Kierzkowska et al., 2013), hydrotalcites (Reijers et al., 2006; van Selow et al., 2009b), lithium zirconates (Nair et al., 2004; Ochoa-Fernandez et al., 2005), lithium orthosilicates (Ebner et al., 2006; Ochoa-Fernandez et al., 2007), etc. Of these, calcium-based sorbents are the most widely proposed in the literature for the SER process due to the following reasons: the carbonation of CaO directly supplies the heat required for the steam reforming of methane; the CaO carbonation rate at 600–700°C is reasonably fast; the presence of CaO minimizes carbon deposition; and CaO-based materials can be obtained from a wide range of natural and low-priced sources (Harrison, 2008). Many works have experimentally demonstrated for a wide range of operating conditions (i.e., at temperatures between 550 and 700°C and pressures between 1 and 10 bar) that mixtures of reforming catalyst and CaO-based sorbents are able to generate a product gas with a H_2 content close to the maximum allowed by the SER equilibrium (Balasubramanian et al., 1999; Li and Cai, 2007; Lopez Ortiz and Harrison, 2001; Lysikov et al., 2008).

However, the rapid loss of CO_2 sorption capacity after multiple carbonation/calcination cycles (Abanades, 2002; Curran et al., 1967; Grasa and Abanades, 2006) is a major problem with these materials. High-performance Ca-based sorbents that maintain a stable CO_2 carrying capacity higher than 30 wt.% have been developed in recent years (Erans et al., 2016; Kierzkowska et al., 2013). A significant number of synthetic CaO-based sorbents have been tested in the SER operation. CaO supported on $Ca_{12}Al_{14}O_{33}$ (mayenite) (75/25 wt.%) showed higher CO_2 sorption capacity after 50 cycles in thermogravimetric analyzer (TGA) than natural limestone under the same test conditions (Li et al., 2006). The uniform distribution of this binder among the CaO particles provided a stable structure that enhances the resistance of the sorbent to sintering. This material mixed with a nickel-based reforming catalyst in a fixed bed reactor generated at 630°C and at atmospheric pressure a product gas containing 95 vol.% of H_2 (on a dry basis) (Li et al., 2006). Martavaltzi and Lemonidou (2010) tested a sorbent with similar composition using the same mixed metal oxide (85/15 wt.%) over consecutive cycles of SER/$CaCO_3$ calcination in a fixed bed. The material exhibited an initial CO_2 sorption capacity of about 0.44 g CO_2/g sorbent with a moderate loss of activity of 15% after five cycles. During the successive SER cycles, the product gas contained H_2 concentrations higher than 92 vol.% (on a dry basis). CaO derived from natural limestone supported on calcium aluminate cement presented stable carrying capacities of almost 30% after 30 cycles of carbonation/calcination in TGA (Manovic and Anthony, 2010). More recent studies have demonstrated the suitability of this sorbent for SER applications (Broda et al., 2013; Garcia-Lario et al., 2015a). Moreover, mixed reforming catalyst/CaO materials that are combined in the same solid matrix have been developed to minimize diffusional effects between particles, to ensure a uniform solids distribution in the reactor, and to reduce the amount of inert. Recent works carried out in microreactors have demonstrated that these bifunctional materials are able to produce H_2-enriched gas with compositions close to that allowed by the SER equilibrium (Aloisi et al., 2017; Garcia-Lario et al., 2015b; Martavaltzi and Lemonidou, 2010).

As mentioned above, another serious drawback of the SER technology (and in general of any calcium-looping system) is the need to find an energy-efficient procedure for the regeneration of the sorbent. The CaO/$CaCO_3$/CO_2 equilibrium requires the regeneration of the solid sorbent to be carried out at temperatures of around 900°C (Baker, 1962) to produce pure CO_2 ready for transport and permanent storage or for other disposal strategies.

Several process schemes have been reported in the literature for this purpose. The only one developed to large-pilot scale is the oxy-fuel combustion of additional fuel in a regenerator, originally proposed by Shimizu et al. (1999), and subsequently tested at TRL5 (Rodríguez et al., 2011; Symonds et al., 2012) and TRL6 (Arias et al., 2013; Diego et al., 2016b; Strohle et al., 2014) for postcombustion applications. Martínez et al. (2013a) carried out the thermal integration of a SER plant involving oxy-combustion for the regeneration of the Ca-based sorbent. The main drawback of the oxy-combustion concept is the need for a costly and energy consuming air separation unit (ASU). In an attempt to circumvent the ASU and increase the overall energy efficiency, different alternatives have been proposed, such as the heating by means of hot combustion gases (Stevens et al., 2007), the indirect heating through heat transfer surfaces (Abanades et al., 2005; Junk et al., 2013) or the use of the waste heat from a coupled solid oxide fuel cell (Boshu et al., 2008; Meyer et al., 2011). However, these may face major economic obstacles respect to the oxy-fired benchmark, as they require very large heat-transfer areas operating at extremely high temperature. On the other hand, the use of high-temperature streams of Ca-rich solids that come from a combustor, as proposed for cement applications (Diego et al., 2016b; Rodríguez et al., 2008) can be ruled out in this case, since the high temperatures involved (around 1000°C) would lead to an unacceptable drop in the activity of the CaO-sorbent during the subsequent SER stage.

Another option for avoiding the need for an ASU to calcine $CaCO_3$ is the use of a coupled chemical looping combustion (CLC) system. In the CLC, the fuel is oxidized thanks to the reduction/oxidation cycle of a solid oxygen carrier (typically a metal oxide) (Lyngfelt et al., 2001). This avoids the dilution of the combustion products with nitrogen from air and generates a product gas that is highly concentrated in CO_2 (without the need for a subsequent costly gas separation step) (Ishida et al., 2002). General electric first proposed the use of the heat released from the oxidation of nickel with air to directly calcine $CaCO_3$ (Lyon and Cole, 2000). Direct heating without the need for intermediate heat-exchange devices (through the coupling of endothermic and exothermic reactions in one single stage) should result in greater theoretical energy efficiency (Dupont et al., 2008). However, an inherent drawback of this system is that this is not a CO_2 capture system. The air which is used for the oxidation/calcination operation generates CO_2 highly diluted with N_2, and as a result additional steps downstream would be necessary to separate and purify the CO_2. Other recent CLC configurations are aimed at reducing the energy penalty in the CO_2-sorbent regeneration step by using the sensible heat of Ni-based oxygen carriers (which also act as catalysts for SMR) that come from an oxidation reactor at temperatures above 1000°C (Wolf and Yan, 2005). The results obtained from the thermodynamic analysis of these combined SER-CLC processes demonstrate their potential for achieving higher H_2 yields and higher CO_2 capture efficiencies than conventional SMR processes that use other CO_2 capture technologies (Ryden and Ramos, 2012). However, the consumption of a significant fraction of the inlet methane due to its oxidation with NiO (during the period in which the oxygen carrier is reduced) and the need for large amounts of steam (used as sweep gas) during the $CaCO_3$ calcination stage has a negative impact on the thermal efficiency of these systems (Antzara et al., 2015). Another alternative could be to use an iron oxide chemical loop based on the scheme proposed by Fernández and Abanades (2016a) to calcine $CaCO_3$ in a cement plant. This process takes advantage of the large heat-transfer capacity of the iron oxides coming at a very high temperature from an air reactor. The operation is carried out by means of an arrangement of interconnected fluidized bed reactors, which facilitates the introduction of a solids separation step between the CaO and reforming catalyst and the denser iron oxide particles. This operation reduces the energy demand for the $CaCO_3$ calcination operation considerably (Fernández and Abanades, 2016b). The need for additional equipment (ASUs, additional reactors) or extremely challenging heat transfer conditions (i.e., very large heat transfer areas or gas or solid flows at very high temperatures) probably explains why none of these processes has yet achieved a large pilot demonstration status.

The Ca—Cu looping process was proposed (Abanades and Murillo, 2009) to enhance the production of H_2 with inherent CO_2 capture (on the basis of the SER concept) and to provide an alternative solution for the regeneration of $CaCO_3$ while delivering a concentrated stream of CO_2. In this process, a Cu/CuO chemical loop supplies directly the heat needed to regenerate the Ca-based sorbent by exploiting the exothermic reduction of CuO with gaseous fuels (CH_4, CO, H_2 present in the system), which produces a highly concentrated stream of CO_2 and H_2O_v. Cu-based materials have a number of favorable features for chemical looping applications. These include high reaction rates for the reduction and oxidation stages, the absence of thermodynamic restrictions for the complete conversion of the fuel to CO_2 and H_2O, a high oxygen transfer capacity (0.5 mol of O_2/mol metal), a high resistance to carbon deposition, and their relatively low cost and low environmental impact. They also show good chemical and mechanical stability under repeated oxidation/reduction cycles, which is of great importance in processes involving fixed bed reactors, where the replacement of the solids can be costly. However, CuO has a relatively low melting point (1085°C) and therefore it cannot be used in the highest temperature range of CLC systems (1000–1200°C). At temperatures over 950°C, Cu-based materials tend to agglomerate and their reactivity is drastically reduced (Adanez et al., 2012).

This chapter reviews the progress made so far in the development of the Ca—Cu looping process. After a detailed description of each stage involved in the Ca—Cu process, the main advances regarding process design, reactor modeling and thermal integration in industrial processes are discussed. Finally, current research in the development of combined Ca—Cu materials is examined.

2. DESCRIPTION OF THE Ca—Cu LOOPING PROCESS

The basic scheme of the Ca—Cu looping process for SER consists of a sequence of three reaction steps as depicted in Fig. 2 (Abanades et al., 2010). The first stage involves the SER of methane to produce a H_2-rich gas. The CaO-based sorbent removes the CO_2 and forms $CaCO_3$. The Cu-based solid, which is in its reduced form, remains unreacted during this SER reaction stage. A reforming catalyst is needed, since Cu does not have sufficient catalytic activity to carry out the reforming reactions. In the second stage, the Cu-based material is oxidized with air. As this reaction is highly exothermic, the operating temperature needs to be controlled to avoid the decomposition of $CaCO_3$ through partial calcination. If the operation is carried out

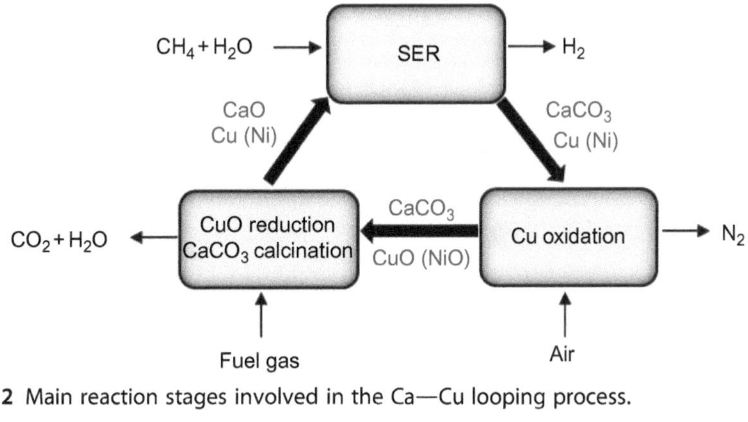

Fig. 2 Main reaction stages involved in the Ca—Cu looping process.

in fluidized bed reactors, the excess of heat can be extracted from the oxidation reactor, as in a typical boiler, to produce steam for power generation. If the process takes place in fixed beds, a more complex heat management strategy is needed. A low oxygen content in the feed, achieved by recycling a large fraction of the product gas (which is virtually pure N_2), and a low temperature operation (i.e., in the solids bed and in the gaseous feed) are the conditions required to limit the increase in temperature in the oxidation front (Fernández et al., 2012a).

In the third stage, a suitable $CuO/CaCO_3$ molar ratio in the reactor is required to ensure that the heat released from the reduction of CuO is sufficient to completely decompose the $CaCO_3$ formed during the SER without the need for an external energy supply. The composition of the fuel gas greatly affects the $CuO/CaCO_3$ ratio required in the reactor to ensure the complete conversion of the solids. Attending to the enthalpies of the reduction reactions (see Eqs. 5–7), a minimum $CuO/CaCO_3$ molar ratio of 3.1 will be needed if methane is used as reducing gas (Eq. 5). In case of using pure H_2, the $CuO/CaCO_3$ molar ratio will decrease to 1.8 (Eq. 6). The higher reduction enthalpy of CuO with CO (Eq. 7) should allow the lowest possible $CuO/CaCO_3$ molar ratio in the reactor (i.e., 1.3). The reduction/calcination operation needs to be carried out at atmospheric pressure to facilitate the calcination reaction at temperatures of around 900°C, which reduces the energy demand of the process and avoids solids sintering (Fernández et al., 2012a).

$$4CuO + CH_4 \rightarrow 4Cu + CO_2 + 2H_2O \quad \Delta H_{298K} = -44.5\ kJ/mol \tag{5}$$

$$CuO + H_2 \rightarrow Cu + H_2O \quad \Delta H_{298K} = -85.8\ kJ/mol \tag{6}$$

$$CuO + CO \rightarrow Cu + CO_2 \quad \Delta H_{298K} = -126.9\ kJ/mol \tag{7}$$

The process can be carried out using three interconnected fluidized bed reactors or an arrangement of fixed beds operating in parallel if the process is aimed at producing H_2 at high pressure (Abanades and Murillo, 2009). Although the Ca—Cu looping concept can be adapted for postcombustion capture systems (Abanades and Murillo, 2009), most of the works on this novel process have been focused on developing systems to produce high-purity H_2 and/or power (Fernández and Abanades, 2017a). A simplified scheme of the process carried out in parallel fixed bed reactors is represented in Fig. 3.

The SER stage must takes place at 600–750°C, at S/C molar ratios of between 2.5 and 5, and at pressures of between 10 and 15 bar in order to

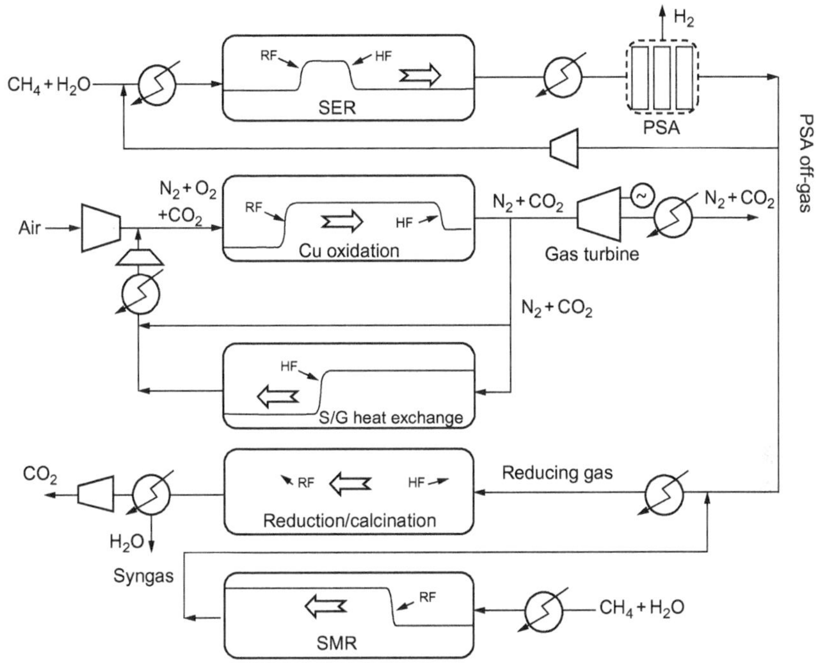

Fig. 3 Illustration of the Ca—Cu looping process in an arrangement of fixed bed reactors for the production of H_2 (RF: reaction front, HF: heat exchange front) (Fernández and Abanades, 2017b).

achieve high H_2 production yields with high CO_2 capture efficiencies. The feed with a S/C molar ratio of three into a reactor performing at a pressure of 10 bar and a temperature of around 700°C would seem a reasonable trade-off to obtain a product gas with about 90 vol.% H_2 (on a dry basis) and very low amounts of CO_2 and CO (Fernández and Abanades, 2017b). The gaseous feed needs to be preheated to temperatures around 700°C to favor the fast conversion of methane and steam to produce H_2 via Eqs. (1)–(3). The part of the bed where the sorbent has already been carbonated acts like a conventional reformer where methane and steam partially react to generate a product gas mainly composed of H_2, CO, and CO_2. The bed is left at a temperature of about 520–550°C, the temperature range allowed by the SER equilibrium. Downstream, the CO_2 resulting from the SMR is almost completely removed from the gas phase in a sharp carbonation front (RF in Fig. 3). The heat released via Eq. (4) increases the temperature in the carbonation front. A heat plateau forms between the carbonation front and the resulting heat exchange front (HE in Fig. 3) that moves ahead. A starting temperature in the bed of around 550–600°C ensures a suitable SER performance. Higher starting temperatures in the solids would lead to higher temperatures in the heat plateau during the SER operation, which greatly reduces the CO_2 capture efficiency. When the carbonation front approaches the reactor exit, the CaO-based sorbent is almost completely carbonated. As a result, the CO_2 capture efficiency starts to decrease, the methane conversion is reduced, and the concentration of H_2 in the product gas falls below 90 vol.%. This point marks the transition to the next reaction stage. A PSA unit can be used downstream to produce almost pure H_2. The resulting PSA off-gas, which is mainly composed of H_2, unconverted CH_4 and CO, can be used as reducing gas in the sorbent regeneration stage (Martinez et al., 2014). This reduces the need for additional methane to calcine the $CaCO_3$ calcination, lowers the Cu/Ca molar ratio required in the bed and highly increases the CO_2 capture efficiency of the process, since the unconverted fuel in the SER is finally emitted after the calcination stage as a CO_2-rich gas suitable for transport and storage.

The Cu oxidation stage needs to be performed at a high pressure to maximize the energy recovery options of the energy released during the oxidation and to minimize the calcination of the CO_2 sorbent. Hence, the product gas can be expanded in a gas turbine to produce electricity. If the Ca—Cu looping process is intended for power generation, the production of electricity can be boosted by burning the H_2 generated in the SER stage (Martínez et al., 2013b). In a system with an oxygen carrier with high Cu content (i.e., higher than 20 wt.%), the temperature profiles resulting from the oxidation of Cu can be moderated by reducing the O_2 content and the temperature of the gaseous feed (<300°C). The recirculation of a large part of the product gas (virtually pure N_2) greatly increases the total inlet flow and dilutes the O_2 in the feed. In these conditions, the heat exchange front, resulting from the Cu oxidation (heat exchange front (HF) in Fig. 3), advances much faster than the oxidation front (RF). As the gas approaches the oxidation front, its temperature rapidly increases due to the heat released from the Cu oxidation (Tegge, 1993). The oxygen reacts completely and almost pure N_2 leaves the oxidation front (a very low content of CO_2 is possible as a result of the calcination of $CaCO_3$). A heat plateau forms (Tegge, 1993), but the heat exchange front advances ahead at a sufficient distance to ensure the maximum temperature remains well below 900°C, which minimizes the partial calcination of the sorbent and prevents side reactions from taking place in the Cu-based oxygen carrier (Fernández et al., 2012a). When the heat exchange front approaches the reactor exit, the product gas is emitted at the maximum temperature (i.e., the temperature of the heat plateau). The Cu oxidation stage ends when the oxidation front reaches the reactor exit. At this point, the oxidized bed has been left at the temperature of the inlet gas (i.e., lower than 300°C), which is too low to carry out the reduction/calcination stage. Therefore, an additional solid/gas heat exchange step is necessary to prepare the bed by heating up the solids just before the calcination step.

Only a small part of the product gas emitted at the maximum temperature can be expanded in a gas turbine for power generation. The main fraction must be recirculated and cooled down to temperatures lower than 300°C. The excess of sensible heat from the recirculated gas is transferred to the oxidized bed during the heat-exchange step. After this operation, most of the solids have reached temperatures of above 800°C, allowing the reduction/calcination stage to be initiated. As the recirculated gas contains a small amount of CO_2, a partial recarbonation of previously calcined particles is possible, as long as the temperature favors this reaction.

The reduction/calcination stage is carried out at atmospheric pressure to favor the calcination of $CaCO_3$ at temperatures of around 900°C. The PSA off-gas resulting from the H_2 purification step after the SER, the syngas obtained in a subsequent SMR stage (explained below) or a mixture of both fuel gases can be used for the reduction of the CuO-based particles to Cu. Under these circumstances, the Cu/Ca molar ratio required in the reactor will be significantly lower than 3.1 (i.e., the ratio needed if pure methane is used, as explained above). A starting temperature in the bed higher than 800°C facilitates the total conversion of the fuel gas to CO_2 and H_2O in a sharp reduction front (RF in Fig. 3). The temperature profile rapidly rises to around 900°C because of the heat generated in the reduction front. A heat plateau forms at this temperature, which favoring the rapid calcination of the Ca-based sorbent. Both the reduction and calcination fronts move forward together during this stage, while the solids that have been totally converted are left behind (Alarcón and Fernández, 2015). The high concentration of reducing gases in the feed (above 90% vol.% of H_2, CH_4, and CO) will make the reduction (and calcination) fronts advance much faster than the heat exchange front resulting from the reduction of CuO to Cu. For this reason, when the bed approaches total reduction, most of the solids will be left at a temperature of around 900°C, which is too high for a new SER stage to be initiated.

An additional SMR stage can be introduced to convert the excess of sensible heat stored in the bed into a H_2/CO-rich product gas for direct use in the CuO reduction stage (Fernández et al., 2012a; Martinez et al., 2014). A S/C molar ratio in the feed of about 1.5 will reduce the production of CO_2 during the steam reforming, thereby minimizing the fraction of the CaO-based sorbent carbonated during the operation (Fernández et al., 2012a). The mixture of methane and steam rapidly reacts at 900°C resulting in a dramatic drop in temperature to values below 600°C (as permitted by the SMR equilibrium). A product gas mainly composed of H_2 and CO (with low contents of CH_4 and CO_2) is generated at 900°C. This high temperature impedes the carbonation of the sorbent. As the bed gradually cools down because of the endothermicity of the SMR reactions (Eqs. 1 and 2), the solids are left at temperatures below 600°C, which are suitable for the subsequent SER stage of the Ca—Cu looping process.

3. PROCESS DESIGN

The first study to take into account the basic mass and energy balances for every stage of the Ca—Cu looping process was carried out by Abanades et al. (2010). The objective of this work was to theoretically demonstrates the potential of this novel concept for a reasonable set of operating conditions. Fernández et al. (2012a) presented the first conceptual design of the complete process for producing H_2 in a system of fixed bed reactors. A simple reactor model assuming sharp reaction and heat exchange fronts was applied to describe the dynamic performance of the process and to capture the main mass and heat transfer mechanisms and reactions that take place at each reaction step. The kinetics were assumed to be sufficiently fast to take place in very narrow reactions fronts that move ahead as the gas passes through the bed and the solids are converted. This assumption was previously adopted by Noorman et al. (2007) to theoretically describe the performance of a CLC systems in fixed bed reactors. In the study by Fernández et al. (2012a), an ideal plug flow model was considered as a reasonable representation of the flow pattern in large-scale packed beds, in which the solid–gas reactions proceed rapidly and axial gas dispersion effects can be neglected. Under these conditions, the velocity of the RFs along the fixed beds could be calculated by means of a mass balance in the RF. Moreover, when the reactions involved are not thermally neutral and/or when the temperatures of the gases and the solids in contact are different, heat exchange fronts arise. The velocity at which the heat fronts move toward the reactor exit can be estimated by means of a heat balance in these heat exchange fronts.

When the operation is carried out at high superficial velocities, there is an intense heat transfer between gases and solids. Under these conditions, the temperature of the gases and the solids is assumed to be very similar along the bed, which makes it possible to explain each cyclic stage using a single axial temperature profile along the entire reactor. Fig. 4 represents a general description of the temperature profiles resulting from an exothermic solid–gas reaction taking place in a fixed bed reactor. The maximum increase in temperature in the RF (ΔT_{max}) can be calculated by means of an energy balance at the RF and it mainly depends on the enthalpy of reduction, the stoichiometry of the reaction, and the concentration of the gas/solid reagents. The formulation of this heat balance depends on the relative positions of the reaction and heat transfer fronts in the bed. The equations used for these calculations are listed in Table 1. This basic model allowed precise operating windows and material requirements to be defined for every stage of the Ca—Cu looping process.

On the basis of the work carried out by Fernández et al. (2012a), Martinez et al. (2014) presented a conceptual design of the Ca—Cu looping for the production of H_2. A reference case with a production of 30,000 Nm^3/h of pure H_2 was examined, which is the typical output of tubular reformers (FTR) installed in refineries (Aasberg-Petersen et al., 2001; Dybkjaer et al., 2007). A CaO-based sorbent with 40% of CO_2 sorption capacity, a Cu-based oxygen carrier with 65 wt.% of active phase, and a reforming catalyst 18 wt.% NiO over Al_2O_3 were assumed to be the

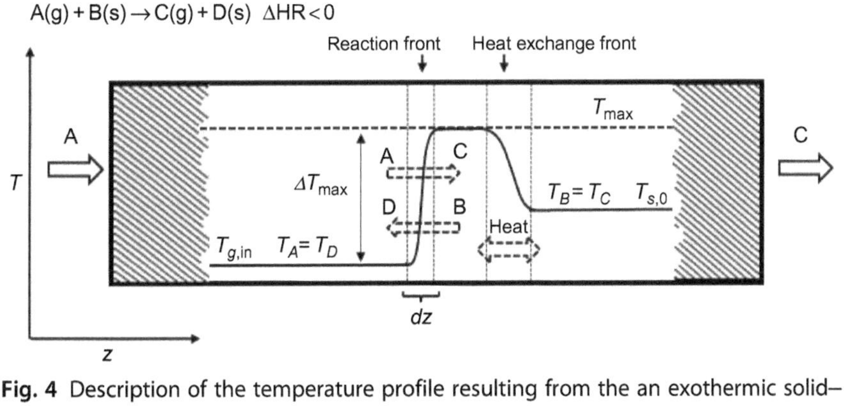

Fig. 4 Description of the temperature profile resulting from the an exothermic solid–gas reaction carried out in a fixed bed reactor (Fernández et al., 2012a).

Table 1 Equations of the Basic Model Assuming Sharp Reaction and Heat Exchange Fronts (Fernández et al., 2012a)

Velocity of the RF (u_r) $u_r = \dfrac{\rho_g u_g x_g M_s}{\varepsilon \rho_s x_s M_g \varphi}$
Velocity of the heat exchange front (u_e) $u_e = \dfrac{\rho_g u_g c_{p,g}}{\epsilon \rho_s c_{p,s}}$
Maximum increase in temperature in the RF ($\Delta T_{\max}$) $\Delta T_{\max} = \dfrac{-\Delta H_r}{\dfrac{C_{ps} M_s}{X_s \varphi} - \dfrac{C_{pg} M_g}{X_g}}$ When $u_r > u_e$ $\Delta T_{\max} = \dfrac{-\Delta H_r}{\dfrac{C_{pg} M_g}{X_g} - \dfrac{C_{ps} M_s}{X_s \varphi}}$ When $u_r < u_e$

composition of the materials, which are typical values reported in the literature for calcium looping (Blamey et al., 2010; Erans et al., 2016; Kierzkowska et al., 2013), CLC (Chuang et al., 2008; Garcia-Lario et al., 2013; Imtiaz et al., 2012), and steam reforming (Rostrup-Nielsen et al., 2002), respectively. A particle size for the solids of 0.01 mm was chosen for the calculation of the pressure drop. Moreover, a CO_2-sorbent/catalyst mass ratio of 3.5 and a Cu/Ca molar ratio of two were selected for the bed composition in order to ensure a production of H_2 close to the SER equilibrium (Fernández et al., 2012b; Rout et al., 2011) and to ensure neutral conditions during the reduction/calcination stage (Fernández et al., 2012a), respectively.

The size of the reactors and the gas velocities chosen for each stage of the process must be able to accommodate the resulting gas flows with sufficient residence time inside the bed and reasonable pressure drops. Martinez et al. (2014) determined the minimum number of reactors that would need to be operating in parallel assuming for the SER stage a maximum gas velocity of 0.5 m/s to ensure the highest possible production of H_2 and for the rest of stages a maximum pressure drop of 0.5 bar, which would limit the gas velocity during the Cu oxidation to around 2 m/s. Under these conditions, a theoretical time for a complete cycle of the Ca—Cu process was estimated to be about 150 min. A total of 15 reactors (including rinse and pressurization/depressurization steps) was shown to be sufficient to carry out the entire operation.

In a subsequent work, Fernández and Abanades (2017b) evaluated new operating conditions with the aim of reducing the number of reactors required so as to achieve a more compact configuration. They used a more rigorous reactor model (explained in Section 4) to describe the dynamic operation of the reactors and adopted heat management strategies similar to those previously developed for CLC (Hamers et al., 2013; Spallina et al., 2013) and SER (Najmi et al., 2016; Riboldi and Bolland, 2016) systems performed in fixed bed reactors. A reference case with an output flow of 30,000 Nm^3/h of pure H_2 was studied, assuming solids compositions similar to those adopted by Martinez et al. (2014) for the purpose of comparison.

Moreover, the SER stage was designed to be performed at a relatively low pressure (10 bar) and S/C ratio (3) in order to avoid CaO hydration. Although some works have reported that the CO_2 sorption capacity of CaO-based materials increases after hydration (Hughes et al., 2004; Manovic and Anthony, 2007), the swelling of the particles due to $Ca(OH)_2$ formation could undermine the mechanical properties of the sorbent (Curran et al., 1967; Fuerstenau et al., 1981). With the new process scheme, only five reactors (i.e., one reactor per stage) was found to be sufficient, larger in volume but permitting a drastic reduction in the number of high-temperature valves required to operate the process.

The effect of the stage duration and the maximum allowable pressure drop along the bed on reactor geometry was evaluated. A reasonable trade-off between both parameters was established taking into account that shorter stage durations allow the conversion of the inlet gas in smaller cross-sectional areas, while higher length/diameter ratios also reduces the area, but at the expense of having to perform with higher gas velocities (and therefore with greater pressure drops). For the reactor design, a minimum length/diameter ratio of two and a maximum pressure drop of about 10% per stage were assumed, which are geometrical constraints and operational limits that have been applied to CLC fixed beds (Spallina et al., 2015). These conditions allow each reaction stage of the Ca—Cu looping process to be accomplished in 15 min using one single reactor 6 m long with an internal diameter of 3 m.

As can be seen in Fig. 5, the flow diagram proposed for the Ca—Cu looping process is similar to other arrangements developed for SER and CLC (Spallina et al., 2015; van Selow et al., 2009a; Wright et al., 2011). Elements common to hydrogen production plants, such as a cooler located downstream of the reformer and a PSA unit for H_2 purification are needed. An air compressor, a gas turbine, and a subsequent heat recovery steam generator, which are commonly used in natural gas combined cycles (NGCCs) are also present in the operational diagram, in addition to the cooler and the

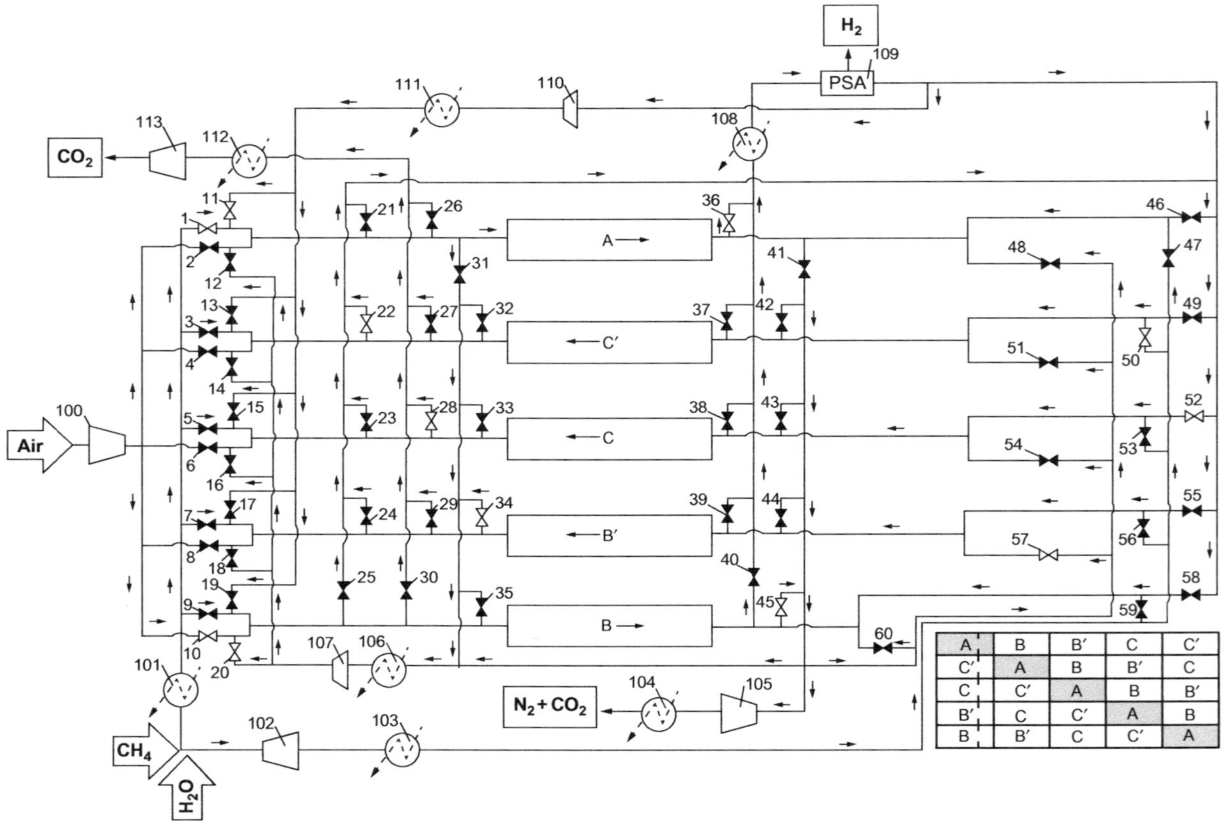

Fig. 5 Operational scheme of the Ca—Cu looping process (*white valves* are open and *black valves* are closed for the specific operational point marked by a *dotted line* in the sequential diagram represented on the *bottom right-hand side*). *Reprinted from Fernández JR, Abanades JC: Optimized design and operation strategy of a Ca—Cu chemical looping process for hydrogen production. Chem Eng Sci 166:144–160, 2017b, copyright (2017), with permission from Elsevier.*

CO_2 compressor commonly used in any CO_2 capture and storage (CCS) system. Finally, a complex valve set-up system synchronizes the operation of the fixed beds. In the reactor performing in the SER stage (A), a valve located upstream needs to be open to feed in the mixture of methane and steam, while the valves responsible for feeding in diluted air remain closed. A valve downstream of the reforming reactor needs to be open to direct the H_2-rich gas to the PSA for purification. Subsequent valves regulate the fraction of the PSA-off gas that is destined for the reduction/calcination stage (C) and the fraction recirculated to the reforming reactor. A valve located upstream of the reactor performing in the Cu oxidation stage (B) is open to allow compressed air to be supplied. Downstream of the oxidation reactor, several valves regulate the flow of the O_2-depleted air directed to the gas turbine for power generation and the fraction that is recirculated to preheat the solids in the reactor performing as B′. Finally, another valve must to open to introduce the mixture of methane and steam into the SMR reactor (C′). Another valve located downstream is open to direct the syngas produced in (C′) (together with part of the PSA off-gas) to reactor (C), where the reduction/calcination stage takes place.

4. REACTOR MODELING AND EXPERIMENTAL VALIDATION AT LABORATORY SCALE

More elaborate reactor models, similar to those experimentally validated in fixed bed CLC systems (Hamers et al., 2014; Noorman et al., 2010), have been developed to obtain more accurate estimations of the temperature and gas composition profiles at every stage of the Ca—Cu looping process (Fernández et al., 2012b, 2013; Martini et al., 2016). The behavior of the fixed bed reactors is usually described using a one-dimensional adiabatic pseudohomogeneous model. This dynamic model simulates unsteady state processes taking into account axial mass and heat dispersion and reaction kinetics, considering both gaseous and solid components. Mass and heat transfer resistances between the gas and solid phases are neglected.

The main model equations used in the modeling studies are listed in Table 2. The kinetics obtained by Xu and Froment (1989) were chosen to describe the SMR and WGS reactions. The empirical equation calculated by Rodríguez et al. (2011) to describe the carbonation of calcium oxide was also taken into account. The carbon dioxide equilibrium pressure, which depends only on temperature, was determined by a correlation from Baker (1962). Although some catalytic activity of copper is known for

Table 2 Mass and Energy Balances Together With Kinetic Equations Typically Used to Model the Ca—Cu Looping Process

Component mass balances $\varepsilon \frac{\partial C_i}{\partial t} = -u_g \frac{\partial C_i}{\partial Z} + \frac{\partial}{\partial Z}\left(D_{eff} \frac{\partial C_i}{\partial Z}\right) + \eta(1-\varepsilon) r_i$
Energy balance $\left((1-\varepsilon)\rho_s C_{ps} + \varepsilon \rho_g C_{pg}\right) \frac{\partial T}{\partial t} = -u_g \rho_g C_{pg} \frac{\partial T}{\partial Z} + \frac{\partial}{\partial Z}\left(\lambda_{eff} \frac{\partial T}{\partial Z}\right) - \Sigma \eta_i (1-\varepsilon) H_{ri}$
Reaction rates
Steam methane reforming and water gas shift (Xu and Froment, 1989) $R_1 = \frac{1}{(DEN)^2} \frac{k_1}{p_{H_2}^{2.5}} \left(p_{CH_4} p_{H_2O} - \frac{p_{CO} p_{H_2}^3}{K_1}\right); \; R_2 = \frac{1}{(DEN)^2} \frac{k_2}{p_{H_2}^{3.5}} \left(p_{CH_4} p_{H_2O}^2 - \frac{p_{CO_2} p_{H_2}^4}{K_2}\right)$
$R_3 = \frac{1}{(DEN)^2} \frac{k_3}{p_{H_2}} \left(p_{CO} p_{H_2O} - \frac{p_{CO_2} p_{H_2}}{K_3}\right) DEN = 1 + K_{CO} p_{CO} + K_{H_2} p_{H_2} + K_{CH_4} p_{CH_4} + K_{H_2O} \frac{p_{H_2O}}{p_{H_2}}$
Steam methane reforming and water gas shift (Numaguchi and Kikuchi, 1988) $r_1 = A_1 e^{\frac{-E_{act,1}}{RT}} \left(\frac{p_{CH_4} p_{H_2O} - \frac{p_{CO} p_{H_2}^3}{K_1}}{p_{H_2O}^{1.596}}\right); \; r_2 = A_2 e^{\frac{-E_{act,2}}{RT}} \left(\frac{p_{CO} p_{H_2O} - \frac{p_{CO_2} p_{H_2}^3}{K_2}}{p_{H_2O}}\right)$
Carbonation and calcination
$\frac{dX_{carb}}{dt} = \frac{k_{carb}}{P^q} (X_{max} - X)\left(v_{CO_2} - v_{CO_2,eq}\right)$ (Rodríguez et al., 2011)
$r_i = \frac{\rho_i}{M_w^i} A_i e^{\frac{-E_{act,i}}{RT}} (1-X)^{2/3} \left(c_{CO_2} - \frac{p_{CO_2}^{eq}}{RT}\right) \left(\frac{p}{p_0}\right)^{-q_p}$ (Lee, 2004)
$\frac{dX_{calc}}{dt} = k_{calc} (1 - X_{calc})^{2/3} \left(C_{CO_2,eq} - C_{CO_2}\right)$ (Martínez et al., 2012)

Table 2 Mass and Energy Balances Together With Kinetic Equations Typically Used to Model the Ca—Cu Looping Process—cont'd

Cu oxidation and CuO reduction
$r_i = \frac{\rho_i 1000}{M_i}\frac{dX_i}{dt}$ (García Labiano et al., 2004)
$\frac{t}{\tau} = X_i;\ \tau = \frac{\rho_{m,i} L_i}{2k_i C_i}$ (García Labiano et al., 2004)
$r = \eta k_0 e^{\frac{-E_{act}}{RT}} c_i^n \left(\frac{p}{p_0}\right)^{-q_p}$ (Hamers et al., 2014)
Axial mass dispersion coefficient (Edwards and Richardson, 1968) $D_{eff} = \left[\frac{0.73}{ReSc} + \frac{0.5}{\varepsilon + \frac{9.7\varepsilon^2}{ReSc}}\right] u_g d_p$
Effective axial heat dispersion (Vortmeyer and Berninger, 1982)
$\lambda_{eff} = \lambda_{0bed} + \frac{RePrk_g}{Pe_{az}} + \frac{Re^2 Pr^2 k_g^2}{6(1-\varepsilon)Nu}$ where λ_{0bed} is calculated according to Krupiczka (1967)
$\lambda_{0bed} = \left(\frac{k_s}{k_g}\right)^{0.28 - 0.757\log(\varepsilon) - 0.057\log\left(\frac{k_s}{k_g}\right)}$
where Pe_{az} is calculated according to Gunn and Misbah (1993)
$Pe_{az} = \frac{2p}{1-p};\ p = 0.17 + 0.29e\left[\frac{-24}{R_e}\right]$
Gas-to-particle heat transfer coefficient (Gunn, 1987)
$Nu = 2 + 1.8Re^{0.5}Pr^{0.33}$

SMR and WGS, only nickel kinetics were used to describe these reaction rates. These works demonstrate that the SER operation is favored by performing at moderate temperatures between 600 and 700°C, low pressures between 5 and 15 bar, high S/C ratios (up to 6), and space velocities up to 3.5 kg/m^2s. Under these conditions, the SER process is able to yield methane conversions of up to 85% and generate an outlet gas with hydrogen purities of above 90% (on a dry basis). A sensitivity analysis of the main operating parameters (i.e., operating temperature, pressure, S/C ratio in the feed, space time, catalyst-to-sorbent ratio in the bed) was carried out for a standard SER stage (Fernández et al., 2012b). In a subsequent work (Fernández et al., 2012c), the SER operation was simulated for the conditions of the Ca—Cu looping process.

An analogous reactor model was constructed to simulate the Cu oxidation stage with nitrogen recycling, as happens in the Ca—Cu looping process (Fernández et al., 2013, 2014). The oxidation kinetics for supported copper particles were taken from García Labiano et al. (2004) assuming a shrinking core model (SCM) mechanism (typically reported in the literature on copper-based materials in CLC applications), where the chemical reaction is the main resistance to the overall process (see Table 2). The amount of nickel in the bed (required for the previous SMR) was considered to have any influence during the oxidation operation and was not incorporated into the model. The calcination of calcium carbonate was described using the SCM taken from Martínez et al. (2012). These works theoretically demonstrated that a recirculation of around 80% of the product gas (which means an O_2 content of about 3–4 vol.% in the feed) and inlet temperatures of about 150°C allow the maximum temperature achieved during the Cu oxidation to remain well below 900°C. Under these conditions, an operating pressure in the fixed bed of more than 15 bar should ensure a minimum calcination of the Ca-based sorbent during this reaction stage.

The simultaneous CuO reduction/$CaCO_3$ calcination stage was simulated in detail by Alarcón and Fernández (2015) for a large fixed bed reactor under adiabatic conditions. A dynamic pseudohomogeneous model was also used for this study in which the kinetics obtained by García Labiano et al. (2004) were implemented to describe the reduction of CuO using different fuel gases and the kinetics calculated by Martínez et al. (2012) for the calcination of $CaCO_3$ (see Table 2). This work, which included a sensitivity analysis of the main parameters of the process (i.e., the temperature, fuel gas composition, and CuO/$CaCO_3$ ratio), theoretically confirmed the viability of this reaction stage. With an appropriate proportion of oxygen carrier and carbonated sorbent, the heat provided by the reduction of CuO should be sufficient to calcine the $CaCO_3$ present in the bed thereby avoiding the need for an additional energy supply. Starting temperatures in the bed above 700°C should permit a fast CuO reduction. The fuel gas would be totally converted in a relatively sharp RF and a stream rich in CO_2 ready for transport, reuse, or permanent storage would be obtained as product gas.

Qin et al. (2015) studied the reaction rates of the reduction of CuO with methane and the calcination of $CaCO_3$ in a TGA for different gas atmospheres and temperatures. A random nucleation model for the CuO reduction and an empirical calcination model based on that obtained by Fang et al. (2008) showed a satisfactory fitting of the experimental data for the operating conditions tested (i.e., temperatures between 800 and 900°C under atmospheres of up to 60 vol.% CO_2 for the calcination reaction and of up to 40 vol.% CH_4 for the CuO reduction). These kinetic equations were incorporated into a dynamic reactor model (see balances included in Table 2) and the combined reduction/calcination stage was simulated in a large-scale fixed bed reactor (i.e., 7 m long with an inner diameter of 0.3 m). It was found that hot spots may form during the operation, although a suitable solids distribution together with the addition of steam and/or CO_2 to the inlet gas can moderate the temperature profiles.

In a recent work (Martini et al., 2016), the three main reaction stages of the Ca—Cu process were also modeled using the packed bed reactor model developed at Eindhoven University (Hamers et al., 2014). The results obtained from a pseudohomogeneous model (see Table 2) and those resulting from a simplified model that assumed sharp fronts (see Table 1) (much faster in computational calculations) were compared. The kinetics obtained by Numaguchi and Kikuchi (1988) were chosen in the pseudohomogeneous model to describe the SMR and WGS reactions. The carbonation of calcium oxide and the subsequent calcination of calcium carbonate were described using the SCM previously obtained by Lee (2004) and Martínez et al. (2012). The oxidation and reduction kinetics for supported copper particles were taken from the SCM obtained by Hamers et al. (2014) and García Labiano et al. (2004), including a factor (q_p) which represents the effect of pressure on the diffusion of species as well as the competitive adsorption of inactive species on the active surface of the Cu-based particles. The maximum and minimum values achieved in both temperature and concentration profiles, as well as the velocities of reaction and heat exchange fronts calculated using both models were in a reasonably good agreement. The operability windows for each reaction stage were identified through sensitivity analyses of the main operating parameters, such as the CaO and Cu content in the bed, the composition of the inlet gases, the temperature, and the pressure. A product gas with a maximum content of 95 vol.% H_2 was found to be if the SER stage was performed at 650°C with a S/C molar ratio of five in the feed. The dilution of the inlet gas below 5 vol.% O_2 allowed the Cu oxidation stage to be accomplished with moderate temperature profiles and a low level of $CaCO_3$ calcination. Several solids bed compositions were tested using fuel gases with different compositions in order to find the best for performing the reduction/calcination under energy neutral conditions.

The dynamic behavior during a complete cycle of the Ca—Cu looping process was simulated by Fernández and Abanades (2017b), assuming that the initial conditions of each reaction step were the result of the previous stage. A dynamic model similar to those developed for CLC and SER applications in fixed beds was used (see Table 2). New operating strategies were evaluated with the aim of reducing the number of reactors required (to achieve a more compact configuration), to increase CO_2 capture efficiency and to avoid possible side reactions that might undermine the mechanical properties of the materials (e.g., the hydration of CaO during the SER stage).

The simulations showed that the feed of inlet gas during the SER stage countercurrent via the part of the reactor that was at the highest temperature (once the previous SMR stage had finished) favored the production of an outlet gas containing above 92 vol.% H_2 on a dry basis, which is close to the maximum value permitted by the SER equilibrium. The SER operation at 10 bar with temperature profiles below 730°C prevented CaO hydration and greatly reduced the emissions of CO_2 (less than 1 vol.% CO_2 on average). A starting temperature profile of around 520°C after the SER stage allowed the total conversion of O_2 during the Cu oxidation operation despite the low O_2 content in the feed (i.e., 2.3 vol.%). The recirculation of about 90% of the product gas and its subsequent cooling down to 300°C limited the maximum temperature in the oxidation front to around 860°C, which minimized the $CaCO_3$ calcination (about 1 vol.% CO_2 in the product gas) and led to the formation of a large heat plateau. In the heat exchange stage, the excess of sensible heat in the recirculated gas increased the temperature of more than 2/3 of the oxidized bed up to 860°C. In the CuO reduction/$CaCO_3$ calcination stage, the feed of the fuel gas via the part of the bed that was at the highest temperature led to the complete conversion of the reducing gases to CO_2 and steam, resulting in a product gas with easily separable CO_2 for its subsequent reuse and/or permanent storage (Fig. 6).

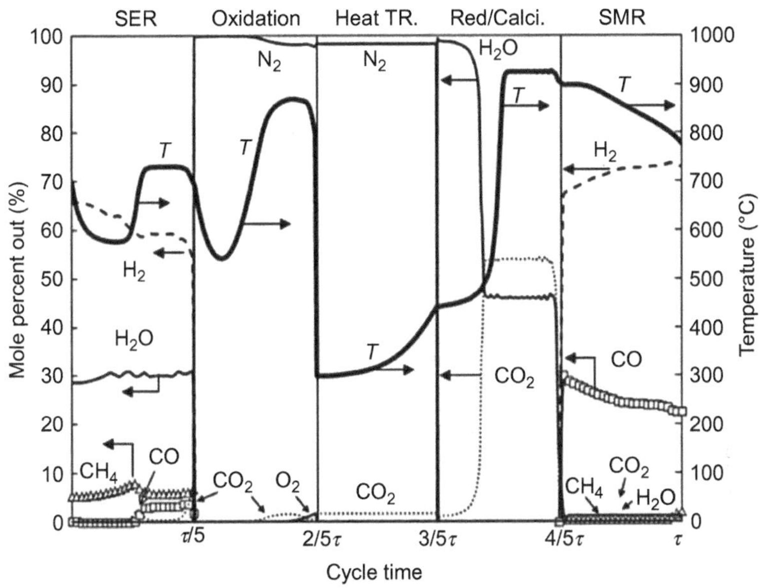

Fig. 6 Theoretical temperature and composition of the product gas in each stage of the Ca—Cu looping process for the conditions evaluated by Fernández and Abanades (2017b). *Reprinted from Fernández JR, Abanades JC: Optimized design and operation strategy of a Ca—Cu chemical looping process for hydrogen production. Chem Eng Sci 166:144–160, 2017b, copyright (2017), with permission from Elsevier.*

The large excess of sensible heat left in the solids after the regeneration of the sorbent was used in the SMR stage to produce a high-temperature syngas containing around 70 vol.% H_2 and 25 vol.% CO, a composition which is highly suitable for the reduction/calcination operation, leaving the fixed bed with a favorable temperature profile of around 600°C for a new SER stage. The use of the PSA off-gas obtained from the H_2 purification step downstream of the SER reactor as a reducing gas avoided the need for an additional supply of methane in the reduction/calcination stage. Moreover, the unconverted carbon from the SER operation was finally emitted as CO_2 during the sorbent regeneration stage, thereby increasing the CO_2 capture efficiency of the process up to 95%, which is about 10 net points higher than those reported for other benchmark CO_2 capture technologies, such as the absorption with MDEA. Fig. 7 shows a summary of the simulation results reported by Fernández and Abanades (2017b) for a complete cycle of the Ca—Cu looping process.

Significant advances have also been made in the performance of the materials used over the last decade, in particular in relation to the reactivity

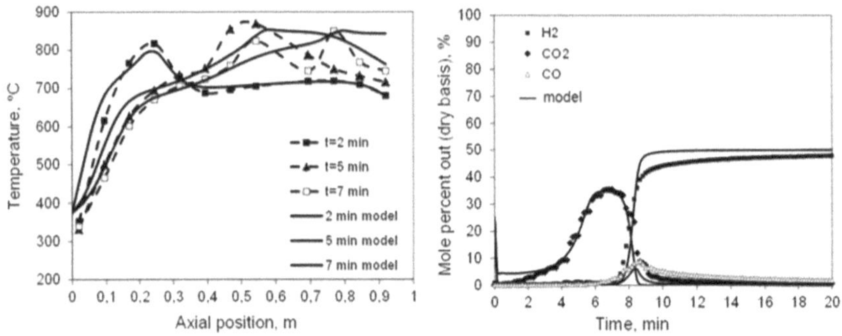

Fig. 7 Evolution of the temperature profiles inside the reactor and of the product gas composition during the reduction/calcination operation carried out using H_2 as reducing gas. *Reprinted from Alarcón JM, Fernández JR, Abanades JC: Study of a Cu–CuO chemical loop for the calcination of $CaCO_3$ in a fixed bed reactor. Chem Eng J 325:208–220, 2017, copyright (2017), with permission from Elsevier.*

of Ca- and Cu-based materials for the carbonation/calcination and oxidation/reduction reactions. In Section 7, an overview about this field of study is reported, so in the next paragraphs only the modeling activities to account for phenomena at particle and/or grain level are discussed.

To date, most of the works on the Ca—Cu looping process have considered particle models previously developed to describe separately the behavior of calcium oxide and copper oxide materials in calcium looping and CLC. Although CaO and CuO can be used in mixtures at particle level, the possibility of synthesizing combined Ca—Cu materials has recently become a topic of investigation, as explained below. The objective of these "all in one" materials is to reduce the amount of inert phase in the solids (thereby decreasing the thermal ballast) and to favor heat transfer during the reduction/calcination stage. Several particle models have been developed to obtain a thorough understanding of the performance of these combined materials.

Qin et al. (2016b) proposed a changing grain size model to describe the calcination of uniformly distributed spherical composites of Ca—Cu. The effect of the initial solids temperature (between 800 and 950°C), operating temperature (between 800 and 950°C), particle porosity (from 0.1 to 0.5), particle diameter (ranging from 0.5 to 2.5 mm), and grain size (from 200 to 800 nm) were studied for the CuO reduction with methane and the simultaneous calcination of $CaCO_3$. Their results showed that the CuO reduction kinetics was not affected by the increase in temperature to above 800°C. However, CuO conversion was favored by a smaller particle and grain size and a larger porosity. With respect to $CaCO_3$ calcination, solid conversion was enhanced by a higher temperature and particle porosity and a lower grain size.

An analogous particle model was later used to show the impact of operating conditions and particle characteristics on the calcination of combined Ca—Cu solids, in which the CuO grains are located in the core whereas the $CaCO_3$ grains are located inside the shell of the particles (Qin et al., 2016a). The results obtained from the simulations demonstrated that operating temperatures close to 900°C favored the cooccurrence of both reactions. In contrast, higher temperatures than 900°C resulted in faster $CaCO_3$ calcination rate compared to that of CuO reduction. In the case of porosity, both reactions took place simultaneously for particle porosities between 0.3 and 0.5. However, when the shell porosity was below 0.3, greater resistance to the diffusion of methane through the shell led to a significant decrease in the CuO reduction rate. The performances of core-in-shell and uniformly distributed particles were also compared. Simulation results showed that a core-in-shell distribution reduces temperature variations in the solids, which favors the coupling of CuO reduction with $CaCO_3$ decomposition.

Perreault and Patience (2016) constructed an empirical particle model based on data obtained in a fixed bed reactor to describe the carbonation of a homogeneously distributed CuO–CaO solid supported on 10wt.% of calcium aluminate. The study was carried out at atmospheric pressure for carbonation temperatures between 425 and 670°C, while the CO_2 mole fraction in the feed was varied from 10% to 20%. The use of a micro fixed bed reactor circumvented the influence of the power density of TGAs on solids sintering during the calcination stage (Alonso et al., 2014) that could lead to erroneous calculations of the kinetic parameters. The authors observed a more pronounced solid crystallinity after multiple carbonation/calcination cycles. This phenomenon increased the resistance to the diffusion of CO_2 through the calcium carbonate phase, causing a drop in the carbonation kinetics of the combined Ca—Cu particles.

Attrition resistance has also become a key operating factor in fluidized bed systems. Asiedu-Boateng et al. (2016) studied the attrition of Ca—Cu composites synthesized using different fractions of cement binder. A much greater fragmentation was observed in combined Ca—Cu solids than in commercial fluidized bed powders, such as a fluid-catalytic cracking catalyst or vanadyl pyrophosphate. However, increasing the amount of cement in Ca—Cu composites significantly improved their resistance to attrition.

Attrition resistance and attrition rates were found to correlate well for wide ranges of particle sizes, gas velocities, and temperatures.

In a recent work, Fernández et al. (2016) demonstrated experimentally the viability of the critical CuO reduction/$CaCO_3$ calcination stage at laboratory scale under pseudoadiabatic conditions. Commercial Ca- and Cu-based materials in pellet form were tested in a fixed bed reactor with a length of 1 m and inner diameter of 38 mm. A fast and complete conversion of CuO with H_2 was observed even at relatively low initial temperatures (i.e., above 400°C). However, only starting temperatures in the solids bed higher than 700°C allowed a simultaneous reduction/calcination operation to be performed. The zones in the fixed bed with lower temperatures were left uncalcined. In a subsequent study (Alarcón et al., 2017), the effect of the fuel gas composition on the reduction/calcination was also assessed. Mixtures of CO and H_2 presented a high reactivity with the CuO-based material, resulting in the complete reduction of CuO to Cu in a sharp RF and the total oxidation of the gaseous fuel to CO_2 and H_2O during the prebreakthrough period. Owing to the high reduction enthalpy of CuO with CO, lower CuO/$CaCO_3$ molar ratios in bed composition were required to maintain neutral conditions in the reduction/calcination front. The Cu-based material catalyzed the reverse WGS reaction, favored by the high temperature and the high CO_2 content in the atmosphere achieved during the prebreakthrough period. Moreover, the multicycle operation at high temperature modified the chemical composition of the functional materials tested in that work and mixed Ca—Cu oxides were formed. These oxides were carbonated in the presence of CO_2 and this affected the CO_2 sorption capacity of the solids bed. The experimental results obtained in this work were validated by a dynamic reactor model incorporating existing knowledge on kinetics and basic assumptions about gas flow and heat transfer (see Table 2). Fig. 7 demonstrates the good agreement between the simulations and the experimental results in the reduction/calcination stage (Alarcón et al., 2017).

Alarcón et al. (2017) also studied the Cu oxidation stage under relevant conditions for the Ca—Cu looping process. The dilution of the O_2 in the feed during the Cu oxidation stage by recirculating a large part of the product gas (virtually pure N_2) allowed to keep the maximum temperature below 800°C, which would minimize the partial calcination of the Ca-based sorbent during the operation. Low starting temperatures in the bed of around 400°C allowed the rapid oxidation of Cu in a narrow RF and the complete conversion of O_2 was achieved in spite of the very low O_2 content in the

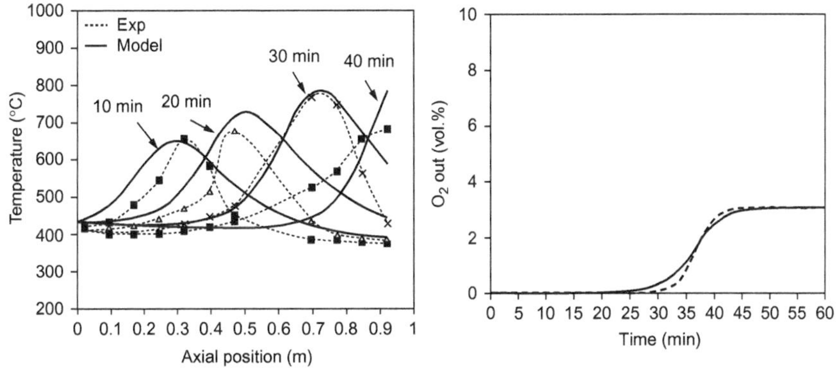

Fig. 8 Temperature profiles and outlet gas composition obtained during the Cu oxidation step carried out with a flow rate of 30 NL/min (3 vol.% O_2) at 400°C. *Reprinted from Alarcón JM, Fernández JR, Abanades JC: Study of a Cu–CuO chemical loop for the calcination of $CaCO_3$ in a fixed bed reactor. Chem Eng J 325:208–220, 2017, copyright (2017), with permission from Elsevier.*

feed (i.e., 3 vol.% O_2). A dynamic reactor model (see Table 2) was validated using the experimental results obtained in this work. The model predicted the evolution of the temperature profiles and the composition of the product gas during both the Cu oxidation stages reasonably well, as can be seen in Fig. 8.

Grasa et al. (2017) focused their investigation on the SER stage using a commercial Ni-based catalyst and a CaO-based sorbent supported on $Ca_{12}Al_{14}O_{33}$. The performance of both materials was evaluated after repetitive oxidation–reduction–reforming cycles, as is common practice for the Ca—Cu looping process. The effect of the space velocity of CH_4 on gas product composition was studied and the operational limits for the system were experimentally found for different operating pressures up to 9 bar. The reforming catalyst (cycled 200 times under the oxidizing/reducing conditions of the Ca—Cu process) and the CaO-based sorbent were able to fulfill the SER equilibrium composition up to a space velocity of 2.5 kg $CH_4\,h^{-1}\,kg\,cat^{-1}$ (i.e., about 0.53 m/s of gas velocity inside the fixed bed reactor), which would be enough to allow the process to be scaled up. Sorbent carbonation reaction rates up to 4.42×10^{-2} kmol h^{-1} kg sorb^{-1} were determined in the experiments. This parameter determines the maximum space velocity at which CH_4 (and steam) can be successfully converted to H_2 through this sorbent/catalyst system. Gas products with H_2 contents higher than 90 vol.% (on a dry basis) were obtained at 650°C for S/C molar ratios from 3 to 4, sorbent to catalysts weight ratios from 4 to 15, and

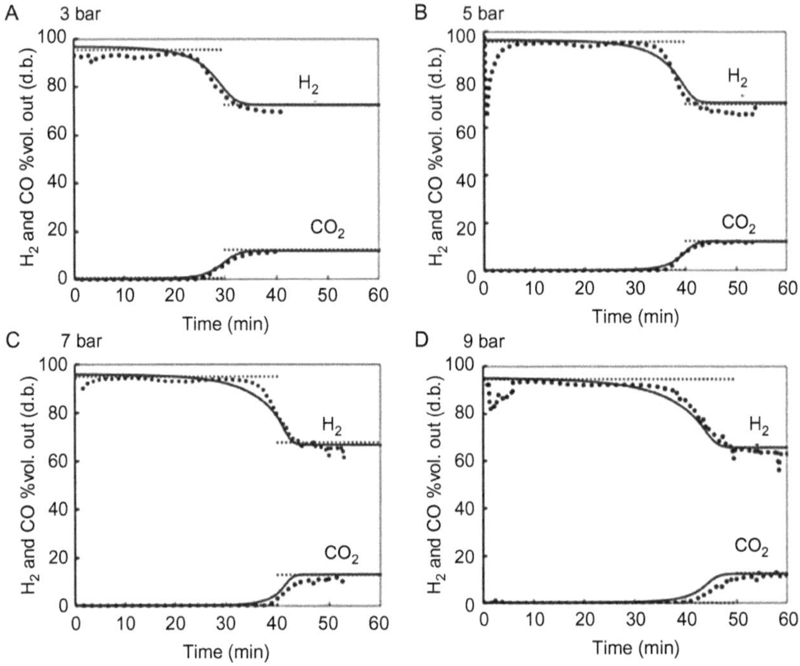

Fig. 9 Evolution of the product gas composition (only represented H_2 and CO_2 contents) during the SER operation at different pressures (on a dry basis and free of N_2). Experimental conditions: S/C molar ratio of 3.2, sorbent/catalyst weight ratio of 5, 650°C. *Reprinted from Grasa G, Navarro MV, López JM, Díez L, Fernández JR, Murillo R: Validation of the SER H_2 production stage under relevant conditions for the Ca/Cu process. Chem Eng J 324:266–278, 2017, copyright (2017), with permission from Elsevier.*

operating pressures between 1 and 9 bar. The dynamic reactor model based on the reforming kinetics obtained by Xu and Froment (1989) and the carbonation kinetics by Rodríguez et al. (2011), listed in Table 2, gave a good description of the transient performance of the SER reaction in a fixed bed reactor. The duration of both the prebreakthrough and breakthrough periods and the product gas composition were accurately predicted, as shown in Fig. 9.

5. THERMAL INTEGRATION OF THE Ca—Cu LOOPING PROCESS IN LARGE-SCALE PLANTS

Published works investigating the thermal integration of the Ca—Cu looping process on a large scale shows that high CO_2 capture efficiencies are achievable with moderate efficiency penalties compared to benchmarking systems without CCS. Martínez et al. (2013b) carried out a first approach integrating a Ca—Cu chemical loop system in a NGCC, employing the simple reactor model (see Table 1) used by Fernández et al. (2012a) for the preliminary conceptual design of the process. The integration of the Ca—Cu process with the combined cycle was carried out by means of the H_2-enriched gas produced in the SER stage (generated at 650°C and subsequently cooled down to 300°C before being introduced in the combustion chamber) and by means of the not-recirculated product gas resulting from the Cu oxidation stage (virtually pure N_2), which is expanded in the gas turbine.

In a subsequent work, Martínez et al. (2013c) carried out a sensitivity analysis in order to determine the operating conditions that would reduce the energy penalty of the process. Particular attention was paid to the effect of the S/C molar ratio used in the SER stage, the maximum temperature allowed during Cu oxidation, and the temperature of the reduction/calcination stage on the overall plant performance. It was concluded that the best conditions for the process were to perform the SER stage with a S/C molar ratio of five in the feed, to limit the temperature during the Cu oxidation stage to below 830°C (with a pressure during the SER and Cu oxidation stages of 20 bar), and to regenerate the Ca-based sorbent at 850°C. Under these conditions, the Ca—Cu looping process would make it possible to achieve an energy efficiency of about 51% (i.e., 8% points of energy penalty compared to a reference NGCC without CO_2 capture) with a reduction in CO_2 emissions of 90%. These results indicate that the process could yield similar energy efficiencies to the highest reported ones of other precombustion systems designed for producing electricity, such as the ATR or the POX. Moreover, the Ca—Cu looping process relies on a more compact design with a smaller number of stages while the costly ASU is avoided. The integration between the Ca—Cu looping process and the combined cycle for the generation of power is depicted in Fig. 10.

The process design of a H_2 production plant based on this novel concept has also been explained in a more recent work (Martinez et al., 2014). Natural gas (rather than pure methane as in previous works) was adopted as a more realistic feedstock for the process. The presence of higher hydrocarbons and sulfur compounds in the natural gas made it necessary to include in the layout of the process prereforming and sulfur removal stages to eliminate these compounds (which would be harmful for the reforming catalyst required in the SER and SMR stages) before feeding the fuel gas into the Ca—Cu looping process. The PSA-off gas obtained from the H_2

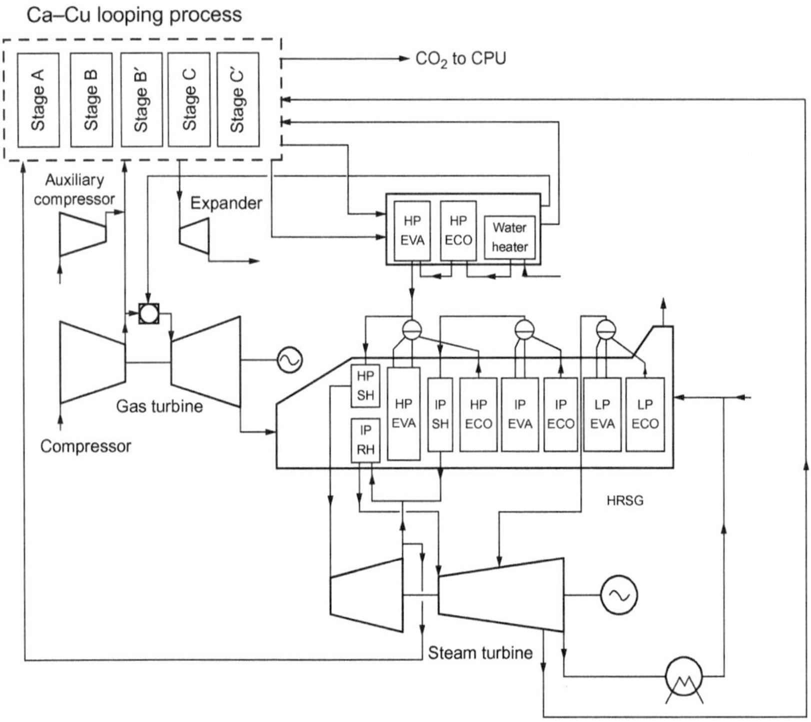

Fig. 10 Layout of the Cu—Ca looping process integrated with a combined cycle for power generation (Martínez et al., 2013c).

purification step was recirculated and used as reducing gas to diminish the thermal input that would be required as well as the Cu/Ca ratio in the bed, as explained above. Mass and energy balances for the Ca—Cu looping process and for benchmark hydrogen production plants with and without CO_2 capture were performed.

In terms of the equivalent hydrogen production efficiency, the Ca—Cu looping process reached a value of 77%, which is six points higher than a steam reforming plant (FTR) that uses amines for CO_2 absorption and capture, but lower than that of a conventional FTR without CO_2 capture (i.e., 83%) due to the higher heat output and the electric power exported of this arrangement. In terms of CO_2 capture efficiency, with the Ca—Cu looping process a value of 93% was feasible, which is eight points higher than the value achievable in a FTR plant with MDEA absorption. Moreover, the value of the specific primary energy consumption for CO_2 avoided (SPECCA) calculated for the Ca—Cu looping process was significantly lower than the benchmark FTR plant with amine absorption (i.e., 1.6

Table 3 Performance Parameters for the Reference H_2 Production Plants With and Without CO_2 Capture and for the Ca—Cu Looping Process (H_2 Output of 30,000 m^3N/h in Every Case) (Martinez et al., 2014)

	FTR Plant	FTR + MDEA	Ca–Cu Process
S/C molar ratio	2.7	4.0	3.0
Thermal input (MW)	121.9	130.8	113.5
Steam turbine output (MW)	3.3	3.8	0.3
ASU consumption (MW)	—	—	—
CO_2 compression (MW)	—	2.2	2.1
H_2 compression (MW)	—	—	1.0
Other plant auxiliaries (MW)	1.0	1.4	0.2
Heat output (MW)	8.6	4.1	4.3
Net electric plant output (MW)	2.4	0.3	−4.1
H_2 production efficiency (%)	73.9	68.8	83.6
Eq. H_2 production efficiency (%)	83.3	71.6	77.1
Equivalent CO_2 emission (gCO_2/MJH_2)	68.4	9.3	6.5
Carbon capture ratio (%)	—	84.9	93.0
SPECCA (MJ/kg CO_2)	—	3.3	1.6

and 3.3 MJ/kg CO_2, respectively). Table 3 summarizes the plant performance indexes obtained by Martinez et al. (2014) for the reference plants with and without CO_2 capture, together with those obtained for the Ca—Cu looping process.

6. THE Ca—Cu LOOPING PROCESS IN POSTCOMBUSTION AND OTHER INDUSTRIAL APPLICATIONS

The application of the Ca—Cu chemical loop in postcombustion capture systems has also been studied by some authors. Can Ozcan et al. (2015) carried out a detailed study of this novel concept designed to capture the CO_2 present in the flue gas of a coal-fired power plant. In their configuration, the thermal power needed for the calcination of $CaCO_3$ was provided by the reduction of CuO with natural gas. A copper–calcium composite with 7.5 wt.% CuO, 74 wt.% CaO, and 18.5 wt.% Al_2O_3 was assumed for the calculations. The reactivity of this functional material was tested in a TGA showing significantly higher CO_2 carrying capacities than the CaO derived from natural limestone during the first 30 carbonation/calcination cycles, although the tendency observed pointed to similar residual activities for both synthetic and natural limestone would be achieved after 60 cycles (i.e., around 0.1 of CaO conversion). A reference power plant without CO_2 capture (with a thermal input from coal combustion of about 623 MWth and a power efficiency of 43.3%) was modified to incorporate three different CO_2 capture systems for the purpose of comparison (i.e., calcium looping, oxy-combustion, and amine scrubbing). A power efficiency of about 36.6% was calculated for the Ca—Cu looping process with a CO_2 capture efficiency of 90%. A slightly lower energy efficiency (i.e., 33%) for a calcium looping system was obtained with the same CO_2 capture efficiency. The power efficiency in the oxy-firing power plant decreased to 31.4% (with a CO_2 capture efficiency of 94%) due to the energy penalty associated with the ASU and the amine scrubbing system presented the lowest energy efficiency (i.e., 29.6%). Despite the promising results obtained by the Ca—Cu looping process, the use of a large amount of natural gas for the calcination of $CaCO_3$ at atmospheric pressure calls into question the feasibility of this postcombustion application, since it comes with a large inherent penalty compared to other natural gas capture technologies performed at high pressure.

Duhoux et al. (2016) extended this study by evaluating the effect of calcination and Cu oxidation temperatures on the performance of the process. Higher oxidation temperatures and larger solids circulation rates between the reactors led to a slightly higher energy efficiency at the expense of a reduction in the CO_2 capture efficiency (due to the partial calcination of $CaCO_3$ during the oxidation stage), an increase in operation costs and a possible loss of solids reactivity due to their exposure to higher oxidation temperatures.

Other recent works have also examined the integration of the Ca—Cu looping process in industrial processes. Fernández et al. (2017) proposed the use of a Ca—Cu chemical loop to decarbonize a substantial part of the blast furnace gas (BFG) generated during the steelmaking process. This configuration, carried out in interconnected fluidized beds, involves a sorption-enhanced water gas shift (SEWGS) fluidized bed reactor fed with the BFG. In this stage, the WGS takes place in the presence of CaO-based solids that act as CO_2 sorbent, and a H_2/N_2 gas stream is produced. The $CaCO_3$ formed in the SEWGS reactor is regenerated in a fuel reactor/calciner fed with coke oven gas (COG) and a stream mainly composed of CO_2 and

H_2O_v is generated. The heat required for the calcination of $CaCO_3$ is provided by the exothermic reduction of CuO with the COG and by the sensible heat of hot Cu-based solids coming from an oxidation stage. The reduced copper-based solids obtained from the fuel reactor/calciner are oxidized and heated up in an air reactor. The separation of the Ca- and Cu-based solids by segregation after the reduction/calcination stage enhances the energy efficiency of the process because the Cu-based solids act as thermal ballast in the SEWGS reactor and in the transition from this stage to the fuel/reactor calciner. Moreover, an efficient segregation of the solids would allow the lime required for the steelmaking process (almost pure CaO) to be obtained downstream of the fuel/reactor calciner (substituting the lime plant). The decarbonized H_2-enriched gas resulting from the SEWGS reactor will contain about 35 vol.% of H_2 and less than 1 vol.% of CO and CO_2 in N_2. This gas could be burnt to supply heat for the coke oven, hot stoves, or boilers (as happens in steel mills with conventional fuel gases); for power generation; or for producing additional steel through a direct reduced iron route. Around 27% of the BFG could be decarbonized via the SEWGS reaction using CaO and about 31% of the CO_2 emitted in the steel mill could be captured. The cold gas efficiency obtained with this process is relatively low (lower than 30%), but a competitive overall performance can be obtained if the heat generated in the process is efficiently recovered to feed a steam cycle and produce electricity. A simplified scheme of the proposed Ca—Cu looping process for steel mills is shown in Fig. 11A.

Martínez et al. (2017) evaluated the application of a Ca—Cu looping process to an ammonia production plant taking into account the inherent advantage of this process for providing almost pure streams of H_2 and N_2 as part of its products. The basic idea is to mix the H_2-rich stream produced in the SER stage with the N_2-rich stream obtained from the Cu oxidation stage in a suitable proportion to feed into an ammonia synthesis unit. There is an excess of N_2 produced in the Cu oxidation stage, so one fraction is directed to the ammonia plant and the rest can be expanded in a gas turbine for the generation of electricity. Apart from a PSA unit needed for the purification of the H_2-rich gas, a second PSA unit located downstream of the Cu oxidation stage must remove the CO_2 generated by the calcination of $CaCO_3$ and then delivers almost pure N_2 (see Fig. 11B).

The proposed system would be able to reduce by up to 14% the natural gas demand with respect to a reference state-of-the-art ammonia plant (i.e., from 28 to 24 GJ_{LHV}/t_{NH_3}). However, if the electric balance is taken into

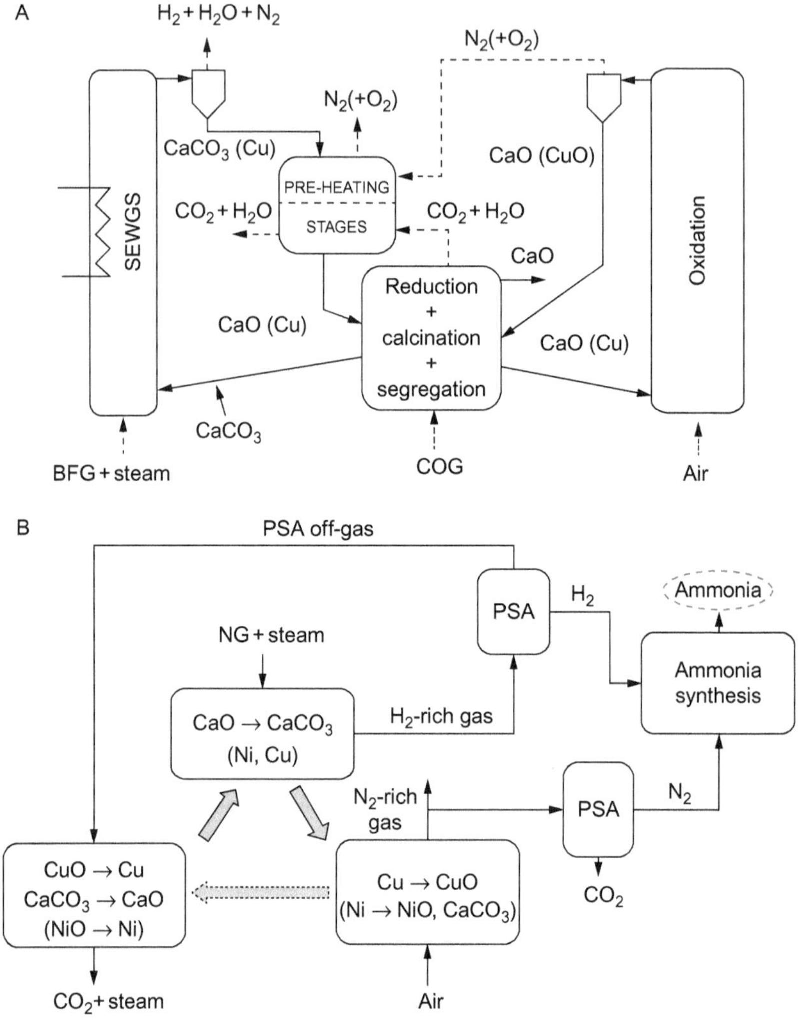

Fig. 11 (A) Ca—Cu looping process for the CO_2 capture in steel mills and (B) Ca—Cu looping process for the production of ammonia with CO_2 capture.

account, the Ca—Cu looping process is less advantageous because a considerably amount of electricity must be imported from the grid. The higher electric consumption in this novel process is mainly associated with the air compressor in the Cu oxidation stage and the H_2 compressor upstream of the PSA unit (H_2 from SER is generated at a lower pressure than H_2 from conventional SMR or ATR, which results in a greater compression duty for the purification of the H_2). Moreover, there is less heat available in the

Ca—Cu looping process for the generation of steam. As a result, the equivalent primary energy consumption is only reduced by 4% in the novel process compared to the benchmark ammonia plant. However, carbon emissions can be reduced by up to 85% with respect to the reference plant. The CO_2 emitted in the Ca—Cu looping process would be that contained in the N_2-rich stream expanded in the gas turbine, the CO_2 present in the N_2 purification PSA off-gas, and the CO_2 associated with the electricity imported from the grid. In a conventional ammonia plant with MDEA absorption, only 75% of the CO_2 generated in the process can be captured because a significant amount of CO_2 is directly emitted into the atmosphere in the FTRs.

7. DEVELOPMENT OF NOVEL MATERIALS FOR THE Ca—Cu LOOPING PROCESS

High-performance materials with direct application to the Ca—Cu looping process have been developed in the last few years. As explained in Section 1, novel CaO-based sorbents obtained by different synthesis routes for other calcium looping and SER applications have shown stable CaO conversions higher than 30%. Combined reforming catalyst/CaO solids have also been synthesized in an attempt to reduce diffusional resistances between particles and minimize the amount of inerts in the SER process. Martavaltzi and Lemonidou (2010) developed several NiO–CaO–$Ca_{12}Al_{14}O_{33}$ hybrid solids. A micro fixed bed reactor was used to evaluate their reforming activity and CO_2 sorption capacity. The synthetic material composed of 70 wt.% CaO, 20 wt.% NiO, and 10 wt.% $Ca_{12}Al_{14}O_{33}$ exhibited the best performance, producing a H_2-enriched gas with 90 vol.% H_2 (on a dry basis) and less than 3 vol.% of CO and CO_2 (i.e., values close to those allowed by the SER equilibrium). This material was subjected to multiple carbonation/calcination cycles in a TGA (with 15 vol.% CO_2 at 650°C and 100 vol.% He at 850°C, respectively), showing a stable CO_2 carrying capacity of about 0.31 gCO_2/g solid after 45 cycles.

Garcia-Lario et al. (2015b) synthesized a combined NiO/CaO material supported over calcium cement aluminate (i.e., 74 wt.% CaO, 18.5 wt. NiO, 7 wt.% cement). The stability of this solid in terms of CO_2 sorption capacity was determined by means of multiple carbonation/calcinations tests carried out in a TGA. After 80 cycles, a stable CO_2 carrying capacity higher than 20 wt.% was observed (for calcination periods performed at 900°C and 70 vol.% CO_2 in air). The catalytic activity of the combined solid was studied in a 4mm-long fixed bed reactor, alternating between SER periods carried out at 650°C with a S/C molar ratio of three and sorbent regeneration periods at 900°C in air. Although the solid showed a small loss of reforming activity after cycling, it was able to generate in every SER stage a product gas with H_2 contents of around 90vol.% (on a dry basis), close to the SER equilibrium composition. Another good example of the progress achieved with hybrid NiO–CaO materials for SER is illustrated in a recent work by Aloisi et al. (2017). They prepared different nickel–calcium aluminates by an impregnation method with increasing CaO and Ni contents up to 65wt.% and 10wt.%, respectively. The solid with the highest content in CaO maintained a CO_2 carrying capacity of around 20% after 50 carbonation/calcination cycles performed in a TGA (with the calcination stages carried out at 900°C in air). Catalytic tests in a microreactor at 650°C and at atmospheric pressure showed that H_2 contents of about 90vol.% in the product gas were achieved for residence times below 8 $NlCH_4\,h^{-1}\,g_{Ni}^{-1}$.

As far as Cu-based oxygen carriers are concerned, a wide variety of synthesis routes and binders have been studied in order to develop solids with high Cu loading that show high resistance to agglomeration and deactivation. De Diego et al. (2004) prepared copper-based materials supported over alumina, silica, titania, sepiolite, and zirconia with CuO loads ranging from 20 to 80wt.%. Their reactivity during oxidation and reduction cycles was tested in a TGA. They observed that the solids obtained by coprecipitation, mechanical mixing, and impregnation exhibited high reactivity and total conversion during the cyclic process. However, the carriers with higher CuO contents showed unacceptable mechanical degradation after several cycles.

In a subsequent work, Chuang et al. (2008) synthesized copper-based carriers up to 82wt.% CuO using alumina as binder. The solids were obtained by applying coprecipitation, mechanical mixing, and impregnation synthesis methods. The reactivity of these materials was assessed in a small fluidized bed reactor. CO and air were used in the reduction and oxidation stages, respectively. All the synthetic carriers showed a great reactivity at temperatures above 800°C. However, only the coprecipitated solids exhibited a reasonably high resistance to attrition and agglomeration after long duration cyclic tests. The agglomeration observed in the rest of the carriers was due to the irregular dispersion of the CuO throughout the alumina. They also found that coprecipitation at pH values higher than 9.7 improved the mechanical properties of the Cu-based carriers. pHs below 7 resulted in solids particularly susceptible to attrition. X-ray diffraction (XRD) analyses demonstrated that $CuAl_2O_3$ was formed at high temperatures. As this mixed oxide proved to be fully reducible with CO, the O_2 carrying capacities of the Cu-based solids were not affected. Moreover, solids experienced O_2 uncoupling (i.e., some of the CuO decomposed to Cu_2O and O_2) during the reduction stages at temperatures higher than 900°C.

Imtiaz et al. (2012) carried out a detailed study on the effect of the pH value on the development of oxygen carriers with high Cu loads (over alumina) using a coprecipitation method. They found that pH values higher than 12 permitted the synthesis of carriers with CuO contents above 80 wt.%. The performance of these materials during the redox cycles was tested in a laboratory-scale fluidized bed reactor at 800°C alternating between methane and air atmospheres. The oxygen carriers with up to 88 wt.% CuO showed a good redox stability after 25 cycles with carrying capacities of up to 0.176 g O_2/g solid and excellent resistance to attrition. However, solids with CuO contents higher than 90 wt.% showed agglomeration and loss of O_2 carrying capacity after the multicycle operation. This study also demonstrated that the incorporation of potassium ions (KOH) into the matrix of the particles impeded carbon deposition. KOH retained part of the steam formed during the oxidation of CH_4, which inhibited the decomposition of methane into carbon and hydrogen.

Garcia-Lario et al. (2013) produced copper-based pellets with an equivalent diameter of 2.5 mm via impregnation technique over a support of alumina with a Cu load of 60 wt.%. The performance of the Cu-based pellets was tested in a TGA showing an excellent reactivity with methane, H_2, and CO (at temperatures between 600 and 800°C with concentrations of the reducing gas up to 50 vol.%). The Cu-based pellets maintained a stable O_2 transport capacity and fast reduction and oxidation rates after 100 redox cycles. The results obtained in TGA demonstrated that the reduction kinetics was well described with a SCM. Fig. 12 includes several cross-sectional images of pellets at different conversion levels, from a completely oxidized pellet on the left to a totally reduced pellet on the right. As can be seen in the center of Fig. 12, the half reduced pellet behaved like a nonporous solid with an inner core still in an oxidized state (black color), which is surrounded by an external zone of reduced pellet (reddish color).

In the case of combined Ca—Cu materials, Manovic and Anthony (2011a) were the first to synthesize CaO/CuO-based composites in pellet form using calcium aluminate as binder (45:45:10 in terms of mass fraction, respectively). Successive reduction/calcination, oxidation, and carbonation cyclic tests were carried out in a TGA. The reduction/calcination step was

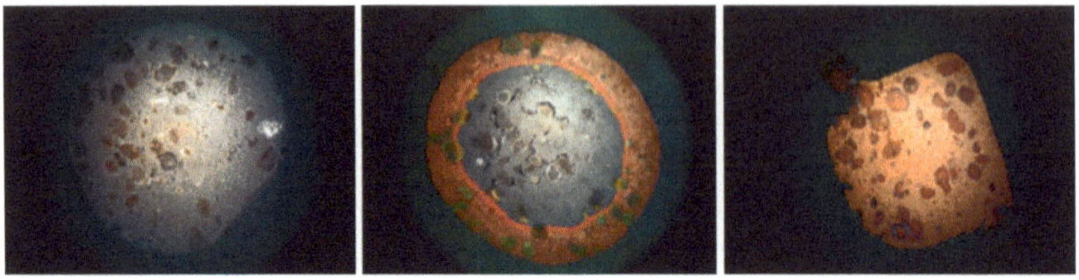

Fig. 12 Cross-sectional images of an oxidized pellet (*left*), a half reduced pellet (*center*), and a totally reduced pellet (*right*), 60 wt.% Cu over alumina. *Reprinted with permission from Garcia-Lario AL, Martinez I, Murillo R, Grasa G, Fernández JR, Abanades JC: Reduction kinetics of a high load Cu-based pellet suitable for Ca/Cu chemical loops. Ind Eng Chem Res 52:1481–1490, 2013. Copyright (2013) American Chemical Society.*

performed at 800°C in an atmosphere of methane. The oxidation of the solid was carried out at 800°C using air and the carbonation step was performed at 600°C with 20 vol.% CO_2. Fast kinetics was observed for each reaction and the copper-based phase was totally converted during both the oxidation and the reduction stages, which confirmed the suitability of this hybrid material for the Ca—Cu looping process.

In a later paper (Manovic and Anthony, 2011b), core-in-shell (with the CuO phase in the inner core) and homogeneous composites with different Ca/Cu ratios were developed. Up to 10 wt.% of calcium aluminate cement was added as binder to improve the mechanical strength of the solids. Mixtures of between 50 and 75 wt.% CuO were granulated. Pellets of about 1.4 mm were prepared from the granulation of the powdered materials by using a mechanical pelletizer with the addition of spray water. All the composites showed a good reactivity during the TGA experiments. However, only the core-in-shell materials supported on calcium aluminate cement exhibited reasonable resistance to attrition after 2 h of fluidization in a laboratory-scale bubbling bed.

Subsequently, Manovic et al. (2011) incorporated manganese oxides (which are inexpensive and nontoxic materials) into CaO-based pellets. The results showed that these oxygen carriers are less suitable than copper oxide for performing the simultaneous reduction/calcination operation as their reduction with methane and/or with a syngas provides insufficient heat for the calcination of $CaCO_3$. Apart from that, the authors incorporated a small amount of nickel oxide (i.e., about 3 wt.%) into the CaO/CuO pellets to evaluate the performance of these solids during the SER step. Higher CaO conversion rates for particles with NiO were observed in the TGA compared to particles without any catalyst when exposed to mixtures of methane and steam at 600°C. This study also showed that bauxite is a suitable support for Ca—Cu composites as it enhances the mechanical strength of the composites and favors the formation of mayenite ($Ca_{12}Al_{14}O_{33}$), which helps to maintain the long-term CO_2 carrying capacity of the sorbent.

Qin et al. (2012) combined MgO with CaO and CuO (25:11:64 in terms of mass fraction, respectively) following different synthesis procedures (i.e., wet mixing, sol–gel, and mechanical mixing). These novel materials showed a good reactivity during the reduction and oxidation tests carried out in a TGA using methane and air, respectively. However, it was observed that high CuO contents hindered the carbonation of the solids, although the presence of MgO enhanced the conversion of carbonate. For example, the carbonation conversion of CaO–CuO was only 18% after the fourth cycle (in a carbonation stage carried out at 650°C using 15 vol.% CO_2 for 10 min). In contrast, the CaO–CuO–MgO composite showed a carbonation value of around 40% in the fourth cycle under the same operating conditions. Magnesium oxide favored the uniform dispersion of both CuO and CaO and prevented the surface of the CaO from being covered by CuO. This greatly reduced the resistance to the diffusion of the CO_2 during the carbonation stage.

In a subsequent work, Qin et al. (2013) proposed different Ca—Cu precursors, supports and reactivation methods to maintain the cyclic stability of the composite materials. Ca—Cu solids were prepared from calcium and copper oxides, acetates, hydroxides, and hydrates by means of a wet mixing method. Their reactivity was studied in a TGA during alternating carbonation, calcination/reduction, and oxidation cycles conducted isothermally at 750°C. The Ca—Cu composite obtained from calcium oxide and copper acetate showed the best reactivity during the Cu/CuO redox cycles and the lowest decay in CO_2 uptake capacity (i.e., a CaO conversion of about 50% was measured after 40 cycles, which is quite a high value due to the mild calcination conditions employed in this work). Moreover, three different supports were tested (i.e., magnesium oxide, alumina, and cement). The sample with MgO as binder exhibited the highest CO_2 sorption capacity, as previously observed by Qin et al. (2012). In contrast, the solid with alumina showed the greatest decay in CO_2 carrying capacity with the number of cycles. XRD analyses demonstrated that side reactions between CaO and alumina during the calcination stage led to the formation of $Ca_{12}Al_{14}O_{33}$, which consumed part of the active CaO present in the Ca precursor. Furthermore, the calcination of the copper acetate at 900°C for 2 h before it was mixed with the CaO and MgO precursors in the synthesis procedure made it possible to reduce the aggregation of CuO and enhanced the CO_2 sorption capacity of the Ca—Cu composite in the TGA tests. The addition of steam (up to 13 vol.% H_2O_v) during all the reaction stages greatly improved CaO conversion in the composites. The Ca—Cu material obtained from calcium oxide and copper acetate and supported on MgO only showed a decay in the CO_2 sorption capacity of about 3% after 10 carbonation–reduction/calcination–oxidation cycles. These results are in agreement with those reported by Donat et al. (2012), who observed that the presence of steam during the carbonation of CaO reduced the resistance to the diffusion of CO_2 through the carbonated product.

Kierzkowska and Müller (2012) developed combined Ca/Cu solids prepared by a coprecipitation method (i.e., 64:36 and 82:18 in terms of mass fraction of CuO and CaO in order to achieve the suitable proportion of solids for the neutral reduction/calcination operation with CO and CH_4, respectively). Successive cycles of carbonation (using 40 vol.% CO_2 for 40 min), reduction/calcination (using 10 vol.% CH_4 for 20 min), and oxidation (using 4.2 vol.% O_2 for 20 min) were carried out in a TGA at 750°C. The combined materials showed excellent CuO conversions (close to 100%) and stable CO_2 carrying capacities after 15 cycles (i.e., values of around 0.2 gCO_2/g for the solid with the highest CaO load and around 0.1 gCO_2/g for the solid with the lowest amount of CaO). XRD analyses of aged samples showed that mixed Ca—Cu oxides were formed after the cyclic operation, which were able to form calcite and other carbonates that tend to improve the CO_2 carrying capacity.

In a later work, Kierzkowska and Müller (2012) analyzed the effect of different supports (e.g., Al_2O_3, MgO, or $MgAl_2O_4$) on solids reactivity. A sol–gel technique was used for the synthesis of the composites. This procedure allowed a better dispersion of the components inside the matrix of the solids. The performance of the Ca—Cu materials during a cyclic operation was studied in a TGA following the same methodology of the previous work during the cyclic tests. Materials doped with Mg^{2+} showed the highest CO_2 uptake (as also reported by other authors in previous works) and minimum carbon deposition. Of the materials developed in this work, the composite with a CuO/CaO molar ratio of 1.3 supported on $MgAl_2O_4$ showed the highest CO_2 carrying capacity (i.e., 0.13 gCO_2/g) after 15 cycles.

Ridha et al. (2015) carried out successive redox cycles in a micro fixed bed reactor in order to study homogeneous and core-in-shell Ca—Cu solids (with a particle diameter of about 0.6 mm). The cyclic tests consisted of an initial oxidation/carbonation step carried out at 650°C using 15 vol.% CO_2 and 4 vol.% O_2 (N_2 balance) for 15 min and a subsequent reduction/calcination step performed at 850°C using a mixture of 67 vol.% CO and 33 vol.% H_2 for 15 min. Very sharp breakthrough curves were observed at the end of each reaction stage, which confirmed the high reactivity of both synthetic materials. Homogeneous materials showed larger surface areas and pore volumes, resulting in higher CaO conversions (up to 32%) and better mechanical strength. In contrast, core-in-shell solids showed CaO conversions of around 10% after 11 cycles and were more susceptible to fragmentation.

Rahman et al. (2015) studied in a TGA the performance of CuO–CaO–cement pellets (50:40:10 in terms of mass fraction, respectively) in different carbonation and calcination environments. The pellets showed the highest CaO conversion when carbonation was carried out with CO_2 and steam together with CO and H_2 (simulating the composition of a syngas). Under these conditions, the CO_2 concentration increased via the enhanced WGS reaction, resulting in an improvement in the carbonation kinetics. Calcination in CO_2 and steam resulted in a greater decay in CaO conversion after 15 cycles in comparison with calcination in pure N_2 (i.e., 40% vs 50%), as a result of sintering effects.

Kazi et al. (2017) used a hydrothermal preparation method to develop low cost Ca–Cu composites supported on mayenite. A material with 53:22 contents of CuO and CaO, respectively, showed stable CO_2 capture and O_2 carrying capacity of 15 g CO_2/100 g and 7 g O_2/100 g after 50 cycles in TGA. Moreover, a satisfactory mechanical stability in this material was observed. A crushing strength of up to 11 N was measured, which is higher than those observed in previous materials developed for the Ca–Cu looping process.

In a recent work, Duan et al. (2016) studied the potential of combined Ca/Cu materials for releasing O_2 at temperatures of up to 940°C. The experiments were carried out in a laboratory-scale bubbling bed. They demonstrated that composites with 75 wt.% CuO were able to provide a stable O_2 concentration of up to 0.6 vol.%, which might allow them to be used in chemical looping with oxygen uncoupling (CLOU) applications. No appreciable attrition and agglomeration were observed in the solids during the cyclic tests. Moreover, XRD analyses of aged samples showed the presence of Ca_2CuO_3. The formation of mixed Ca—Cu oxides may reduce the

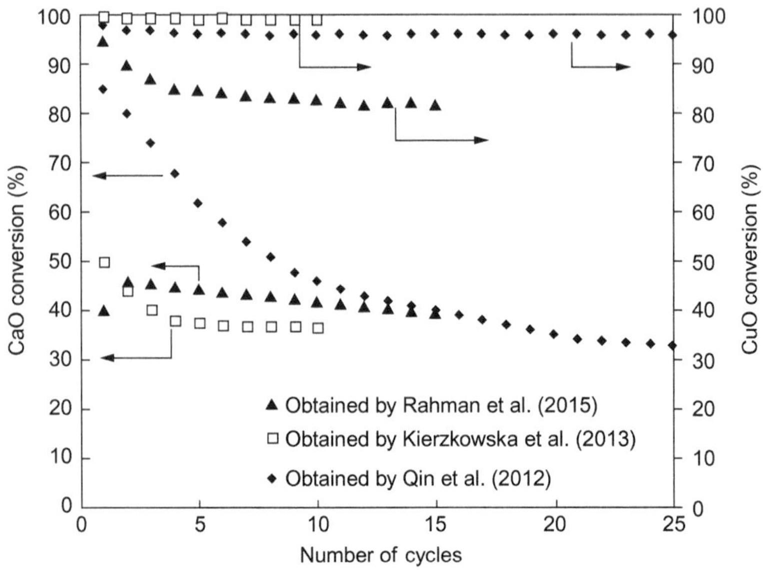

Fig. 13 Evolution of carbonation and oxidation conversions of synthetic Ca/Cu-based materials over multiple cycles. *Reprinted from Fernández JR, Abanades JC: Overview of the Ca-Cu looping process for hydrogen production and/or power generation. Curr Opin Chem Eng 17:1–8, 2017a, copyright (2017), with permission from Elsevier.*

availability of CuO and undermine the potential of these materials to release oxygen over a long period of time, although further investigation on this matter is required.

Fig. 13 shows the performance over multiple cycles of three combined Ca—Cu materials developed separately by different groups. As can be seen, reasonably high solids conversions can be achieved, although more detailed studies on their stability over hundreds of cycles are needed in order to ensure their viability on a commercial scale.

8. CONCLUSIONS

The Ca—Cu looping process combined with the SER is a promising precombustion CO_2 capture technology, since it allows the energy-intensive sorbent regeneration stage to be carried out with a moderate energy penalty. Although this novel process can be adapted to several configurations, most authors favor the parallel fixed bed reactor arrangement. State-of-the-art reactor models have been developed to simulate the dynamic behavior of every reaction stage and to obtain accurate estimations of the temperature and gas composition profiles along the fixed beds during the process. The operability windows have been identified through sensitivity analyses of the main operating parameters, such as the CaO and Cu content in the bed, the composition of the inlet gases, temperature, and pressure. The SER stage must takes place at 600–750°C, at S/C molar ratios of between 2.5 and 4, and at pressures up to 15 bar in order to achieve high H_2 production yields with high CO_2 capture efficiencies. In the Cu oxidation stage, oxygen contents in the feed below 5 vol.% and low temperature operation (i.e., in the solids bed and in the gaseous feed) are the conditions required to keep the temperature in the oxidation front below 830°C. A suitable $CuO/CaCO_3$ molar ratio in the bed, which depends on the composition of the reducing gas, ensures that the heat released during the reduction of CuO is sufficient to completely decompose the $CaCO_3$ formed during the SER.

Different operating strategies have been evaluated to reduce the number of reactors required, to increase the CO_2 capture efficiency, and to prevent side reactions that might undermine the characteristics of the materials. Several works have experimentally confirmed the viability of all the reaction stages at laboratory scale. However, more detailed experimental studies are required before the process can be scaled up to large pilot scale demonstration. Recent integration works have demonstrated that this concept can be implemented in large-scale NGCCs and H_2 production plants. High CO_2 capture efficiencies are achievable with moderate efficiency penalties compared to benchmarking systems without CCS. An important issue that requires independent development is the need for high temperature switching valves (water cooled or with special materials) to operate these systems. Significant advances have also been made in the development of synthetic Ca- and Cu-based materials. Novel CaO-based sorbents for CO_2 capture obtained by different synthesis routes have shown stable CaO conversions higher than 30%. Combined reforming catalyst/CaO solids have yielded product gases with more than 90 vol.% H_2 on a dry basis (close to the SER equilibrium). Oxygen carriers with CuO loads of up to 80 wt.% have shown high resistance to agglomeration and deactivation. Novel Ca—Cu composites have also been designed to reduce the amount of inerts and favor heat transfer during the reduction/calcination operation. Future work needs to focus on improving the long-term reactivity and mechanical stability of these materials as a necessary step for the scaling-up of this technology.

REFERENCES

Aasberg-Petersen K, Bak Hansen JH, Christensen TS, et al: Technologies for large-scale gas conversion, *Appl Catal Gen* 221(1–2):379–387, 2001.

Abanades JC: The maximum capture efficiency of CO_2 using a carbonation/calcination cycle of $CaO/CaCO_3$, *Chem Eng J* 90:303–306, 2002.

Abanades JC, Murillo R: Method of capturing CO_2 by means of CaO and the exothermic reduction of a solid, US8506915 B2. Priority date Sept 16, 2009.

Abanades JC, Anthony EJ, Wang J, Oakey JE: Fluidized bed combustion systems integrating CO_2 capture with CaO, *Environ Sci Technol* 39:2861–2866, 2005.

Abanades JC, Murillo R, Fernández JR, Grasa G, Martínez I: New CO_2 capture process for hydrogen production combining Ca and Cu chemical loops, *Environ Sci Technol* 44:6901–6904, 2010.

Abanades JC, Arias B, Lyngfelt A, et al: Emerging CO_2 capture systems, *Int J Greenhouse Gas Control* 40:126–166, 2015.

Adanez J, Abad A, Garcia-Labiano F, Gayan P, de Diego LF: Progress in chemical-looping combustion and reforming technologies, *Prog Energy Combust Sci* 38:215–282, 2012.

Alarcón JM, Fernández JR: $CaCO_3$ calcination by the simultaneous reduction of CuO in a Ca/Cu chemical looping process, *Chem Eng Sci* 137:254–267, 2015.

Alarcón JM, Fernández JR, Abanades JC: Study of a Cu-CuO chemical loop for the calcination of $CaCO_3$ in a fixed bed reactor, *Chem Eng J* 325:208–220, 2017.

Aloisi I, Di Giuliano A, Di Carlo A, Foscolo PU, Courson C, Gallucci K: Sorption enhanced catalytic steam methane reforming: experimental data and simulations describing the behaviour of bi-functional particles, *Chem Eng J* 314:570–582, 2017.

Alonso M, Criado Y, Abanades JC, Grasa G: Undesired effects in the determination of CO_2 carrying capacities of CaO during TG testing, *Fuel* 127:52–61, 2014.

Antzara A, Heracleous E, Bukur DB, Lemonidou AA: Thermodynamic analysis of hydrogen production via chemical looping steam methane reforming coupled with in situ CO_2 capture, *Int J Greenhouse Gas Control* 32:115–128, 2015.

Arias B, Diego ME, Abanades JC, et al: Demonstration of steady state CO_2 capture in a 1.7MWth calcium looping pilot, *Int J Greenhouse Gas Control* 18:237–245, 2013.

Asiedu-Boateng P, Legros R, Patience GS: Attrition resistance of calcium oxide-copper oxide-cement sorbents for post-combustion carbon dioxide capture, *Adv Powder Technol* 27:786–795, 2016.

Baker EH: The calcium oxide-calcium dioxide system in the pressure range 1-300 atmospheres, *J Chem Soc* 70:464–470, 1962.

Balasubramanian B, Lopez Ortiz A, Kaytakoglu S, Harrison DP: Hydrogen from methane in a single-step process, *Chem Eng Sci* 54:3543–3552, 1999.

Barrelli L, Bidini G, Gallorini F, Servili S: Hydrogen production through sorption-enhanced steam methane reforming and membrane technology: a review, *Energy* 33:554–570, 2008.

Blamey J, Anthony EJ, Wang J, Fennell PS: The calcium looping cycle for large-scale CO_2 capture, *Prog Energy Combust Sci* 36:260–279, 2010.

Boot-Handford ME, Abanades JC, Anthony EJ, et al: Carbon capture and storage update, *Energ Environ Sci* 7:130–189, 2014.

Boshu H, Mingyang L, Xin W, Ling Z: Chemical kinetics-based analysis for utilities of ZEC power generation system, *Int J Hydrogen Energy* 33:4673–4680, 2008.

Broda M, Manovic V, Imtiaz Q, Kierzkowska A, Anthony E, Muller C: High-purity hydrogen via the sorption-enhanced steam methane reforming reaction over a synthetic CaO based sorbent and a Ni catalyst, *Environ Sci Technol* 47:6007–6014, 2013.

Brun-Tsekhovoi AR, Zadorin AN, Katsobashvili YR, Kourdyumov SS: The process of catalytic steam-reforming of hydrocarbons in the presence of a carbon dioxide acceptor. Hydrogen energy progress VII, proceedings of the seventh world hydrogen energy conference, Moscow, Russia, September 25-29, vol.2. New York: Pergamon; P. 885, 1988.

Can Ozcan D, Macchi A, Lu D, et al: Ca-cu looping process for CO_2 capture from a power plant and its comparison with ca-looping, oxy-combustion and amine-based CO_2 capture processes, *Int J Greenhouse Gas Control* 43:198–212, 2015.

Chuang SY, Dennis JS, Hayhurst AN, Scott SA: Development and performance of Cu-based oxygen carriers for chemical-looping combustion, *Combust Flame* 154:109–121, 2008.

IPCC Climate Change 2014: Mitigation of climate change. Contribution of working group III to the fifth assessment report of the intergovernmental panel on climate change. Cambridge University Press, Cambridge, United Kingdom and New York, NY, USA 2014.

Curran GP, Fink CE, Gorin E: CO_2 acceptor gasification process. Fuel gasification, *Am Chem Soc*: 141–165, 1967.

De Diego LF, Garcia-Labiano F, Adanez J, et al: Development of Cu-based oxygen carriers for chemical-looping combustion, *Fuel* 83:1749–1757, 2004.

Di Carlo A, Bocci E, Zuccari F, Dell'Era A: Numerical investigation of sorption enhanced steam methane reforming process using computational fluid dynamics Eulerian-Eulerian code, *Ind Eng Chem Res* 49:1561–1576, 2010.

Diego ME, Arias B, Mendez A, et al: Experimental testing of a sorbent reactivation process in La Pereda 1.7 MWth calcium looping pilot plant, *Int J Greenhouse Gas Control* 50:14–22, 2016b.

Donat F, Florin NH, Anthony EJ, Fennell PS: Influence of high-temperature steam on the reactivity of CaO sorbent for CO_2 capture, *Environ Sci Technol* 46:1262–1269, 2012.

Duan L, Godino D, Manovic V, Montagnaro F, Anthony EJ: Cyclic oxygen reléase characteristics of bifunctional copper oxide/calcium oxide composites, *Energ Technol* 4:1171–1178, 2016.

Duhoux B, Mehrani P, Lu D, Symonds RT, Anthony EJ, Macchi A: Combined calcium looping and chemical looping combustion for post-combustion carbon dioxide capture: process simulation and sensitivity analysis, *Energ Technol* 4:1–14, 2016.

Dupont V, Ross AB, Knight E, Hanley I, Twigg MV: Production of hydrogen by unmixed steam reforming of methane, *Chem Eng Sci* 63:2966–2979, 2008.

Dybkjaer I: Tubular reforming and autothermal reforming of natural gas—an overview of available processes, *Fuel Process Technol* 42:85–107, 1995.

Dybkjaer I, Rostrup-Nielsen JR, Aasberg-Petersen K: Hydrogen and synthesis gas. In *Encyclopedia of hydrocarbons*, Rome, 2007, Treccani, pp 469–500.

Ebner AD, Reynolds SP, Ritter JA: Understanding the adsorption and desorption behavior of CO2 on a K-promoted HTLC through non-equilibrium dynamic isotherms, *Ind Eng Chem Res* 45:6387–6392, 2006.

Edwards MF, Richardson JF: Gas dispersion in packed beds, *Chem Eng Sci* 23:109–123, 1968.

Erans M, Manovic V, Anthony EJ: Calcium looping sorbents for CO_2 capture, *Appl Energy* 180:722–742, 2016.

Fang F, Li ZS, Cai NS: Experiment and modeling of CO2 capture from flue gases at high temperature in a fluidized bed reactor with Ca-based sorbents, *Energy Fuel* 23:207–216, 2008.

Fernández JR, Abanades JC: CO_2 capture from the calcination of $CaCO_3$ using iron oxide as heat carrier, *J Clean Prod* 112:1211–1217, 2016a.

Fernández JR, Abanades JC: Novel process for hydrogen production through the sorption enhanced reforming of methane combined with chemical looping combustion, *Chem Eng Trans* 52:535–540, 2016b.

Fernández JR, Abanades JC: Overview of the Ca-Cu looping process for hydrogen production and/or power generation, *Curr Opin Chem Eng* 17:1–8, 2017a.

Fernández JR, Abanades JC: Optimized design and operation strategy of a Ca-Cu chemical looping process for hydrogen production, *Chem Eng Sci* 166:144–160, 2017b.
Fernández JR, Abanades JC, Murillo R, Grasa G: Conceptual design of a hydrogen production process from natural gas with CO_2 capture using a Ca–Cu chemical loop, *Int J Greenhouse Gas Control* 6:126–141, 2012a.
Fernández JR, Abanades JC, Murillo R: Modeling of sorption enhanced steam methane reforming in an adiabatic fixed bed reactor, *Chem Eng Sci* 84:1–11, 2012b.
Fernández JR, Abanades JC, Grasa G: Modeling of sorption enhanced steam methane reforming—Part II: simulation within a novel Ca/Cu chemical loop process for hydrogen production, *Chem Eng Sci* 84:12–20, 2012c.
Fernández JR, Abanades JC, Murillo R: Modeling of Cu oxidation in adiabatic fixed-bed reactor with N_2 recycling in a Ca/Cu chemical loop, *Chem Eng J* 232:442–452, 2013.
Fernández JR, Abanades JC, Murillo R: Modeling of Cu oxidation in adiabatic fixed-bed reactor with N_2 recycling, *Appl Energy* 113:945–1951, 2014.
Fernández JR, Alarcón JM, Abanades JC: Investigation of a fixed-bed reactor for the calcination of CaCO3 by the simultaneous reduction of CuO with a fuel gas, *Ind Eng Chem Res* 55:5128–5132, 2016.
Fernández JR, Martínez I, Abanades JC, Romano MC: Conceptual design of a ca-cu chemical looping process for hydrogen production in integrated steelworks, *Int J Hydrogen Energy* 42:11023–11037, 2017.
Fuerstenau MC, Shen CM, Palmer BR: Liquidus temperatures in the calcium carbonate-calcium hydroxide-calcium oxide and calcium carbonate-calcium sulfate-calcium sulfide ternary systems, *Ind Eng Chem Process Des Dev* 20(3):441–443, 1981.
García Labiano F, de Diego LF, Adánez J, Abad A, Gayan P: Reduction and oxidation kinetics of a copper-based oxygen carrier prepared by impregnation for chemical-looping combustion, *Ind Eng Chem Res* 43:8168–8177, 2004.
Garcia-Lario AL, Martinez I, Murillo R, Grasa G, Fernandez JR, Abanades JC: Reduction kinetics of a high load cu-based pellet suitable for ca/cu chemical loops, *Ind Eng Chem Res* 52:1481–1490, 2013.
Garcia-Lario AL, Aznar M, Martínez I, Grasa G, Murillo R: Experimental study of the application of a $NiO/NiAl_2O_4$ catalyst and a CaO-based synthetic sorbent on the sorption enhanced reforming process, *Int J Hydrogen Energy* 40:219–232, 2015a.
Garcia-Lario AL, Grasa G, Murillo R: Performance of a combined CaO-based sorbent and catalyst on H_2 production, via sorption enhanced methane steam reforming, *Chem Eng J* 264:697–705, 2015b.
Gorin E, Retallick WB: Method for the production of hydrogen. US patent 3.108.857, 1963.
Grasa G, Abanades JC: CO_2 capture capacity of CaO in long series of carbonation/calcination cycles, *Ind Eng Chem Res* 45:8846–8851, 2006.
Grasa G, Navarro MV, López JM, Díez L, Fernández JR, Murillo R: Validation of the SER H_2 production stage under relevant conditions for the Ca/Cu process, *Chem Eng J* 324:266–278, 2017.
Gunn DJ: Axial and radial dispersion in fixed beds, *Chem Eng Sci* 42:363–373, 1987.
Gunn DJ, Misbah MMA: Bayesian estimation of heat transport parameters in fixed beds, *Int J Heat Mass Transf* 36:2209–2221, 1993.
Hamers HP, Gallucci F, Cobden PD, Kimball E, van Sint Annaland M: A novel reactor configuration for packed bed chemical-looping combustion of syngas, *Int J Greenhouse Gas Control* 16:1–12, 2013.
Hamers HP, Gallucci F, Cobden PD, Kimball E, van Sint Annaland M: CLC in packed beds using syngas and CuO/Al_2O_3: a model description and experimental validation, *Appl Energy* 119:163–172, 2014.
Harrison DP: Sorption-enhanced hydrogen production: a review, *Ind Eng Chem Res* 47:6486–6501, 2008.

Hufton JR, Mayorga S, Sircar S: Sorption-enhanced reaction process for hydrogen production, *AICHE J* 45(2):248–256, 1999.
Hughes RW, Lu D, Anthony EJ, Wu Y: Improved long-term conversion of limestone-derived sorbents for in situ capture of CO2 in a fluidised bed combustor, *Ind Eng Chem Res* 43(18):5529–5539, 2004.
IEA: *Hydrogen production and storage. R&D priorities and gaps,* Paris, France, 2006, IEA Publications.
IEA: *Technology roadmap. Hydrogen and fuel cells,* Paris, France, 2015, IEA Publications.
Imtiaz Q, Kierzkowska AM, Broda M, Müller CR: Synthesis of Cu-rich, Al_2O_3-stabilized oxygen carriers using a coprecipitation technique: redox and carbon formation characteristics, *Environ Sci Technol* 46:3561–3566, 2012.
Ishida M, Yamamoto M, Ohba T: Experimental results of chemical-looping combustion with $NiO/NiAl_2O_4$ particle circulation at 1200 degrees C, *Energ Conver Manage* 43:1469–1478, 2002.
Johnsen K, Ryu HJ, Grace JR, Lim CJ: Sorption-enhanced steam reforming of methane in a fluidised bed reactor with dolomite as –acceptor, *Chem Eng Sci* 61(4): 1195–1202, 2006.
Junk M, Reitz M, Strohle J, Epple B: Thermodynamic evaluation and cold flow model testing of an indirectly heated carbonate looping process, *Chem Eng Technol* 36: 1479–1487, 2013.
Kazi SS, Aranda A, di Felice L, Mayer J, Murillo R, Grasa G: Development of cost effective and high performance composite for CO_2 capture in Ca–Cu looping process, *Energy Procedia* 114:211–219, 2017.
Kierzkowska AM, Müller CR: Development of calcium-based, copper-functionalised CO2 sorbents to integrate chemical looping combustion into calcium looping, *Energ Environ Sci* 5:6061–6065, 2012.
Kierzkowska AM, Pacciani R, Müller CR: CaO-based CO2 sorbents: from fundamentals to the development of new, highly effective materials, *ChemSusChem* 6(7):1130–1148, 2013.
Krupiczka R: Analysis of thermal conductivity in granular materials, *Int Chem Eng*: 122–144, 1967.
Lee DK: An apparent kinetic model for the carbonation of calcium oxide by carbon dioxide, *Chem Eng J* 100:71–77, 2004.
Li Z, Cai N: Modeling of multiple cycles for sorption-enhanced steam methane reforming and sorbent regeneration in fixed bed reactor, *Energy Fuel* 21:2909–2918, 2007.
Li Z, Cai N, Yang J: Continuous production of hydrogen from sorption-enhanced steam methane reforming in two parallel fixed-bed reactors operated in a cyclic manner, *Ind Eng Chem Res* 45:8788–8793, 2006.
Lopez Ortiz A, Harrison DP: Hydrogen production using sorption-enhanced reaction, *Ind Eng Chem Res* 40:5102–5109, 2001.
Lyngfelt A, Leckner B, Mattisson T: A fluidized-bed combustion process with inherent CO2 separation; application of chemical-looping combustion, *Chem Eng Sci* 56:3101–3113, 2001.
Lyon RK, Cole JA: Unmixed combustion: an alternative to fire, *Combust Flame* 121:249–261, 2000.
Lysikov AI, Trukhan SN, Okunev AG: Sorption enhanced hydrocarbons reforming for fuel cell powered generators, *Int J Hydrogen Energy* 33:3061–3066, 2008.
Manovic V, Anthony EJ: Steam reactivation of spent CaO-based sorbent for multiple CO_2 capture cycles, *Environ Sci Technol* 41(4):1420–1425, 2007.
Manovic V, Anthony E: CO_2 carrying behavior of calcium aluminate pellets under high-temperature/high-CO2 concentration calcination conditions, *Ind Eng Chem Res* 49:6916–6922, 2010.

Manovic V, Anthony EJ: Integration of calcium and chemical looping combustion using composite CaO/CuO-based materials, *Environ Sci Technol* 45:10750–10756, 2011a.

Manovic V, Anthony EJ: CaO-based pellets with oxygen carriers and catalysts, *Energy Fuel* 25:4846–4853, 2011b.

Manovic V, Wu Y, He I, Anthony EJ: Core-in-Shell CaO/CuO-based composite for CO_2 capture, *Ind Eng Chem Res* 50:12384–12391, 2011.

Martavaltzi CS, Lemonidou AA: Hydrogen production via sorption enhanced reforming of methane: development of a novel hybrid material-reforming catalyst and CO2 sorbent, 65:4134–4140, 2010.

Martavaltzi C, Pefkos TD, Lemonidou AA: Operational window of sorption enhanced steam reforming of methane over CaO-Ca12Al14O33, *Ind Eng Chem Res* 50:539–545, 2011.

Martínez I, Grasa G, Murillo R, Arias B, Abanades JC: Kinetics of calcination of partially carbonated particles in a Ca-looping system for CO_2 capture, *Energy Fuel* 26:1432–1440, 2012.

Martínez I, Romano MC, Chiesa P, Grasa G, Murillo R: Hydrogen production through sorption enhanced steam reforming of natural gas: thermodynamic plant assessment, *Int J Hydrogen Energy* 38:15180–15199, 2013a.

Martínez I, Murillo R, Grasa G, Fernández JR, Abanades JC: Design of a hydrogen production process for power generation based on a Ca-Cu chemical loop, *Energy Procedia* 37:626–634, 2013b.

Martínez I, Murillo R, Grasa G, Fernández JR, Abanades JC: Integrated combined cycle from natural gas with CO_2 capture using a Ca-Cu chemical loop, *AIChE J* 59:2780–2794, 2013c.

Martinez I, Romano MC, Fernandez JR, Chiesa P, Murillo R, Abanades JC: Process design of a hydrogen production plant from natural gas with CO2 capture based on a novel Ca/Cu chemical loop, *Appl Energy* 114:192–208, 2014.

Martínez I, Armaroli D, Gazzani M, Romano MC: Integration of the Ca-Cu process in ammonia production plants, *Ind Eng Chem Res* 56:2526–2539, 2017.

Martini M, van den Berg A, Gallucci F, van Sint Annaland M: Investigation of the process operability windows for Ca–Cu looping for hydrogen production with CO_2 capture, *Chem Eng J* 303:73–88, 2016.

Meyer J, Mastin J, Bjornebole TK, Ryberg T, Eldrup N: Techno-economical study of the zero emission gas power concept, *Energy Procedia* 4:1949–1956, 2011.

Meyer J, Mastin J, Pinilla CS: Sustainable hydrogen production from biogas using sorption-enhanced reforming, *Energy Procedia* 63:6800–6814, 2014.

Nair BN, Yamaguchi T, Kawamura H, Nakao S: Processing of lithium zirconate for applications in carbon dioxide separation: structure and properties of the powders, *J Am Ceram Soc* 87:68–74, 2004.

Najmi B, Bolland O, Colombo KE: A systematic approach to the modeling and simulation of a sorption enhanced water gas shift (SEWGS) process for CO_2 capture, *Sep Purif Technol* 157:80–92, 2016.

Noorman S, van Sint Annaland M, Kuipers H: Packed bed reactor technology for chemical-looping combustion, *Ind Eng Chem Res* 46:4212–4220, 2007.

Noorman S, Gallucci F, van Sint Annaland M, Kuipers H: Experimental investigation of a CuO/Al_2O_3 oxygen carrier for chemical-looping combustion, *Ind Eng Chem Res* 49:9720–9728, 2010.

Numaguchi T, Kikuchi K: Intrinsic kinetics and design simulation in a complex reaction network: steam-methane reforming, *Chem Eng Sci* 43:2295–2301, 1988.

Ochoa-Fernandez E, Rusten HK, Jakobsen HA, Rønning M, Holmen A, Chen D: Sorption enhanced hydrogen production by steam methane reforming using Li2ZrO3 as sorbent: sorption kinetics and reactor simulation, *Catal Today* 106:41–46, 2005.

Ochoa-Fernandez E, Haugen G, Zhao T, et al: Process design simulation of H2 production by sorption enhanced steam methane reforming: evaluation of potential CO2 acceptors, *Green Chem* 9:654–662, 2007.

Perreault P, Patience GS: Carbonation and deactivation kinetics of a mixed calcium oxide-copper oxide sorbent/oxygen carrier for post-combustion carbon dioxide capture, *Chem Eng J* 306:726–733, 2016.

Qin C, Yin J, Liu W, An H, Feng B: Behavior of CaO/CuO based composite in a combined calcium and copper chemical looping process, *Ind Eng Chem Res* 51:12274–12281, 2012.

Qin C, Yin J, Luo C, An H, Liu W, Feng B: Enhancing the performance of CaO/CuO based composite for CO_2 capture in a combined Ca–Cu chemical looping process, *Chem Eng J* 28:75–86, 2013.

Qin C, Feng B, Yin J, Ran J, Zhang L, Manovic V: Matching of kinetics of $CaCO_3$ decomposition and CuO reduction with CH4 in Ca-Cu chemical looping, *Chem Eng Sci* 262:665–675, 2015.

Qin C, Manovic V, Ran J, Feng B: Simulation of the calcination of a core-in-shell CuO/$CaCO_3$ particle for Ca-Cu chemical looping, *Fuel* 181:522–530, 2016a.

Qin C, Yin J, Feng B, Ran J, Zhang L, Manovic V: Modelling of the calcination behaviour of a uniformly-distributed CuO/$CaCO_3$ particle in Ca-Cu chemical looping, *Appl Energy* 164:400–410, 2016b.

Rahman RA, Mehrani P, DY L, Anthony EJ, Macchi A: Investigating the use of CaO/CuO sorbents for in situ CO2 capture in a biomass gasifier, *Energy Fuel* 29:3808–3819, 2015.

Reijers HTJ, Valster-Schiermeier SEA, Cobden PD, VandenBrink RW: Hydrotalcite as CO2 sorbent for sorption-enhanced steam reforming of methane, *Ind Eng Chem Res* 45:2522–2530, 2006.

Riboldi L, Bolland O: Pressure swing adsorption for coproduction of power and ultrapure H_2 in an IGCC plant with CO_2 capture, *Int J Hydrogen Energy* 41:10646–10660, 2016.

Ridha FN, Lu D, Macchi A, Hughes RW: Combined calcium looping and chemical looping combustion cycles with CaO-CuO pellets in a fixed bed reactor, *Fuel* 153:202–209, 2015.

Rodríguez N, Alonso M, Grasa G, Abanades JC: Process for capturing CO_2 arising from the calcination of the $CaCO_3$ used in cement manufacture, *Environ Sci Technol* 42:6980–6984, 2008.

Rodríguez N, Alonso M, Abanades JC: Experimental investigation of a circulating fluidized-bed reactor to capture CO_2 with CaO, *AIChE J* 57:1356–1366, 2011.

Rostrup-Nielsen JR: Steam reforming and chemical recuperation, *Catal Today* 145:72–75, 2009.

Rostrup-Nielsen JR, Sehested J, Nørskov JK: Hydrogen and synthesis gas by steam- and CO_2 reforming, *Adv Catal* 47:65–139, 2002. Academic Press.

Rout KR, Jakobsen HA: Reactor performance optimization by the use of a novel combined pellet reflecting both catalyst and adsorbent properties, *Fuel Process Technol* 99:13–34, 2012.

Rout KR, Solsvik J, Nayak A, Jakobsen HA: A numerical study of multicomponent mass diffusion and convection in porous pellets for the sorption-enhanced steam reforming and desorption processes, *Chem Eng Sci* 66:4111–4126, 2011.

Ryden M, Ramos P: H_2 production with CO_2 capture by sorption enhanced chemical-looping reforming using NiO as oxygen carrier and CaO as CO_2 sorbent, *Fuel Process Technol* 96:27–36, 2012.

Shimizu T, Hirama T, Hosoda H, Kitano K, Inagaki M, Tejima K: A twin fluid-bed reactor for removal of CO_2 from combustion processes, *Chem Eng Res Des* 77:62–69, 1999.

Solieman AA, Dijkstra JW, Haije WG, Cobden PD, van den Brink RW: Calcium oxide for CO_2 capture: operational window and efficiency penalty in sorption-enhanced steam methane reforming, *Int J Greenhouse Gas Control* 3:393–400, 2009.

Spallina V, Gallucci F, Romano MC, Chiesa P, Lozza G, van Sint Annaland M: Investigation of heat management for CLC of syngas in packed bed reactors, *Chem Eng J* 225:174–191, 2013.

Spallina V, Chiesa P, Martelli E, et al: Reactor design and operation strategies for a large-scale packed-bed CLC power plant with coal syngas, *Int J Greenhouse Gas Control* 36:34–50, 2015.

Stevens JF, Krishnamurthy B, Atanassova P, Spilker K: Development of 50 kW fuel processor for stationary fuel cell applications, Final Report, DOE/GO/13102-11, 2007.

Strohle J, Junk M, Kremer J, Galloy A, Epple B: Carbonate looping experiments in a 1 MWth pilot plant and model validation, *Fuel* 127:13–22, 2014.

Symonds RT, DY L, Manovic V, Anthony EJ: Pilot-scale study of CO_2 capture by CaO-based sorbents in the presence of steam and SO_2, *Ind Eng Chem Res* 51:7177–7188, 2012.

Tavares MT, Alstrup I, Bernardo CA, Rostrup-Nielsen JR: Carbon formation and CO methanation on silica-supported nickel and nickel-copper catalysts in $CO + H_2$ mixtures, *J Catal* 158:402–410, 1996.

Tegge G: Ullmann's encyclopedia of industrial chemistry. In *Fifth, completely revised edition. Volume B 4: principles of chemical reaction engineering and plant design*, Germany, 1993, VCH Verlagsgesellschaft Weinheim.

van Selow ER, Cobden PD, van den Brink RW, Hufton JR, Wright A: Performance of sorption-enhanced water-gas shift as a pre-combustion CO_2 capture technology, *Energy Procedia* 1:689–696, 2009a.

van Selow ER, Cobden PD, Verbraeken PA, Hufton JR, van den Brink RW: Carbon capture by sorption-enhanced water-gas shift reaction process using Hydrotalcite-based material, *Ind Eng Chem Res* 48:4184–4193, 2009b.

Voldsun M, Jordal K, Anantharaman R: Hydrogen production with CO_2 capture, *Int J Hydrogen Energy* 41:4969–4992, 2016.

Vortmeyer D, Berninger R: Comments on the paper, theoretical prediction of effective heat transfer parameters in packed beds by Anthony Dixon and D.L. Cresswell [AIChE J, 25, 663 (1979)], *AIChE J* 28:508–510, 1982.

Wang Y, Chao Z, Chen D, Jakobsen HA: SE-SMR process performance in CFB reactors: simulation of the CO_2 adsorption/desorption processes with CaO based sorbents, *Int J Greenhouse Gas Control* 5:489–497, 2011.

Williams R. Hydrogen production. US patent 1.938.202, 1933.

Wolf J, Yan J: Parametric study of chemical looping combustion for tri-generation of hydrogen, heat, and electrical power with CO_2 capture, *Int J Energy Res* 29:739–753, 2005.

Wright AD, White V, Hufton JR, Quinn R, Codben PD, van Selow PD: CAESAR: development of a SEWGS model for IGCC, *Energy Procedia* 4:1147–1154, 2011.

Xiu G, Li P, Rodriguez AE: Adsorption-enhanced steam-methane reforming with intraparticle-diffusion limitations, *Chem Eng J* 95:83–93, 2003.

Xu J, Froment GF: Methane steam reforming, methanation and water-gas shift: I. Intrinsic kinetics, *AIChe J* 35:88–96, 1989.

FURTHER READING

Diego ME, Arias B, Abanades JC: Analysis of a double calcium loop process configuration for CO_2 capture in cement plants, *J Clean Prod* 117:110–121, 2016a.

Kierzkowska AM, Müller CR: Sol–gel-derived, calcium-based, copper-functionalised CO2 sorbents for an integrated chemical looping combustion-calcium looping CO2 capture process, *ChemPlusChem* 78:92–100, 2013.

CHAPTER FOUR

Sorption-Enhanced Reaction With Simulated Moving Bed Reactor and PermSMBR Technologies

Dânia S.M. Constantino, Rui P.V. Faria, Alírio E. Rodrigues[1]
Laboratory of Separation and Reaction Engineering—Laboratory of Catalysis and Materials (LSRE-LCM), Faculdade de Engenharia, Universidade do Porto, Porto, Portugal
[1]Corresponding author: e-mail address: arodrig@fe.up.pt

Contents

Abstract

Two of the most recent and promising multifunctional reactors, the Simulated Moving Bed Reactor and the Simulated Moving Bed Membrane Reactor, for the synthesis of oxygenated compounds, were addressed focusing on the methodology for the development and implementation of these technologies based on experimental data and suitable mathematical models. Moreover, different case studies, involving acetals and esters, using the different reactor configurations were illustrated. The two types of reactors were compared under the same operating conditions and, after that, individually optimized in order to find the optimal operating conditions for which the maximum performance is achieved.

The Simulated Moving Bed Reactor proved to be a very competitive process for the synthesis that involves equilibrium-limited reactions. This technology, which combines reaction with chromatographic separation, revealed a strong ability to increase the equilibrium conversion with high productivities and reasonable desorbent consumptions by the continuous removal of water through selective adsorption. However, diluted products are obtained and subsequent units for the products purification are required.

Advances in Chemical Engineering, Volume 51
ISSN 0065-2377
https://doi.org/10.1016/bs.ache.2017.07.005

Combining the Simulated Moving Bed Reactor with water-selective membranes, a new technology was generated, the Simulated Moving Bed Membrane Reactor also designated by PermSMBR. This recent concept that comprises reaction and two different separation techniques exhibited excellent performance for acetals and esters synthesis, even improving the previous reactor productivities with significant reduction on the desorbent demand. Furthermore, its dimensions can be reduced without affecting the productivity, leading to higher savings on the desorbent consumption and lower associated desorbent recovery costs.

1. INTRODUCTION

Process Intensification (PI) has deserved special attention in chemical engineering research in the last decades due to the market challenges. Generally, this subject represents substantial impact on the associated operating and production costs at industrial scale, mainly in the case of the products that are obtained from limited reactions, either by thermodynamic or kinetic issues. A very large number of products of industrial interest like pharmaceutics, solvents, flavors, intermediates, and others involve, generally, reactions strongly limited by equilibrium. The most challenging step is then to study these systems in order to make them capable of answering to the markets demand. Although PI can be achieved by different ways, the focus is typically to improve the performance of the synthesis process by increasing the productivity and purity with the minimal associated costs, making it more sustainable in relation to conventional process. For that, reaction and separation steps have been integrated in multifunctional reactors (single equipment), where the two processes occur simultaneously, commonly designated as reactive separation, presenting substantial advantages in relation to the conventional processes. This way, the equilibrium-limited reaction conversion is overtaken by removing at least one of the products continuously, leading to more energy-efficient processes. Continuous chromatographic reactors like Simulated Moving Bed Reactors (SMBRs) offer many advantages from the PI point of view, particularly for oxygenated compounds production that involves equilibrium-limited reactions, since combining reaction and separation processes leads to higher yields, less desorbent consumptions, and lower capital investments (Rodrigues et al., 2012; Stankiewicz, 2003) compared with the traditional processes (reactor + separator). Moreover, this kind of technology allows to operate at low temperatures leading to a reduction of the energy required, as mentioned previously, and taking advantage over other reactive separation processes such as Reactive Distillation (RD) (Silva et al., 2011). Several studies have been published concerning PI of oxygenated compounds and green solvents synthesis by using SMBR technology successfully (Constantino et al., 2015b; Graça et al., 2011b; Gyani et al., 2014; Mazzotti et al., 1996; Pereira et al., 2009; Silva and Rodrigues, 2005). The optimal design and operating parameters must be studied in detail for each system, in order to achieve the best SMBR performance, since it depends on several factors like the reaction equilibrium and kinetics, the adsorption capacity and selectivity of the stationary phase as well. However, even operating at optimal conditions, the SMBR presents some limitations regarding the product dilution with the desorbent used to regenerate the solid for the continuous operation. This way, the product is usually purified in a complementary separation unit allowing the desorbent recovery for a more sustainable process. Furthermore, the presence of water, generally obtained as by-product in the synthesis of oxygenated compounds, limits the reaction conversion and consequently leads to a higher desorbent consumption for solid regeneration since this species is commonly the most retained compound in the stationary phases typically applied in these processes.

Meanwhile, a new SMBR-based hybrid technology emerged to minimize the SMBR drawbacks, the Simulated Moving Bed Membrane Reactor (PermSMBR). This technology is considered a hybrid reactive process that embraces the process reintensification paradigm (Silva et al., 2011) because it results from the integration of an SMBR with membranes. Thus, the reactor combines two different separation processes, adsorption (chromatography) and pervaporation (membranes), in a single equipment, taking advantage over other intensified processes. This multifunctional reactor is able to reach conversions of 100% with low desorbent consumptions due to the displacement of the equilibrium by the continuous removal of the products: one through the raffinate port and the other through the extract port being one of them additionally eliminated by pervaporation through selective membranes. PermSMBR technology is especially important for applications that involve the synthesis of temperature-sensitive compounds or products with small volatility differences since its principle of separation is based on the differences of adsorptivity and on the membranes permeance. The feasibility of PermSMBR for the synthesis of different compounds like acetals and esters (Graça and Rodrigues, 2017; Pereira et al., 2012; Silva et al., 2011) has been accessed where excellent performance with high productivities and low desorbent consumptions have been obtained, proving that it is a potential technology for the sustainable production of oxygenated compounds.

In this chapter, different case studies of sorption-enhanced reactions, involving synthesis of acetals and esters, with SMBR and PermSMBR will be addressed. The performances of two different technologies for each system will be presented and compared as well as the respective advantages. The mathematical models developed to predict the behavior of these multifunctional continuous reactors will be also described, and the respective design and operating parameters to achieve the optimal performance for each system will be presented.

2. SMBR TECHNOLOGY

Simulated Moving Bed (SMB) is one of the most relevant technologies for the separation of either complex molecules with low volatilities like important biomolecules studied in pharmaceutical industry or temperature-sensitive compounds with high risk of polymerization. It consists in several chromatographic columns operating in series continuously where a stationary and a mobile phase are in contact permanently. The columns are packed with a solid material (stationary phase) with strong adsorption properties, being highly selective for one of the compounds involved in the multicomponent mixture which is carried by a desorbent/eluent (mobile phase) through the stationary phase. The contact between the solid and the liquid phases is performed by simulating the solid countercurrent movement by shifting the inlets (feed and eluent) and the outlets (extract and raffinate) streams periodically at regular time intervals, designated by switching time (t^*), in the fluid flow direction. Indeed, the countercurrent movement maximizes the mass transfer driving force, reducing the costs associated to the eluent and the adsorbent consumptions (Nicoud et al., 1993). Furthermore, the fact that the solid is not actually moving in the SMB avoids the drawbacks associated with the solid motion expected in the True Moving Bed technology (TMB), such as equipment abrasion, mechanical erosion of the adsorbent, and difficulty to keep the plug flow for the solid (Rodrigues et al., 2015a). The separation principle of SMB is based on the differences of affinity of each compound with the solid phase (the adsorbent), leading to different propagation velocities. Hence, the compound for which the stationary phase has more affinity will be more retained, allowing to collect the other compound earlier together with the eluent in the raffinate outlet.

Meanwhile, the concept of the SMB has also been successfully applied to combine reaction with chromatographic separation (Kawase et al., 1996; Mazzotti et al., 1996; Rodrigues and Silva, 2009). Integrating the reaction with this kind of separation technology generated a new technology, the SMBR, which provides advantages regarding the operating costs associated to the conventional reactor–separator sequences. Similar to the SMB a stationary and a mobile phase are also present in SMBR; however, the stationary phase will work as adsorbent and catalyst simultaneously. This way, the SMBR technology is able to increase the equilibrium-limited reaction conversions by the continuous removal of one product, at least, from the reaction media by selective adsorption leading to a cleaner, safer, smaller, and faster process with higher yields and purities of final products. For more efficient processes the eluent recycle is mandatory. Therefore besides the adsorbent, the eluent should be carefully selected in order to regenerate the respective adsorbent for continuous operation, leading to achieve high-purity products. Usually, one of the reactants is used as eluent in most of the SMBR applications in order to favor the reaction rate and, consequently, improve its performance. However, this scenario is not always possible like in systems in which both reactants present peculiar characteristics, such as high viscosities, among others. In these cases, a methodology based on reaction, adsorption, and environmental variables should be adopted to find the most suitable eluent for each system. For that, a screening procedure should be followed taking into account mainly three process variables: the eluent reactivity, the eluent physical state at the reaction conditions, and a hypothetical phase split promoted by the use of the respective eluent, as described by Faria et al. (2013). The solid material should be selected based on its properties through rigorous screening. Although a mixture of materials with adsorptive and catalytic properties, respectively, can be used as stationary phase in SMBR processes, a hybrid material with both adsorptive and catalytic properties is used in most of the SMBR applications, simplifying the implementation. For that, kinetic and adsorption studies are required. The first study might be carried out in batch conditions comparing the respective reaction rates and selecting the most active material. The second one can be performed by using simple elution (frontal) chromatography with the eluent previously selected in a single fixed-bed column packed, preferably, with the most active catalyst observed in the first screening. However, first of all non-reactive binary mixtures should be studied in order to acquire information about adsorption equilibrium by finding the affinity of the adsorbent for each compound in the absence of reaction. After that, adsorption and kinetic data will allow the prediction of the dynamic behavior of the fixed-bed adsorptive reactor (FBAR) packed with the same catalyst/adsorbent when a reactive mixture is fed to this unit by complex mathematical models based on the information previously gathered. A set of different experimental conditions, as temperature and feed ratios for instance, must be considered in both adsorption and catalytic studies in order to provide a broader perspective of the final SMBR process performance, where adsorption and reaction will occur simultaneously in set of FBAR columns that constitutes it.

Regarding SMBR operation, basically, it comprises four streams: two inlet streams (feed and desorbent) and two outlet streams (extract and raffinate), according to Fig. 1. Considering a reaction $A+B \leftrightarrow C+D$, for instance, the extract stream is composed mainly by the eluent (A) and the most retained compound (D), while the raffinate is composed mostly by the less retained compound (C) with the eluent. The streams are distributed by four operating sections. In section 1 (between eluent and extract nodes), the adsorbent is regenerated by desorption of the most retained product from the solid, while in section 2 (between extract and feed nodes) and in section 3 (between feed and raffinate nodes) the reaction occurs, together with the separation of the products. The most adsorbed product in sections 2 and 3 is transported with the solid phase to the extract port. The less adsorbed product in sections 2 and 3 is carried with the liquid in the raffinate port direction. In section 4 (between raffinate and eluent nodes), the less retained product is adsorbed, allowing the regeneration of the eluent before being recycled to the next section (1) (Mazzotti et al., 1997; Rodrigues et al., 2015b). A cycle is finished when the number of switches equals the total number of columns. The cyclic steady state is achieved when no significant differences are

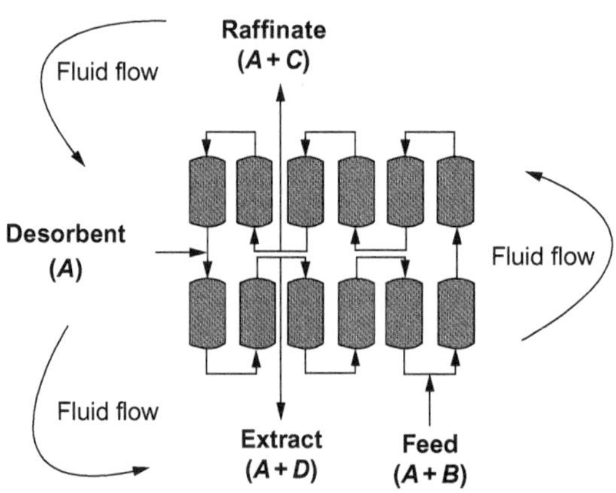

Fig. 1 Schematic representation of an SMBR with configuration 2-4-4-2.

observed in the compounds concentration compared with the profiles and histories from one cycle to the other (deviations less or equal to 1%, typically).

2.1 Modeling an SMBR Process

The proof of concept is fundamental to ensure the reliability of any process. Nevertheless, besides the experimental investigations of several systems using SMBR technology numerical simulations are required to proper process development. Two different strategies can be considered for modeling of SMBR processes: (i) one by modeling a TMBR, making use of the analogy between a TMBR and an SMBR unit and considering an infinite number of columns; and (ii) another by modeling the real SMBR, considering the periodic switch of the inlet and outlet streams. Although the first methodology leads to faster results, a considerable number of columns per section have to be considered to get satisfactory results, leading most of the authors to take into account the second one. This way, the mathematical model here described to predict the performance of SMBR processes just considers the second strategy. It will be able to assess the variations in concentration profiles all over the unit as well as the extract and raffinate streams compositions within a set period assuming constant bed volume and packing porosity, plug flow with axial dispersion, linear driving force (LDF) approximation for the inter- and intraparticle mass transfer rates, velocity variations due to changes in the bulk compositions, and isothermal operation. Hence, the dynamic behavior of the SMBR is described by the following equations:

Bulk fluid mass balance to component i in column k:

$$\frac{\partial C_{ik}}{\partial t} + \frac{\partial (u_k C_{ik})}{\partial z} + \frac{(1-\epsilon)}{\epsilon}\frac{3}{r_p} K_{L,ik}\left(C_{ik} - \overline{C}_{p,ik}\right) = D_{ax,k}\frac{\partial}{\partial z}\left(C_T \frac{\partial x_{ik}}{\partial z}\right) \quad (1)$$

where C_{ik} and $\overline{C}_{p,ik}$ correspond to the liquid-phase concentration and the average of liquid-phase concentration inside the particle of the compound i in the column k, respectively. ϵ represents the bulk porosity, u_k is the interstitial velocity, $D_{ax,k}$ is the axial dispersion coefficient, r_p is the particle radius, and $K_{L,ik}$ corresponds to the global mass transfer coefficient of the compound i in the column k.

Particle mass balance to component i in column k:

$$\frac{3}{r_p} K_{L,ik}\left(C_{ik} - \overline{C}_{p,ik}\right) = \epsilon_p \frac{\partial \overline{C}_{p,ik}}{\partial t} + \left(1-\epsilon_p\right)\frac{\partial \overline{q}_{ik}}{\partial t} - \frac{\nu_i \rho_b}{1-\epsilon} r\left(\overline{C}_{p,ik}\right) \quad (2)$$

in which $\overline{q}_{ik}$ represents the average solid-phase concentration, in equilibrium with $\overline{C}_{p,ik}$, of compound i in the column k, ϵ_p is the catalyst/adsorbent particle porosity, ν_i, ρ_b, and r are the stoichiometric coefficient, bulk density, and the reaction rate, respectively.

From the total mass balance it is possible to determine the interstitial fluid velocity variation, which is given by:

$$\frac{du_k}{dz} = -\frac{(1-\epsilon)}{\epsilon}\frac{3}{r_p}\sum_{i=1}^{NC} K_{L,ik} V_{mol,i}\left(C_{ik} - \overline{C}_{p,ik}\right) \tag{3}$$

where $V_{mol,i}$ is the molar volume of the compound i.

A global mass transfer coefficient is considered combining external and internal mass transfer coefficients, k_e and k_i, respectively, according to the resistances-in-series model given by the following equation:

$$\frac{1}{K_L} = \frac{1}{k_e} + \frac{1}{\varepsilon_p k_i} \tag{4}$$

Although the internal and external mass transfer coefficients can be obtained experimentally, there are correlations to get it. For instance, the internal mass transfer coefficient can be estimated by Glueckauf (1955) (Eq. 5), while the external mass transfer coefficient can be estimated by the Wilson and Geankoplis correlation (Ruthven, 1984), expressed by Eq. (6).

$$k_i = \frac{5D_m/\tau}{r_p} \tag{5}$$

where D_m represents the molecular diffusivity.

$$Sh_p = \frac{1.09}{\varepsilon}\left(Re_p Sc\right)^{0.33} \quad 0.0015 < Re_p < 55 \tag{6}$$

where $Sh_p = k_e d_p / D_m$ (Sherwood number), $Re_p = \rho d_p u/\eta$ (Reynolds number), and $Sc = \eta/(\rho D_m)$ (Schmidt number).

The infinite dilution diffusivities are estimated by the Scheibel correlation (Scheibel, 1954):

$$D^0_{A,B} = \frac{8.2\times 10^{-8}\,T}{\eta_B V_{M,A}^{1/3}}\left[1 + \left(\frac{3V_{M,B}}{V_{M,A}}\right)^{2/3}\right] \tag{7}$$

where $D^0_{A,B}$ is the diffusion coefficient for a dilute solute A into a solvent B and η_B is the viscosity of pure eluent B.

Vignes equation (Vignes, 1966), based on coefficients at infinite dilution, was used to predict the diffusion coefficient in concentrated solutions for binary systems:

$$D_{B,A} = D_{A,B} = \left(D^0_{A,B}\right)^{x_2}\left(D^0_{B,A}\right)^{x_1} \tag{8}$$

The diffusion coefficient for multicomponent concentrated solutions was determined by the Perkins and Geankoplis mixing rule (Perkins and Geankoplis, 1969):

$$D_{A,m}\eta_m^{0.8} = \sum_{\substack{i=1 \\ i \neq A}}^{n} x_i D^0_{Aj}\eta_i^{0.8} \tag{9}$$

where η_m is the viscosity of the mixture and η_i is the viscosity of the component i.

Initial and Danckwerts boundary conditions:

$$t = 0: \quad C_{ik} = \overline{C}_{p,ik} = C_{ik,0} \tag{10}$$

$$z = 0: \quad u_j C_{ik} - D_{axj} C_T \left.\frac{\partial x_{ik}}{\partial z}\right|_{z=0} = u_j C_{ik,F} \tag{11}$$

$$z = L: \quad \left.\frac{\partial C_{ik}}{\partial z}\right|_{z=L} = 0 \tag{12}$$

where the indices F and 0 refer to the feed and initial states, respectively.

Mass balances at the nodes of the inlet and outlet lines of the SMBR:

Desorbent node:

$$C_{i(j=4,\, z=L)} = \frac{u_1}{u_4} C_{i(j=1,\, z=0)} - \frac{u_D}{u_4} C_i^D \tag{13}$$

Extract (j = 2) and raffinate (j = 4) nodes:

$$C_{i(j-1,\, z=L)} = C_{i(j,\, z=0)} \tag{14}$$

Feed node:

$$C_{i(2,\, z=L)} = \frac{u_3}{u_2} C_{i(3,\, z=0)} - \frac{u_F}{u_2} C_i^F \tag{15}$$

where:

$$u_1 = u_4 + u_D \quad \text{(Desorbent node);} \tag{16}$$

$$u_2 = u_1 - u_X \quad \text{(Extract node);} \tag{17}$$

$$u_3 = u_2 + u_F \quad \text{(Feed node);} \tag{18}$$

$$u_4 = u_3 - u_{Raff} \quad \text{(Raffinate node).} \tag{19}$$

The ratio between the fluid interstitial velocity, u_j, and the simulated solid velocity can be defined for each section using the following parameter:

$$\gamma_j = \frac{u_j}{U_s} \tag{20}$$

The relation between the solid interstitial velocity, U_s, and the switching time, t^*, in SMB operation is provided in Eq. (21) where L is the column length.

$$U_s = L/t^* \tag{21}$$

The kinetic parameters and equilibrium constant as well as the multicomponent adsorption parameters are provided from the adsorption and catalytic preliminary studies required for the selection of the hybrid material to use as stationary phase in the SMBR unit, as mentioned before. Afterward, all conditions are gathered to carry out simulations of SMBR performance at specific operating conditions.

The SMBR process performance can be evaluated over a complete cycle according to the following equations and considering, once again as example, a reaction $A + B \leftrightarrow C + D$:

$$PR(\%) = 100 \int_t^{t+N_c t^*} C_{R,C}dt \Bigg/ \left(\int_t^{t+N_c t^*} (C_{R,B} + C_{R,C} + C_{R,D})dt \right) \tag{22}$$

$$PX(\%) = 100 \int_t^{t+N_c t^*} C_{X,D}dt \Bigg/ \left(\int_t^{t+N_c t^*} (C_{X,B} + C_{X,C} + C_{X,D})dt \right) \tag{23}$$

$$X(\%) = 100 \left[1 - \left(Q_X \int_t^{t+N_c t^*} C_{X,B}dt + Q_R \int_t^{t+N_c t^*} C_{R,B}dt \right) \Big/ (Q_F C_{F,B} N_c t^*) \right] \tag{24}$$

$$Prod(\text{kg}_C/\text{L}_{ads}\,\text{day}) = Q_R \int_t^{t+N_c t^*} C_{R,C}dt/((1-\varepsilon)N_c V_{unit} t^*) \tag{25}$$

$$DesC(\mathrm{L}_A/\mathrm{kg}_C) = N_c t^*(Q_D C_{D,A} + Q_F(C_{F,A} - \upsilon_A X C_{F,B})) / \left(Q_R \int_t^{t+N_c t^*} C_{R,C} dt \right) \quad (26)$$

where *PR* and *PX* are the purity of the raffinate and extract streams, respectively. N_c represents the total number of columns and t corresponds to the operating time. X is the conversion of component B and the subscripts F, D, X, and R represent the following streams: feed, desorbent, extract, and raffinate, respectively. The productivity, *Prod*, is defined considering the component C as the target product which is collected in the raffinate stream. The desorbent used (component A) in the reaction is not taken into account to calculate the desorbent consumption parameter (*DesC*), being just considered the amount of the component A used as desorbent/eluent.

2.2 Design and Methodology

The optimization of SMBR-based processes is achieved by studying the effect of the design and operating variables on the performance. Thereby, relevant design variables like the sectional flow rates, fixed-bed columns arrangement by section (configuration), and the switching time ought to be assigned. For that, an approach followed for SMB separations is performed in order to estimate a suitable set of SMBR operating flow rates that would potentially lead to complete separation of the two products of the reaction. This approach makes use of the concept of the TMB, considering the equilibrium theory. The following constraints (Eqs. 27–29) should be defined and fulfilled for the ratio between the fluid and the solid interstitial velocities, γ_j, for each section, j, of the TMB:

$$\gamma_1 > \frac{(1-\varepsilon)}{\varepsilon}\left(\varepsilon_p + (1-\varepsilon_p)\frac{q_{D_1}}{\overline{C}_{p,D_1}}\right) \quad (27)$$

$$\frac{(1-\varepsilon)}{\varepsilon}\left(\varepsilon_p + (1-\varepsilon_p)\frac{q_{C_2}}{\overline{C}_{p,C_2}}\right) < \gamma_2 < \gamma_3 < \frac{(1-\varepsilon)}{\varepsilon}\left(\varepsilon_p + (1-\varepsilon_p)\frac{q_{D_3}}{\overline{C}_{p,D_3}}\right) \quad (28)$$

$$\gamma_4 < \frac{(1-\varepsilon)}{\varepsilon}\left(\varepsilon_p + (1-\varepsilon_p)\frac{q_{C_4}}{\overline{C}_{p,C_4}}\right) \quad (29)$$

The guaranty that the compound (D) is preferentially carried by the liquid phase in section 1 and by the solid phase in the other sections, while the compound (C) is preferentially carried by the solid phase in section 4 and by the liquid phase in the other sections given by the constraints above described, Eqs. (27) and (29), simultaneously. Both sections 1 and 4 have a significant role in SMBR operation. In section 1, the solid phase is regenerated by removing the most adsorbed compound (*D*), while in section 4 the desorbent is cleaned by the adsorption of compound (*C*) and the desorbent is returned to the system through the recycle stream. Thereby, the limiting flow rates in sections 1 and 4 are determined by calculating the ratios between the liquid and solid interstitial velocities in those sections, according to the equilibrium theory, in order not to have contaminations. The equivalence between SMBR and TMBR systems is made by keeping constant the relative liquid velocities:

$$u_j^{SMBR} = u_j^{TMBR} + U_s \tag{30}$$

The ratio between the liquid and solid interstitial velocities in each section, γ_j, is given by Eq. (20). From Eqs. (20) and (30) it is possible to obtain the following equation (31) which allows to determine the limiting ratio values for the critical sections (1 and 4) during an SMB/SMBR process and to find the ideal operating flow rates from that.

$$\gamma_j^{SMBR} = \gamma_j^{TMBR} + 1 \tag{31}$$

The solid velocity is determined according to Eq. (21) by setting the switching time. However, the effect of the switching time on the performance parameters must be studied in order to know what is the value for the optimal operating conditions leading to minimal desorbent consumption (*DesC*) and maximum productivity (*Prod*), ensuring significant conversion (usually set $X \geq 95\%$) with the required raffinate purity (*PR*) for each case.

The successful design and operation of countercurrent separation units depends upon the correct selection of the operating conditions, particularly of the flow rates in each section (Storti et al., 1993). For that, separation (SR) and reactive separation regions (RSRs) are usually obtained by numerical solution according to the mathematical model above described. The main difference is that SR is obtained by simulating an SMB unit to separate the products, *C* and *D*, in absence of reaction, while the RSR is attained by considering an SMBR unit where a specific ratio of the reactants, *A* and *B*, is fed, taking into account the reaction kinetics and the multicomponent adsorption data. The procedure used to define the SR/RSR consists of keeping constant simultaneously the values of the switching time and of the flow rates in sections 1 and 4 for the same configuration of the columns per section. It means that keeping the flow rates ratio in sections 1 and 4 constant (γ_1 and γ_4), obtained from the equilibrium theory, leads to constant desorbent and recycle flow rates. Hence, starting from a low feed flow rate, the SMBR operation is simulated for different extract flow rates determining the range of extract flow rates for which the separation required is successfully accomplished (setting γ_2 and γ_3, according to Fig. 2). Thus, the procedure is repeated increasing the feed flow rate until the maximum value of feed flow rate that ensures the required separation and reaction conversion; usually, this point corresponds to the maximum process performance: maximum productivity and minimum desorbent consumption, since above that maximum feed flow rate value the requirements cannot be fulfilled. At this stage, *PX* is not an important parameter since water is a secondary product. However, in a more detailed study for industrial application it will be an important factor in order to reduce the equipment size and the energy costs for the treatment of the respective extract stream.

Furthermore, it is important to refer that before this kind of study, the miscibility ranges of each system should be studied at the desired operating temperature, through thermodynamic models and the databases in order to

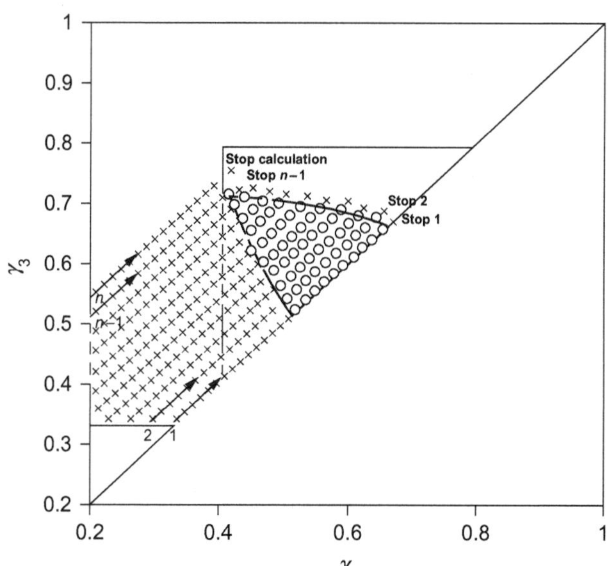

Fig. 2 Procedure to determine a reactive/separation region. *Reprinted from Minceva M, Rodrigues AE: Simulated moving-bed reactor: reactive–separation regions.* AIChE J *51(10): 2737–2751, 2005 by permission of John Wiley & Sons.*

avoid problems in the chromatographic columns when the experimental operation will be performed.

2.3 SMBR for Sorption-Enhanced Reactions

The greatest advantage of SMBR is to increase the equilibrium-limited reaction conversions combining reaction and continuous separation process by selective adsorption of the products without drawbacks associated to the solid movement since it is simulated by shifting the inlets/outlets streams a distance column in the fluid direction by periodic intervals (switching times). This concept leads to higher conversions being more feasible in terms of energy and productivity than the conventional processes where the separation is carried out after the reactor unit. The first applications of SMBR were for gas-phase oxidative coupling of methane (Tonkovich and Carr, 1994) and the hydrogenation of 1,3,5-trimethylbenzene (Ray et al., 1994). However, in the last years, many applications for the liquid phase involving systems with equilibrium-limited reactions have arisen such as: production of bisphenol-A from acetone and phenol (Kawase et al., 1999), isomerization and separation of *p*-xylene (Minceva et al., 2008), esterification of acetic acid with methanol and b-phenethyl alcohol (Lode et al., 2001), and the synthesis of MTBE (Zhang et al., 2001). Other interesting applications are the production of acetals like 1,1-diethylacetal (Silva and Rodrigues, 2005), 1,1-dimethylacetal (Pereira et al., 2007), and 1,1-dibutoxyethane (Graça et al., 2011b); and important esters like ethyl lactate (Pereira et al., 2009) and butyl acrylate (Constantino et al., 2015b).

This document will focus on the systems that involve the synthesis of acetals and esters as case studies, analyzing each system regarding the main steps required for the detailed study of SMBR. For all systems, a common procedure is followed to attain the best performance of the SMBR, which is summarized in Fig. 3.

2.3.1 *Synthesis of Acetals*

Acetals have been considered as oxygenated additives to diesel fuel since the reduction of particles seems to be linearly dependent on the oxygen content (Nord and Haupt, 2005). Its interesting properties lead to the ability to reduce significantly the emission of particles and NO_x without decreasing the ignition quality. Furthermore, some of acetals like 1,1-diethoxyethane have been used in formulation of perfumes and in the flavoring of drinks. In agriculture and pharmaceutical field, this kind of compounds have been used also as intermediate for the synthesis of

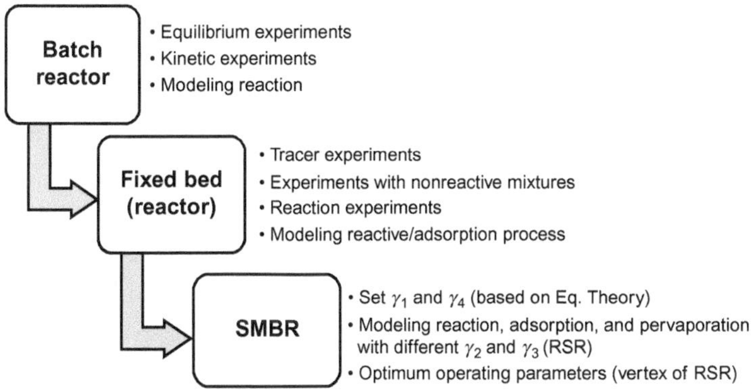

Fig. 3 Summary scheme of the procedure required to study an SMBR process.

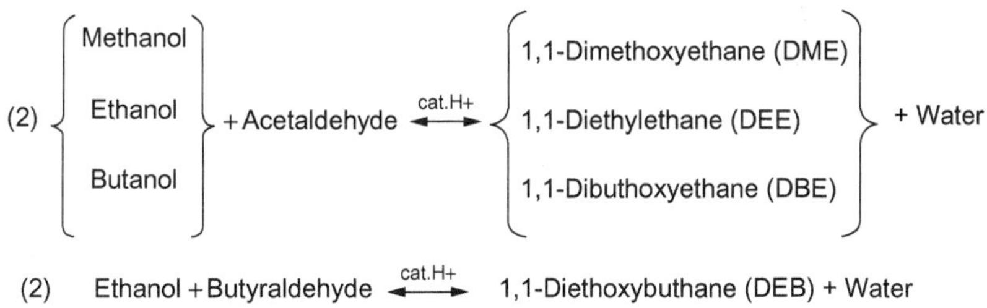

Fig. 4 Overall reaction for synthesis of different acetals.

several industrial chemicals (Boennhoff and Obenaus, 1980; Boennhoff and Sausen, 1983; Laborde, 2003).

The synthesis of acetals is usually performed by reacting two molecules of an alcohol and one molecule of an aldehyde, according to Fig. 4, catalyzed by a strong acid medium. Generally, acid heterogeneous catalysts like ion exchange resins are used in SMBR for acetalization reactions because this kind of materials are selective adsorbents for water, helping removing it from the reaction medium as intended for the extension of the reaction conversion.

In the present section some case studies of acetals production by SMBR will be presented and discussed. All simulation results were acquired considering the Licosep 12-26 SMB pilot-scale unit (by Novasep, France) which is shown in Fig. 5, and it is available at Laboratory of Separation and Reaction Engineering (LSRE) of the Faculty of Engineering of University of Porto (Portugal). The columns installed in the experimental setup can withstand up to 60°C and 60 bar, and its principal characteristics are presented in Table 1.

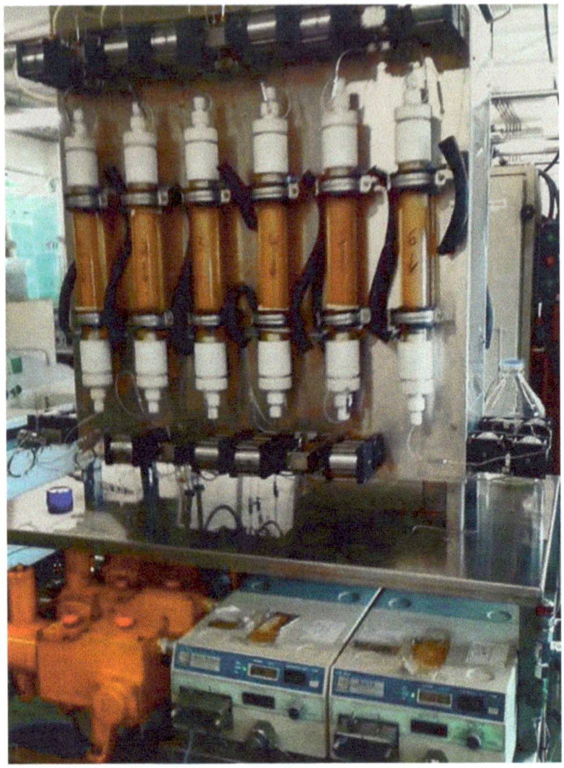

Fig. 5 Licosep 12-26 SMB pilot-scale unit.

Table 1 Characteristics of the Pilot-Scale SMBR Columns Packed With Amberlyst-15

Length of each column (L)	23 cm
Internal diameter of each column (D_i)	2.6 cm
Solid weight (A15)	47.6 g
Bulk density (ρ_b)	390.0 kg/m^3
Particle radius (r_p)	375 μm
Particle porosity (ϵ_p)	0.4
Peclet number	300

2.3.1.1 Case Study (S.1): Synthesis of 1,1-Diethoxyethane or 1,1-Diethylacetal (1,1-DEE)

After the preliminary studies of reaction kinetics, equilibrium, and multicomponent adsorption isotherms, the data required for the detailed study

Table 2 Kinetic Results From Batch Reactor for 1,1-DEE Synthesis (Silva and Rodrigues, 2001)

Parameter	Experimental Conditions	Result
Equilibrium constant	20–60°C; 10 bar	$K_{eq} = \exp\left(\frac{1270}{T(\mathrm{K})} - 3.07\right)$
Reaction rate	16–26°C; 6 bar	$r(\mathrm{mol/g/min}) = k_c \frac{a_A a_B - \frac{a_C a_D}{K_{eq} a_A}}{(1 + K_{s,D} a_D)^2}$
Kinetic constant		$k_c(\mathrm{mol/g/min}) = 3.3 \times 10^{11} \exp\left(\frac{-7824}{T(\mathrm{K})}\right)$
Adsorption constant		$K_{s,D} = 6.4 \times 10^5 \exp\left(\frac{-4003}{T(\mathrm{K})}\right)$

Table 3 Multicomponent Adsorption Parameters for Langmuir Isotherms of 1,1-DEE System Over Amberlyst-15 at 15°C

Component	Q_{ads} (mol/L$_{ads}$)	K (L/mol)
Ethanol (A)	14.3	1.43
Acetaldehyde (B)	20.1	1.58
1,1-DEE (C)	11.45	0.09
Water (D)	44.7	2.01

of the SMBR are gathered in Tables 2 and 3, respectively. The adsorption data were obtained by frontal chromatography in a fixed-bed adsorptive columns over ion exchange resin Amberlyst-15 at 15°C for binary nonreactive mixtures and then were used to predict the dynamic behavior of an FBAR for the synthesis of 1,1-DEE through the mathematical model for the adsorptive reactor. More details about this study can be found in the literature (Silva and Rodrigues, 2002, 2005). The adsorption equilibrium of component i is described by the multicomponent Langmuir adsorption equilibrium isotherm:

$$q_i = \frac{Q_i K_i \overline{C}_{p,i}}{1 + \sum_{j=1}^{NC} K_j \overline{C}_{pj}} \tag{32}$$

where Q_i is the monolayer capacity and K_i is the equilibrium constant for component i.

2.3.1.1.1 Reactive Separation Region The numerical procedure used to build the RSR consists of keeping the fluid/solid interstitial velocity ratio constant in sections 1 and 4 determined from the Equilibrium Theory as mentioned in Section 2.2, for a set switching time and progressively increase the feed and extract flow rates, determining the set of operating conditions that allow to obtain the products within the specified requirements. Accordingly the RSR for the synthesis of 1,1-DEE with SMBR process was determined, where the product specifications are attained (raffinate purity and reaction conversion >95%), which is presented in Fig. 6. TMBR (dashed line) and SMBR (full line) models were used being very similar in terms of size and shape; however, slight difference is observed mainly from the border of impure raffinate. This fact is due to the higher liquid velocities in the SMBR unit, so the residence time is smaller in section 3 of that unit

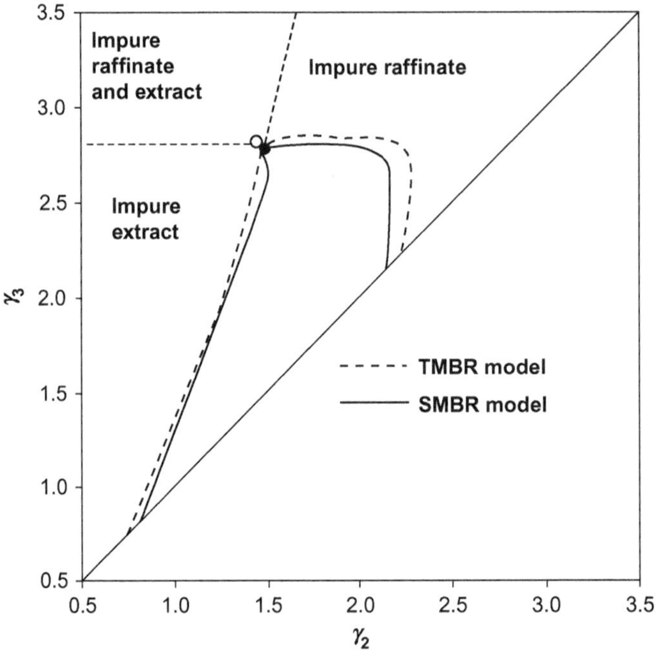

Fig. 6 Reactive separation regions for 1,1-DEE synthesis at 10°C with the TMBR (*solid line*, γ_1=4.30; γ_4=0.515) and the SMBR (*dashed line*, 5.30; 1.52) models using Amberlyst-15 as stationary phase and a configuration 3-3-3-3 (t^* =3.7 min). *Reprinted from Silva VMTM, Rodrigues AE: Novel process for diethylacetal synthesis.* AIChE J *51(10):2752–2768, 2005 by permission of John Wiley & Sons.*

Table 4 Operating Conditions and Performance Parameters for the Reference Case (at the Middle of the Switching Time (3.7 min) of SMBR)

Parameters	Reference Case
Q_F (mL/min)	10
Q_X (mL/min)	35
Q_D (mL/min)	50
Q_{Rec} (mL/min)	20
$C_{Eth,F}$ (mol/L)	8.57
$C_{Acet,F}$ (mol/L)	8.96
X (%)	99.7
PR (%)	98.3
PX (%)	99.7
$Prod$ ($kg_{DEE}/(L_{ads}$ day))	4.02
$DesC$ (L_{Eth}/kg_{DEE})	18.2

Reprinted from Silva VMTM, Rodrigues AE: Novel process for diethylacetal synthesis. *AIChE J* 51(10):2752–2768, 2005 by permission of John Wiley & Sons.

resulting in lower reaction conversion and accordingly the acetaldehyde unreacted will contaminate the raffinate stream. The vertex of each RSR ($\gamma_2=2.78$ for SMBR and $\gamma_3=2.82$ for TMBR) concerns the respective optimal operating conditions to get the maximum performance. Thus, the black point is considered the reference case for the synthesis of 1,1-DEE with SMBR and the respective conditions are shown in Table 4. Above the feed flow rate corresponding to the maximum performance, it is not possible fulfill the purity requirement for any pair of values of (γ_2, γ_3). However, further optimizations of SMBR performance can be investigated by analyzing the RSR for different switching times, configurations of columns per section, or different feed molar compositions of acetaldehyde.

2.3.1.1.2 Effect of Switching Time and Feed Composition Different runs were performed for the synthesis of 1,1-DEE by using SMBR process considering different switching time values, according to Table 5. All runs were carried out at 10°C using three columns per section. A feed mixture of ethanol with 51% of acetaldehyde molar fraction was used, and the feed and raffinate flow rates were 3.0 and 8.0 mL/min, respectively. The other

Table 5 Operating Conditions and Performance Parameters of SMBR for 1,1-DEE Synthesis

Parameters	Run 1	Run 2	Run 3
t^* (min)	3.5	3.75	4.00
Q_D (mL/min)	23.5	22.7	24.0
Q_{Rec} (mL/min)	20.5	21.3	20.0
$C_{Acetal,R}$ (mol/L)	2.37	3.00	2.90
X (%)	97.7 (98.6)	98.1 (98.7)	98.2 (99.0)
PR (%)	79.1 (87.3)	84.1 (90.2)	87.2 (90.0)
PX (%)	97.8 (95.2)	95.2 (97.4)	96.1 (98.3)
$Prod$ ($kg_{Acetal}/(L_{ads}$ day))	3.66 (4.21)	4.65 (4.61)	4.48 (4.05)
$DesC$ (L_{Eth}/kg_{Acetal})	9.81 (8.53)	7.45 (7.50)	8.20 (9.06)

Reprinted from Silva VMTM, Rodrigues AE: Novel process for diethylacetal synthesis. *AIChE J* 51(10):2752–2768, 2005 by permission of John Wiley & Sons.

experimental conditions are presented in Table 5 as well as the SMBR experimental and simulated (inside brackets) performance parameters. According to the results, the conversion for all runs was practically 98%. The complete regeneration of section 1 was not possible due to limitations of the experimental setup requiring higher desorbent flow rate. Analyzing the effect of the switching time it is possible to conclude that increasing this parameter improves the raffinate purity, while the extract purity decreases. However, the best performance is obtained in run 2 (Table 5) in which less desorbent was consumed for the highest productivity observed. This way, the effect of switching time was further investigated for different feed ratios of reactants. Thus, the process performance was evaluated for different values of switching times by feeding 51% (Fig. 7) and 70% (Fig. 8) mole fraction of acetaldehyde, taking into account the minimal extract and purity required (95%). Comparing both figures (Figs. 7 and 8) it is possible to conclude that increasing the acetaldehyde feed composition until 70 mol% a maximum performance is reached with the highest productivity (23.5 $kg_{Acetal}/(L_{ads}$ day)) and the lowest desorbent consumption (2.71 L_{Eth}/kg_{Acetal}) for a switching time of 3.9 min, while for 51 mol% of acetaldehyde feed composition just it is achieved 17.3 $kg_{Acetal}/(L_{ads}$ day) spending 4.22 L_{Eth}/kg_{Acetal} of eluent for the switching time range 3.7–4.1 min. This is related with the residence time of acetaldehyde in each section leading to the adsorbent to work more efficiently as catalyst for an optimum

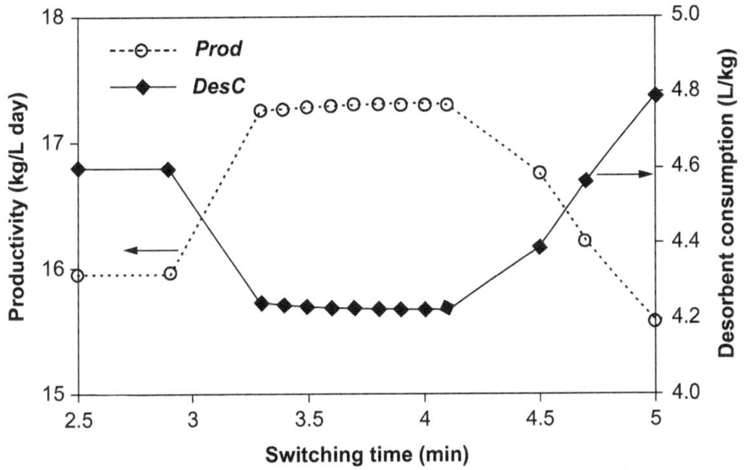

Fig. 7 Effect of switching time feeding 51 mol% of acetaldehyde. Configuration 3-3-3-3 and $T = 10°C$. *Reprinted from Silva VMTM, Rodrigues AE: Novel process for diethylacetal synthesis.* AIChE J *51(10):2752–2768, 2005 by permission of John Wiley & Sons.*

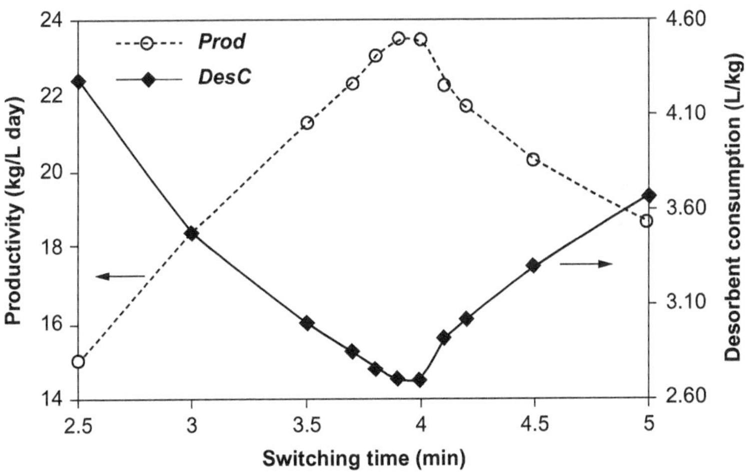

Fig. 8 Effect of switching time feeding 70 mol% of acetaldehyde. Configuration 3-3-3-3 and $T = 10°C$. *Reprinted from Silva VMTM, Rodrigues AE: Novel process for diethylacetal synthesis.* AIChE J *51(10):2752–2768, 2005 by permission of John Wiley & Sons.*

switching time. Indeed, when equivalent molar fraction of reactants is fed to SMBR the acetaldehyde is totally consumed appearing just around the feed port, since the ethanol, used as eluent, is present in large excess along the unit. On the other hand, when acetaldehyde is fed in a ratio of 2.33 in relation to ethanol, this compound has a higher residence time in sections 2 and

3, where the reaction takes place, leading to higher conversions and consequently better performance (higher productivity and lower desorbent consumption). However, higher acetaldehyde concentrations will lead to higher excess of unreacted acetaldehyde due to the lack of ethanol in sections 2 and 3 being the raffinate consequently contaminated. In contrast, for feed compositions below 70% of acetaldehyde, it will be totally consumed and SMBR unit will operate like an SMB where there is just separation of water and acetal, so the productivity and desorbent consumption are the same for a set range of switching time (Fig. 7).

In summary, the maximum performance was achieved increasing the acetaldehyde feed molar fraction to 70% and operating with a configuration of 3-3-3-3; a switching time of 3.9 min; and Q_D=50 mL/min, Q_F=10 mL/min, Q_X=35 mL/min, and Q_{Rec}=25 mL/min. A productivity of 23.5 $kg_{Acetal}/(L_{ads}\,day)$ and a desorbent consumption of 2.71 L_{Eth}/kg_{Acetal} were obtained with 99% of acetal purity (solvent-free).

2.3.1.1.3 Effect of Mass Transfer Resistance A sensitivity analysis to the mass transfer resistance should be always performed since it is a critical factor for the determination of RSR and most of the times these data are obtained from correlations or from experimental methods with large associated errors. The effect of mass transfer resistance in the purity (of extract and raffinate) and in the acetaldehyde conversion is shown in Figs. 9 and 10, respectively. Increasing the global mass transfer coefficient by a factor of 2,

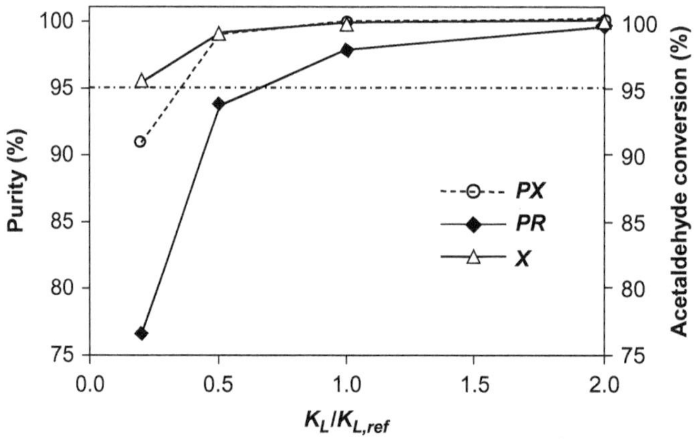

Fig. 9 Effect of mass transfer resistance in the raffinate and extract purities and reaction conversion (51% of acetaldehyde in feed and t^* = 3.7 min). *Reprinted from Silva VMTM, Rodrigues AE: Novel process for diethylacetal synthesis.* AIChE J *51(10):2752–2768, 2005 by permission of John Wiley & Sons.*

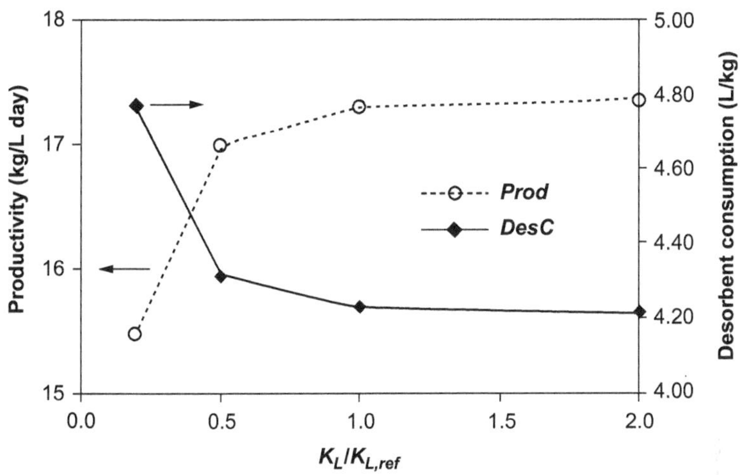

Fig. 10 Effect of mass transfer resistance in the productivity and desorbent consumption (51% of acetaldehyde in feed and $t^* = 3.7$ min). *Reprinted from Silva VMTM, Rodrigues AE: Novel process for diethylacetal synthesis.* AIChE J *51(10):2752–2768, 2005 by permission of John Wiley & Sons.*

regarding the reference case ($K_L/K_{L,ref}=1$), it is observed that the raffinate purity increases from 97.5% (reference case) to 99.5%. Contrarily, decreasing of global mass transfer coefficient significantly affects the performance. Therefore, when higher acetal purity is required smaller particle size should be considered. However, no significant differences are observed in the remaining performance parameters.

2.3.1.1.4 Proof of Concept The mathematical model described in Section 2.1 can be used to assess the feasibility of an SMBR process after including the multicomponent adsorption, equilibrium, and kinetic data for each system. On the other hand, the validation of these data and the assumptions considered along the mathematical model description ought to be performed. For that, experimental investigations at pilot scale were carried out (Silva and Rodrigues, 2005). Fig. 11 shows the experimental and simulated concentration profiles for the synthesis of 1,1-DEE with SMBR, where the run 2 (from Table 5) is presented as an example. This is an essential analysis for the study of the process feasibility and, in this case, revealed a very good agreement between the experimental and the simulated data. This way, SMBR is a very competitive technology for the 1,1-DEE synthesis taking advantage over reactive distillation process since it allows overcoming the drawbacks associated to the complex thermodynamic

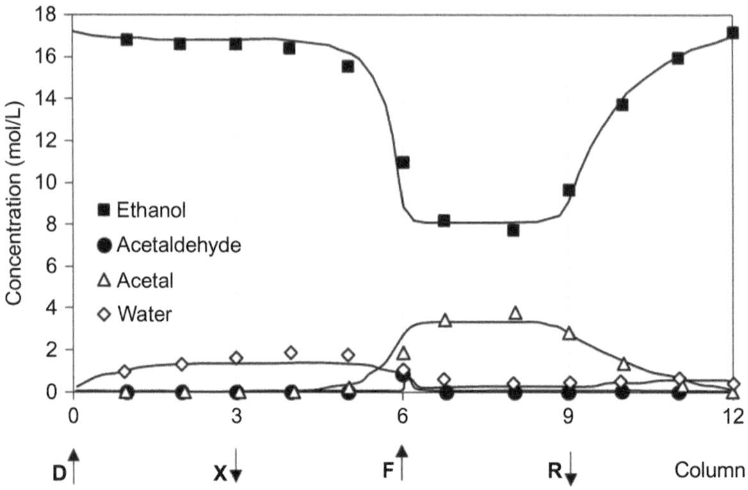

Fig. 11 Experimental (*points*) and simulated (*lines*) concentration profiles in an SMBR at the middle of a switching time at steady state (10th cycle) with $C_{A,F}$=8.57 mol/L and $C_{B,F}$=8.96 mol/L: Run 2. *D*, desorbent; *F*, feed; *R*, raffinate streams; *X*, extract (Silva and Rodrigues, 2005). *Adapted from Silva VMTM, Rodrigues AE: Novel process for diethylacetal synthesis.* AIChE J *51(10):2752–2768, 2005.*

behavior of this system (one ternary and three binary azeotropes) and the small differences between the boiling points of the products (1,1-DEE and water).

2.3.2 Summary of SMBR Application for Synthesis of Other Acetals

The enhancement of several other acetals synthesis has been investigated by SMBR technology, among them are 1,1-dibutoxyethane (1,1-DBE) (Graça et al., 2011b), 1,1-diethoxybutane (1,1-DEB) (Graça and Rodrigues, 2017), and 1,1-dimethoxyethane (1,1-DME) (Pereira et al., 2007). The main operating conditions and the performance parameters reached for each system are summarized in Table 6. All systems were studied considering the LICOSEP unit with 12 columns packed with Amberlyst-15 ion exchange resin with three columns per section (3-3-3-3). The experimental kinetic (Gandi et al., 2005; Graça et al., 2010; Rahaman et al., 2015) and the multicomponent adsorption studies (Gandi et al., 2006; Graça et al., 2011a; Regufe et al., 2016) required for the SMBR application are available in the literature for all systems. The synthesis of 1,1-DME by SMBR was investigated by theoretical and experimental SMBR studies just considering the flow rates limits of the pilot-scale unit LICOSEP, while the other acetals were also studied theoretically, through the mathematical model previously described (see Section 2.1), considering higher flow rates than the limits

Table 6 Summary of Operating and Performance Parameters for the Synthesis of Different Acetals by SMBR Over Amberlyst-15 and Using a Configuration of 3-3-3-3

Parameters/Acetal	1,1-DME	1,1-DEE	1,1-DEB	1,1-DBE
Eluent/desorbent	Methanol	Ethanol	Ethanol	Butanol
Temperature (°C)	20	10	50	50
t^* (min)	3.0	3.9	3.9	3.7
Q_D (mL/min)	25	50	44.4	105
Q_{Rec} (mL/min)	25	25	21	21
Q_X (mL/min)	20	35	46.3	82.5
Q_F (mL/min)	3	10	25	29
Aldehyde feed composition (mol%)	60	70	51	50
X (%)	99.6	99.8	96.3	>95.0
PR (%)	97.0	99.0	95.1	>95.0
PX (%)	99.6	99.0	99.8	>95.0
$Prod$ ($kg_C/(L_{ads}\,day)$)	5.19	23.5	32.2	53.2
$DesC$ (L_A/kg_C)	7.11	2.71	1.86	2.69

imposed by the pilot-scale setup. Hence, the productivity and the desorbent consumption values presented for the other acetals are better than the values presented for the 1,1-DME, mainly due to high desorbent flow rates used to regenerate section 1 which is not possible at experimental conditions, since the limit of the desorbent flow rate is 30 mL/min being not stable above 25 mL/min. The reaction conversions achieved were above 95% for all acetals. However, the aldehyde conversion was particularly high for the synthesis of 1,1-DME even though a higher feed composition of this reactant was used (60% instead of equimolar composition as in the others).

Summarizing, the SMBR process showed to be a competitive, efficient, and environmentally friendly route to produce all acetals.

2.3.3 Synthesis of Esters

Esters like ethyl lactate (EL) and butyl acrylate (BAc) are considered very important materials in Chemical Industry owing to their great market demand in the last years. EL is a biodegradable green solvent which is present in a wide range of applications, as food additive, in perfumery, among others. Besides that, it is a particularly attractive solvent for the coatings industry due to its high solvency power, vapor pressure, surface tension, and high boiling point. This ester also works as a very effective paint stripper and graffiti remover having replaced less safe solvents as toluene and xylene.

Butyl acrylate is a very interesting organic compound being used as a precursor for several products as adhesives, varnishes, finishers of papers, and textiles. This acrylate monomer has been also applied as intermediate in the synthesis of inks, plastics, sealants, coatings, and mostly to produce copolymers.

The synthesis of both compounds involves equilibrium-limited reactions having water as a by-product, so the SMBR process has been proposed as an alternative way to produce these esters. An independent study for each one is presented below.

2.3.3.1 Case Study (S.2): Synthesis of Ethyl Lactate (EL)

EL is usually obtained from the esterification of lactic acid with ethanol in acidic medium, according to the following equation:

$$\text{Ethanol} + \text{Lactic Acid} \xleftrightarrow{\text{Cat.H}^+} \text{ethyl lactate} + \text{water}$$

The kinetic and equilibrium data over Amberlyst-15 were measured and gathered in Table 7, while the multicomponent adsorption data also determined using the Amberlyst-15 as adsorbent are presented in Table 8.

Table 7 Kinetic Results From Batch Reactor for the Synthesis of EL (Pereira et al., 2008)

Parameter	Experimental Conditions	Result
Equilibrium constant	50–90°C; 6 bar	$K_{eq} = 19.35 \exp\left(-\dfrac{515.13}{T(K)}\right)$
Reaction rate	50–90°C; 6 bar	$r(\text{mol/g/min}) = k_c \dfrac{a_A a_B - \dfrac{a_C a_D}{K_{eq}}}{\left(1 + \sum_{i=A}^{D} K_{s,i} a_i\right)^2}$
Kinetic constant		$k_c(\text{mol/g/min}) = 2.7 \times 10^7 \exp\left(\dfrac{-6011.55}{T(\text{K})}\right)$
Adsorption constants		$K_{s,A} = 1.22 \exp\left(\dfrac{359.63}{T(\text{K})}\right)$ $K_{s,D} = 15.19 \exp\left(\dfrac{12.01}{T(\text{K})}\right)$

Table 8 Multicomponent Adsorption Parameters for Langmuir Isotherms of EL System Over A15 at 50°C

Component	Q_{ads} (mol/L_{res})	K (L/mol)
Ethanol (A)	6.30	3.07
Lactic acid (B)	4.94	4.09
EL (C)	3.24	1.82
Water (D)	20.6	7.06

Reprinted from Pereira CSM, Silva VMTM, Rodrigues AE: Fixed bed adsorptive reactor for ethyl lactate synthesis: experiments, modelling, and simulation, *Sep Sci Technol* 44:2721–2749, 2009 by permission of Elsevier.

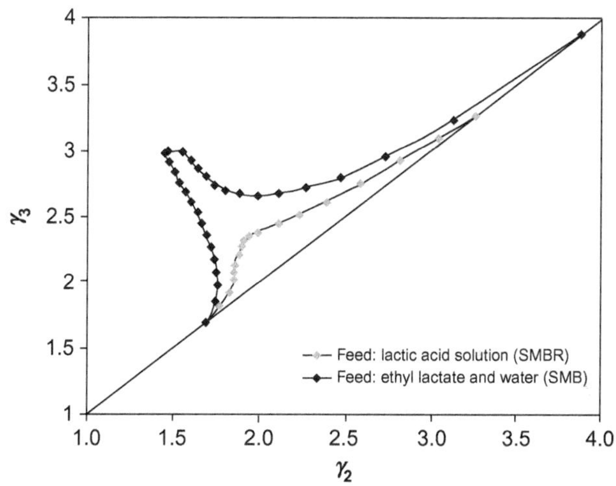

Fig. 12 Separation region (SMB) vs the reactive separation region (SMBR): $C_{B,F}=10.75$ mol/L and $C_{D,F}=9.48$ mol/L as SMBR feed and $C_{C,F}=6.47$ mol/L and $C_{D,F}=12.17$ mol/L as SMB feed; $T=50$°C; $\gamma_1=4.698$ and $\gamma_4=1.492$ (95% of purity). *Reprinted from Pereira CSM, Zabka M, Silva VMTM, Rodrigues AE: A novel process for the ethyl lactate synthesis in a simulated moving bed reactor (SMBR).* Chem Eng Sci *64(14):3301–3310, 2009 by permission of Elsevier.*

2.3.3.1.1 Reactive Separation Regions The processes performance can be compared through the respective RSRs. Fig. 12 evidences the difference between the separation region (SR) for one SMB for the separation of EL and water and the RSR for one SMBR in which EL is synthesized showing that the process is limited by the reaction. In the SMB process a feed solution of EL (C_C = 6.47 mol/L) with water (C_D = 12.17 mol/L) was taken into account, corresponding to the complete conversion of the lactic acid solution (C_B = 10.75 mol/L and C_D = 9.48 mol/L) into EL and water.

A configuration of 3-3-4-2 with a switching time of 2.7 min and purity criteria of 95% for the extract and raffinate was considered for both regions. Nevertheless, the SR is much larger than RSR because the esterification reaction of lactic acid with ethanol is slow, and therefore, the reactants require longer residence times in sections 2 and 3 to attain the desired conversion. Otherwise, unreacted lactic acid will contaminate the raffinate stream.

2.3.3.1.2 Effect of Configuration and Switching Time Three different runs were performed at 50°C by changing either the configuration (arrangement of the columns per section of the setup) or the switching time. A lactic acid solution (85% in water) was used as feed in all runs. The feed, raffinate, desorbent, and recycle flow rates were 1.8, 8.8, 27.4, and 24.7 mL/min, respectively. Table 9 shows the respective operating and performance parameters obtained experimentally and theoretically.

The results reveal that the introduction of more columns into section 3 by taking off section 4, with the same total number of columns, does not improve significantly the process performance since the ethanol is the limiting reactant (just lactic acid solution is fed) and even that the reaction zone is increased it is not possible to fully convert the lactic acid, leading to raffinate stream contamination. As the kinetic of EL synthesis is very slow, it is required to feed to the SMBR a lactic acid concentration that does not lead to a lack of ethanol in the reaction zone in order to allow full reaction conversion.

Table 9 Operating Conditions and Performance Parameters of SMBR Experiments for Synthesis of EL

Parameters	Run 1	Run 2	Run 3
t^* (min)	2.9	2.9	3
Configuration	3-3-3-3	3-3-4-2	3-3-4-2
X (%)	99.1 (97.5)	99.9 (97.9)	100 (98.0)
PR (%)	72.8 (79.2)	73.5 (79.8)	87.2 (90.0)
PX (%)	95.5 (99.9)	95.2 (99.8)	97.8 (99.9)
$Prod$ ($kg_C/(L_{ads}\,day)$)	3.20 (3.66)	3.31 (3.68)	3.36 (3.68)
$DesC$ (L_A/kg_C)	13.4 (11.7)	13.0 (11.7)	12.8 (11.7)

Simulation values are inside brackets.

Reprinted from Pereira CSM, Silva VMTM, Rodrigues AE: Fixed bed adsorptive reactor for ethyl lactate synthesis: experiments, modelling, and simulation, *Sep Sci Technol* 44:2721–2749, 2009 by permission of Elsevier.

The switching time effect was evaluated for the optimal operating point by two different ways: one varying the switching time and keeping constant remaining operating conditions (Fig. 13) and the other varying also the desorbent and recycle flow rates in order to keep constant $\gamma_1 = 4.698$ and $\gamma_4 = 1.492$ (Fig. 14). According to Fig. 13, decreasing the switching time below 2.1 min, the restriction of γ_1 is violated (Eq. 27), so is no longer possible to achieve the minimum purity required (95%), since the solid is not completely regenerated in section 1 leading to raffinate stream contamination with water. Operating the SMBR with a switching time of 2.1 min instead of 2.8 min the productivity is improved 22% and decreases 19.6%, reducing the associated costs due to the products are less diluted. Fig. 14 shows the results for which the interstitial velocity ratios, in sections 1 and 4, were kept by changing the desorbent and recycle flow rates. Similar to the previous case, the productivity increases by decreasing the switching time; however, the desorbent consumption increases with the desorbent flow rate. The productivity also increases by changing the switching time from 2.8 to 2.1; nevertheless, the desorbent consumption also increases. Reducing the switching time from 2.8 to 1 min leads to a significant increase in the productivity by 114.4% and the desorbent consumption is increased 33.3%.

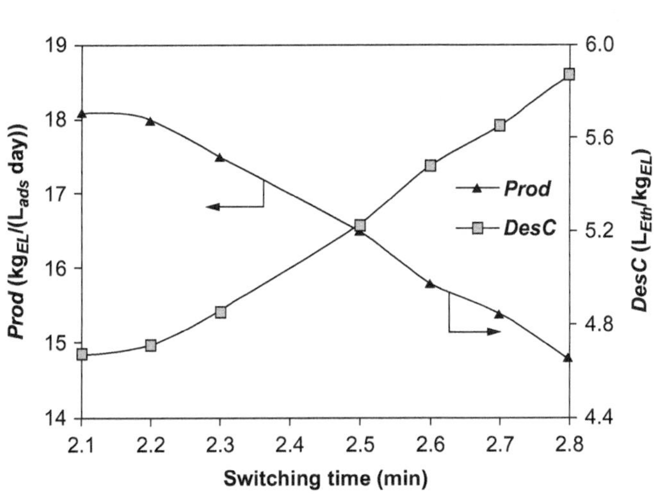

Fig. 13 Effect of switching time on the SMBR performance for the optimal operating points. *Adapted from Pereira CSM, Zabka M, Silva VMTM, Rodrigues AE: A novel process for the ethyl lactate synthesis in a simulated moving bed reactor (SMBR).* Chem Eng Sci *64(14):3301–3310, 2009.*

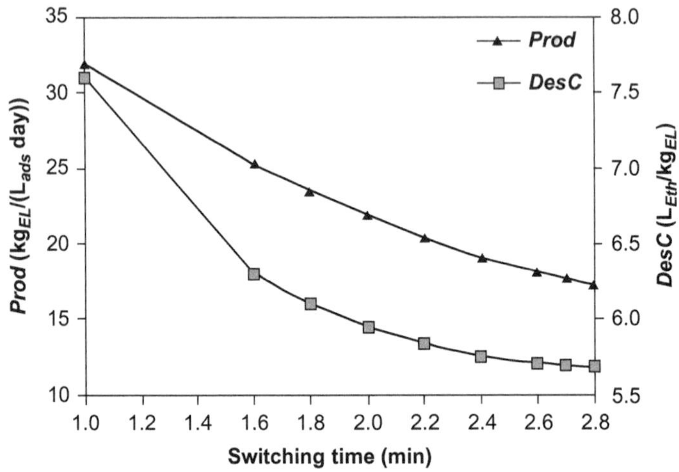

Fig. 14 Effect of switching time on the SMBR performance for the optimal operating points keeping $\gamma_1=4.698$ and $\gamma_4=1.492$. *Adapted from Pereira CSM, Zabka M, Silva VMTM, Rodrigues AE: A novel process for the ethyl lactate synthesis in a simulated moving bed reactor (SMBR).* Chem Eng Sci *64(14):3301–3310, 2009.*

2.3.3.1.3 Effect of Feed Composition A sensitivity analysis to the performance parameters was performed by feeding different concentrations of lactic acid solution (given by Fig. 15) in order to obtain the maximum performance. According to that figure, increasing the lactic acid concentration in the feed until 72% of feed composition of lactic acid solution (85% in water) increases the productivity and the desorbent consumption decreases achieving a productivity value of 17.3 $kg_C/(L_{ads}$ day) and desorbent consumption of 5.36 L_A/kg_C. For lactic acid feed compositions above that value (72%) the inverse is observed. This fact can be associated to the lower amount of ethanol in sections 2 and 3, where the reaction takes place, limiting the reaction conversion. Therefore, higher lactic acid concentration leads to higher amount of unreacted compound and, accordingly, the raffinate stream will be contaminated more easily.

2.3.3.1.4 Proof of Concept As it can be observed the mathematical model proposed (in Section 2.1) predicts the experimental concentration profiles presented in Fig. 16 with reasonable accuracy as well as the performance parameters summarized in Table 9.

2.3.3.2 Case Study (S.3): Synthesis of Butyl Acrylate (BAc)

In the following case, a more detailed study is presented comprising the integration of SMBR with downstream separation units. Moreover, its viability

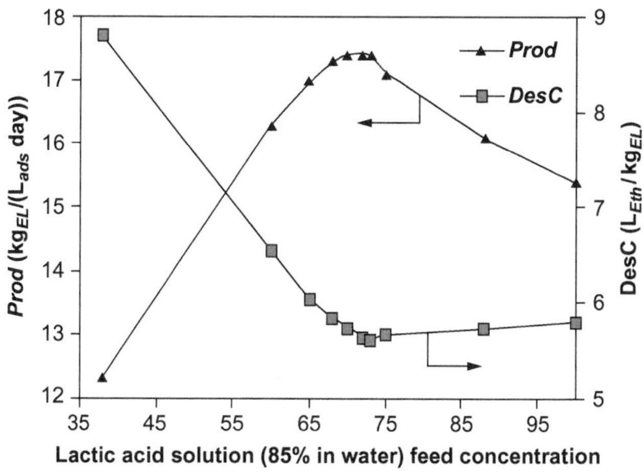

Fig. 15 Influence of the lactic acid solution feed composition in the productivity and desorbent consumption. *Adapted from Pereira CSM, Zabka M, Silva VMTM, Rodrigues AE: A novel process for the ethyl lactate synthesis in a simulated moving bed reactor (SMBR).* Chem Eng Sci *64(14):3301–3310, 2009.*

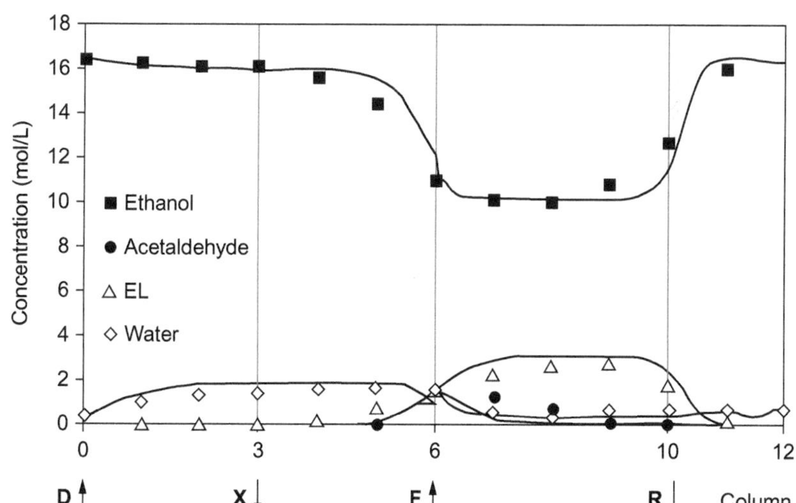

Fig. 16 Experimental (*points*) and simulated (*lines*) concentration profiles for the synthesis of EL by SMBR at the middle of a switching time at steady state (13th cycle) with a feed solution of 85% of lactic acid (in water): Run 3 at $T = 50°C$ ($\gamma_1 = 4.698$ and $\gamma_4 = 1.492$). *Adapted from Pereira CSM, Zabka M, Silva VMTM, Rodrigues AE: A novel process for the ethyl lactate synthesis in a simulated moving bed reactor (SMBR).* Chem Eng Sci *64(14):3301–3310, 2009.*

Table 10 Kinetic Results From Batch Reactor for the Synthesis of BAc (Ostaniewicz-Cydzik et al., 2014)

Parameters	Experimental Conditions	Result
Equilibrium constant	50–90°C	$K_{eq} = \exp\left(\frac{-1490}{T(K)} + 7.21\right)$
Reaction rate	50–90°C	$r(\text{mol}/(\text{g min})) = k_c \frac{a_A a_B - \frac{a_C a_D}{K_{eq}}}{(1 + K_{s,D} a_D)^2}$
Kinetic constant		$k_c(\text{mol}/(\text{g min})) = 1.52 \times 10^7 \exp\left(\frac{-66,988}{RT}\right)$
Adsorption constant		$k_{s,D}(\text{mol}/(\text{g min})) = 1.589$

Table 11 Multicomponent Adsorption Parameters for Langmuir Isotherms of BAc System Over A15 at 90°C (Constantino et al., 2015a)

Component	Q_{ads} (mol/L_{res})	K (L/mol)
Butanol (A)	4.55	7.07
AAc (B)	6.09	1.90
BAc (C)	2.91	1.90
Water (D)	24.18	22.7

and competitiveness at industrial scale are assessed through of an energetic analysis and comparing it with other reactive separation processes.

BAc is produced from the esterification of acrylic acid (AAc) with butanol in an acidic medium, according to the following equation:

$$\text{Butanol} + \text{Acrylic Acid} \xleftrightarrow{\text{Cat.H}^+} \text{BAc} + \text{water}$$

This system presents a complex thermodynamic behavior with five azeotropes (two homogeneous and three heterogeneous). Moreover, there is a high risk of polymerization of AAc and BAc mainly at high temperatures. According to the literature (Schwarzer and Hoffmann, 2002), the conventional industrial process is a complex multistage process using two reactors catalyzed homogeneously and three distillation columns for the recovery of the reactants and purification of the final product. Therefore, a large number of researches have been conducted in order to propose alternative processes for the BAc synthesis to overcome all drawbacks associated to this system and decrease the cost associated to its production by the conventional process. The most of them concerns RD processes which is the most common reactive separation process applied in the industry. Nonetheless, chromatographic reactors (CRs) take advantage over RD columns for complex molecules that are difficult to separate by evaporation process since in CR the separation is made through an adsorption process which can occur at lower temperatures avoiding polymerization issues. This way, more recently, SMBR technology was proposed for the BAc synthesis (Constantino et al., 2015b).

The kinetic and multicomponent adsorption data over Amberlyst-15 were previously investigated and the data are presented in Tables 10 and 11, respectively.

2.3.3.2.1 Reactive/Separation Region The fluid/solid ratio interstitial velocities in sections 1 and 4 were first determined based on the Equilibrium Theory for TMB process. In order to compare the SMB process for BAc and water separation with SMBR for producing and purifying BAc, the respective values for SMB/SMBR process were found ($\gamma_1=9.10$ and $\gamma_4=1.52$) and the reactive/separation regions were determined according to the methodology previously described (Section 2.2), being both regions represented in Fig. 17. The same criteria of raffinate purity (99.5%) were considered. Although, a large difference is verified between the regions, being the SR much larger than RSR. Indeed, the reaction rate is very slow

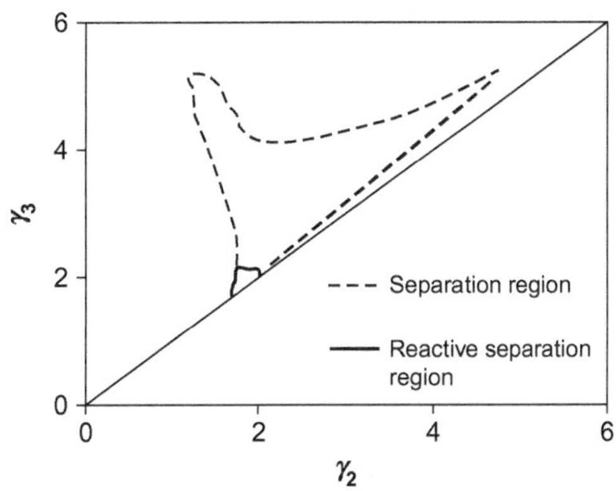

Fig. 17 RSR by feeding an equimolar ratio of AAc and butanol vs SR by feeding an equimolar ratio of BAc and water (90°C; configuration 2-4-4-2; $\gamma_1=9.10$ and $\gamma_4=1.52$). *Adapted from Constantino DSM, Pereira CSM, Faria RPV, Loureiro JM, Rodrigues AE: Simulated moving bed reactor for butyl acrylate synthesis: from pilot to industrial scale.* Chem Eng Process *97:153–168, 2015.*

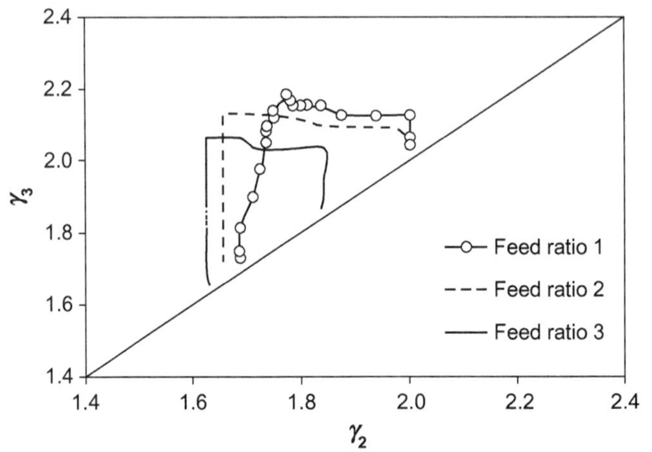

Fig. 18 Effect of feed ratio (butanol/AAc) in RSR. Feed ratio 1 means equimolar ration between the reactants. Feed ratios 2 and 3 consider excess of AAc with ratios of 2 and 3, respectively. *Adapted from Constantino DSM, Pereira CSM, Faria RPV, Loureiro JM, Rodrigues AE: Simulated moving bed reactor for butyl acrylate synthesis: from pilot to industrial scale.* Chem Eng Process *97:153–168, 2015.*

(limiting step), leading to the contamination on the raffinate stream with unreacted AAc.

2.3.3.2.2 Effect of Feed Composition Different reactants molar ratio (butanol/AAc) was simulated for the feed stream of the SMBR resulting in different RSR. Fig. 18 shows that increasing the feed molar ratio (from 1 to 3) the vertex of the region that regard the optimal point moves down and to the left, resulting in more desorbent consumption and less productivity because of the small amounts of the limiting reactant (AAc) in sections 2 and 3. So, the optimal operating conditions fulfilling the required performance criteria with the largest productivity (7.84 $kg_{BAc}/(L_{ads}$ day)) and the lowest desorbent consumption (25.5 L_{But}/kg_{BAc}) are the following: $Q_F = 6.5$ mL/min, $Q_X = 117.6$ mL/min, $Q_D = 122$ mL/min, and $Q_{Rec} = 24$ mL/min, using an equimolar feed ratio of the reactants.

In order to study the effect of different parameters like configuration and switching time in the SMBR performance, a reference case was selected inside the RSR and the respective parameters are given in Table 12.

2.3.3.2.3 Effect of Configuration and Switching Time First, three different runs were simulated by changing the arrangement of columns per each section (configuration): 2-4-4-2, 3-3-3-3, and 2-3-5-2. All conditions presented in the previous table (Table 12) were considered. The

Table 12 Operating and Performance Parameters for a Reference Case of BAc Synthesis by SMBR (90°C, t^* = 3.1 min, Configuration 2-4-4-2)

Parameters	Reference Case
Q_F (mL/min)	5
Q_X (mL/min)	118
Q_D (mL/min)	122
Q_{Rec} (mL/min)	24
$C_{But,F}$ (mol/L)	5.76
$C_{AAc,F}$ (mol/L)	5.76
X (%)	99.7
PR (%)	99.9
PX (%)	99.2
$Prod$ ($kg_{BAc}/(L_{ads}$ day))	6.0
$DesC$ (L_{But}/kg_{BAc})	33.3

different configurations change the contact time between the reactants/products and the catalyst/adsorbent in each section, causing different impact on the process. The results can be seen in Table 13 allowing to set the configuration for which it is possible to get the best performance parameters for the system in study. According to the simulated results, the most suitable configuration is 2-4-4-2 or 2-3-5-2 (very similar results), which can be explained by the larger residence time in the reaction section, enabling higher reaction conversion toward higher raffinate purity. As the first arrangement (2-4-4-2) shows slightly higher productivity and lower desorbent consumption, this configuration will be considered in the subsequent simulations.

Similarly, a sensitivity analysis to the switching time was performed keeping constant $\gamma_1 = 9.10$ and $\gamma_4 = 1.52$. The results (Table 14) show that 3.1 min is the most appropriate switching time presenting the best performance parameters taking into account the required separation criteria of raffinate purity (99.5%). Therefore, this switching time will be used in all further simulations.

2.3.4 From the SMBR to the Final Product: Process Integration

Although the SMBR has shown great versatility and ability to overcome equilibrium-limited reactions toward higher conversions and productivities with low desorbent consumptions, the extract and raffinate streams are

Table 13 Performance Parameters for Different Configurations ($t^* = 3.1$ min)

Run	Run 1	Run 2	Run 3
Config.	2-4-4-2	3-3-3-3	2-3-5-2
PX (%)	99.2	99.2	99.2
PR (%)	99.9	99.8	99.9
X (%)	99.7	99.5	99.7
Prod ($kg_{BAc}/(L_{ads}\,day)$)	6.00	5.96	5.99
DesC (L_{But}/kg_{BAc})	33.3	33.6	33.4

Reprinted from Constantino DSM, Pereira CSM, Faria RPV, Loureiro JM, Rodrigues AE: Simulated moving bed reactor for butyl acrylate synthesis: from pilot to industrial scale. *Chem Eng Process* 97:153–168, 2015 by permission of Elsevier.

Table 14 Performance Parameters for Different t^* (Configuration 2-4-4-2)

Parameter	Run 1	Run 2	Run 3
t^*	3.0	3.1	3.2
PX (%)	99.2	99.2	59.1
PR (%)	97.4	99.9	99.9
X (%)	97.2	99.7	96.2
Prod ($kg_{BAc}/(L_{ads}\,day)$)	5.91	6.00	2.03
DesC (L_{But}/kg_{BAc})	34.9	33.3	95.4

Reprinted from Constantino DSM, Pereira CSM, Faria RPV, Loureiro JM, Rodrigues AE: Simulated moving bed reactor for butyl acrylate synthesis: from pilot to industrial scale. *Chem Eng Process* 97:153–168, 2015 by permission of Elsevier.

diluted in the desorbent used. Hence, the purification of those streams must be investigated in order to find a more sustainable process by recovering the eluent stream and treating the output streams in suitable equipment, since the eluent represents the largest operating cost. Furthermore, an energy consumption estimation is required for the global process to validate if the productivity enhancement overcomes the increase of the cost associated with the desorbent separation and recycle to the SMBR unit. For that, a large-scale SMBR unit was considered based on the ratio between the diameter and the length ($D/L = 3.6$) of each column presented by industrial SMBR units, such as the Parex unit (Minceva and Rodrigues, 2005).

2.3.4.1 Case Study (PInt.1): Synthesis of Butyl Acrylate

Two different process integrations based on SMBR processes have been proposed for the BAc synthesis (Constantino et al., 2015b, 2016). Starting by feeding an SMBR an equimolar amount of reactants to the unit and using butanol as eluent, the respective output streams, extract and raffinate, will comprise essentially butanol/water (96.7/3.3 (mol%)) and butanol/BAc (99.7 mol% of BAc purity (solvent-free basis)), respectively. According to Fig. 19, the authors (Constantino et al., 2015b) suggest a pervaporation membrane module for the dehydration of butanol of the extract stream, since this unit is a very effective and economical technique for the separation of water from organic solvents and solvent mixtures mainly when low water content is observed (10 wt.% or less), as in this case. Additionally, a distillation column was considered for the downstream separation of the target product, allowing the production of BAc 99.5 mol.% pure as required commercially. Butanol obtained in the top stream of the distillation column is returned to the process as feed of the SMBR avoiding contamination of the eluent stream by BAc, while butanol recovered from a pervaporation unit is returned to the process through the eluent stream. Implementing this design for the process integration it is possible to reach 51,500 tons of BAc per year with a purity of 99.5 mol.% consuming just 1 L of butanol per kg of BAc since it allows to recover about 96% of the butanol used which is a major advantage since this compound represents the largest operating cost due to its use as reactant and mainly as eluent of the SMBR unit.

The other design presented in Fig. 20 includes a Fixed-Bed Reactor (FBR) before the SMBR unit, aiming to improve the overall performance (Constantino et al., 2016). Accordingly, reaction takes place first in the FBR, and after the thermodynamic equilibrium is reached, the outlet stream

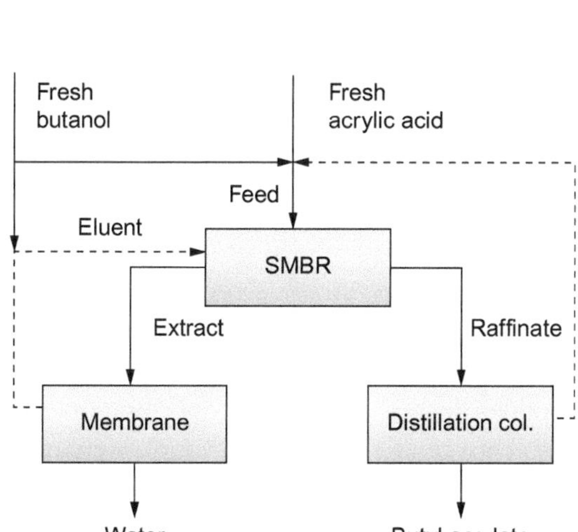

Fig. 19 Process integration proposed based on a conventional SMBR process for BAc synthesis (design A).

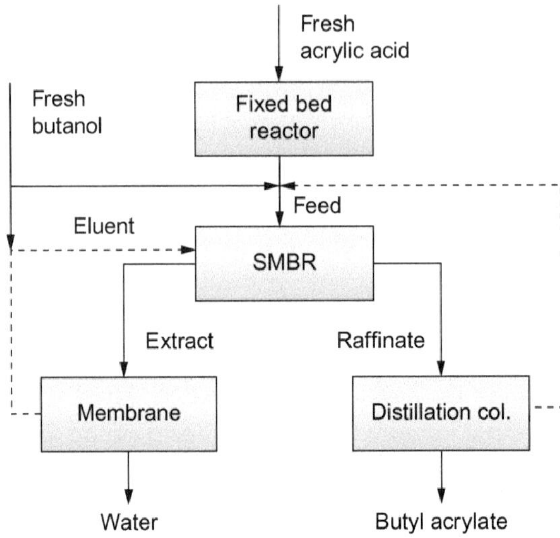

Fig. 20 Process integration proposed based on an enhanced SMBR process for BAc synthesis by coupling an FBR with an SMBR unit (design B).

of FBR is introduced in the SMBR unit which works, essentially, as separation unit turning it into a more efficient process (with higher productivity and less desorbent consumption) according to the SR (Fig. 17) determined previously for this system. Thus, it is possible to achieve complete conversion of unreacted AAc from the FBR, by extending the reaction with the butanol that is present along the SMBR, while the separation occurs simultaneously. Moreover, this design enables to operate under higher feed flow rates than in the previous case (conventional SMBR), leading to higher production capacities and lower desorbent consumptions. Indeed, this new approach has capacity to achieve more 30% of the annual production of BAc (67,000 tons/year) with very similar energy costs associated, according to Table 15, where a summary of the principal operating and performance parameters can be found.

Fig. 21 shows the concentration profiles obtained for both SMBR-based integrated processes according to the operating parameters described in Table 15. Comparing both profiles, it can be concluded that there was an evident extension of the reaction when an FBR is used before the SMBR unit leading to a sudden decrease in butanol concentration in section 3, where the reaction occurs between the AAc and butanol, yielding BAc and water.

In summary, the versatility and the competitiveness of the SMBR were highlighted through all examples given in this section, revealing that this multifunctional reactor is a very promising alternative for the synthesis of several compounds that involve equilibrium-limited reactions, mainly acetals

Table 15 The Optimal Operating and Performance Parameters for the SMBR-Based Integrated Processes at Industrial Scale and 90°C

Parameter/Process	Conventional SMBR	FBR Plus SMBR
Configuration	2-4-4-2	2-3-5-2
Q_F (L/min)	118	130
Q_X (L/min)	2315	2148
Q_D (L/min)	2397	2280
Q_{Rec} (L/min)	475	478
$C_{A,F}$ (mol/L)	5.76	0.57
$C_{B,F}$ (mol/L)	5.76	4.56
$C_{C,F}$ (mol/L)	—	3.47
$C_{D,F}$ (mol/L)	—	3.47
BAc purity (%)	99.5	99.5
SMBR Prod. capacity ($\times 10^3$) (t_{BAc}/year)	51.5	67.0
Energy consumption ($\times 10^3$) (kJ/kg$_{BAc}$)	1.80	1.87

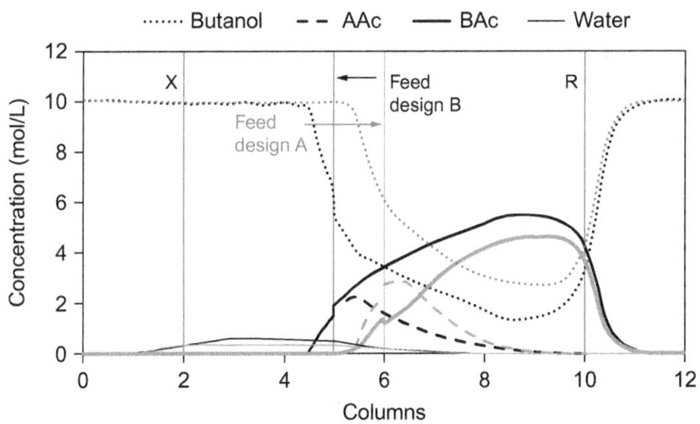

Fig. 21 Concentration profiles in SMBR unit at the middle of the switching time (3.1 min) at steady state (21th cycle) for both integrated processes design at 90°C. *Adapted from Constantino DSM, Faria RPV, Pereira CSM, Loureiro JM, Rodrigues AE: Enhanced simulated moving bed reactor process for butyl acrylate synthesis: process analysis and optimization.* Ind Eng Chem Res *55(40):10735–10743, 2016.*

and esters, having great ability to improve the reaction conversion producing high-purity products with considerably high productivity. Hence, it was possible to provide some examples of the production of these compounds at pilot scale, which can be considered a proof of concept, as it was demonstrated that, with the appropriated process integration strategy, the SMBR has the potential to be the core technology of highly efficient and sustainable processes.

3. PermSMBR TECHNOLOGY

The combination of a separation technique with reaction in the same unit, like in chromatographic reactors, reactive distillation, and membrane reactors, has brought significant advantages to numerous processes by extending the conversion of equilibrium-limited reactions, as illustrated along the previous section where SMBR was successfully applied. Notwithstanding, the process intensification can be considerably strengthened through integration of multifunctional reactors in the same equipment. This will enhance the performance of conventional multifunctional reactors since one or more products will be removed through different ways at the same time, and, consequently, this will lead to higher productivities and lower desorbent consumption. As the SMBR is one of the most commonly used technology in the process intensification for oxygenated compounds production, recently, a new hybrid technology emerged integrating the SMBR with membranes, the Simulated Moving Bed Membrane Reactor (PermSMBR) (Silva et al., 2011). Hence, this novel multifunctional reactor combines reaction with adsorption and permeation processes which can be accomplished with two different configurations, as follows:

(i) Integrated, encompassing a set of modules connected in series which are comprised by a set of tubular selective permeation membranes packed inside or outside with a catalyst/adsorbent, depending on the membrane active layer is placed, according to Fig. 22.

(ii) Coupled, encompassing a set of chromatographic columns packed with a catalyst/adsorbent intercalated with permeation selective membranes modules, according to Fig. 23.

However, independently of its configuration, the PermSMBR operating principle is very similar to the SMB/SMBR technologies where the countercurrent between the fluid and solid phases is performed by simulating the solid motion by shifting periodically all the streams by one column in the direction of the fluid at regular time intervals (switching times). The process

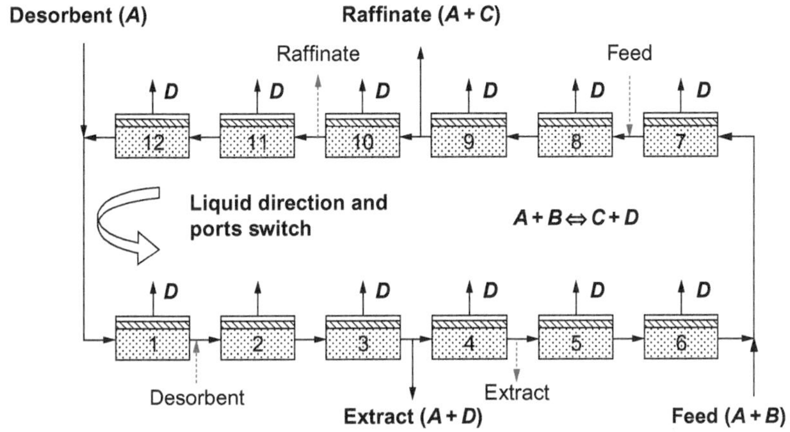

Fig. 22 Schematic diagram of an integrated PermSMBR unit with four sections and three columns per section (3-3-3-3). *Reprinted from Silva VMTM, Pereira CSM, Rodrigues AE: PermSMBR—a new hybrid technology: application on green solvent and biofuel production.* AIChE J *57(7):1840–1851, 2011 by permission of John Wiley & Sons.*

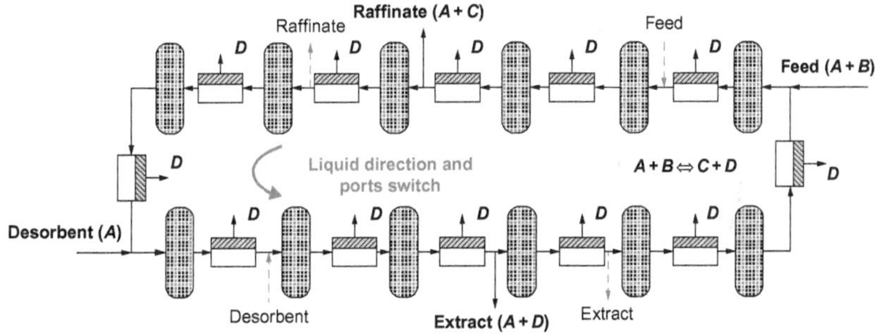

Fig. 23 Schematic diagram of a coupled PermSMBR unit with four sections and three columns and three membranes intercalated per section (3-3-3-3). *Reprinted from Pereira CSM, Silva VMTM, Rodrigues AE: Coupled PermSMBR—process design and development for 1,1-dibutoxyethane production.* Chem Eng Res Des *92(11):2017–2026, 2014 by permission of Elsevier.*

also comprises two input streams, feed and eluent, as the SMB/SMBR; nevertheless, in addition to the two outlet streams, the extract and raffinate, another stream is collected, the permeate stream that comprises all the flows removed through the membrane being rich in the product for which the membrane is selective to. It is important to notice that a vacuum source or sweep gas flow is required to get the permeate streams. Moreover, the membrane flux is easily deactivated in any column if required by switching off the vacuum or cut-off the sweep gas flow.

The membrane can be selective to the most adsorbed product (for which the adsorbent has higher affinity) or to the less adsorbed product (for which the adsorbent has lower affinity). Depending on the membrane selectivity and, consequently, the permeable product, it is possible to eliminate the extract or the raffinate stream if the membrane is selective to the most adsorbed product or to the less one, respectively. Thereafter, the setup can be simplified to three sections (PermSMBR-3s) instead of four, as described for SMBR (in Section 2), which represents another advantage of the PermSMBR over the SMBR besides the extension of the separation of the products by two different separation techniques. Considering a scenario where the extract stream is eliminated, the reactants are fed continuously between sections 1 and 2, where the reaction and the separation of the products that are being formed take place. The most adsorbed product is removed through the membranes, while the less adsorbed product is collected at the outlet of section 2 through the raffinate stream. The desorbent is introduced continuously at the beginning of section 1 being regenerated in section 3 before it returns to section 1. Hence, the complete adsorbent regeneration in section 1 is not required since the most adsorbed product can be removed through the membranes at the end of section 3, preventing the raffinate stream contamination. Regarding the scenario that excludes the raffinate stream, the reactants are fed continuously between sections 2 and 3 where the reaction takes place and the products are separated while are being formed. The most adsorbed product is collected at the end of section 1, while the less adsorbed product is collected through the permeate streams. The full regeneration of the desorbent, which is fed continuously in section 1, is not also required due to removal of the less adsorbed product through the membranes in the beginning of section 1, avoiding the extract contamination (Silva et al., 2011). This way, those configurations enable to decrease the desorbent consumptions significantly compared with the SMBR process where the complete regeneration of the adsorbent is essential to obtain the maximum process performance. Moreover, the associated costs to treat the extract and raffinate streams are considerably reduced once one of that streams can be excluded. Despite the fact that this technology has not been experimentally tested until the moment, it was already investigated theoretically for the synthesis of ethyl lactate and 1,1-diethoxyethane (Silva et al., 2011), 1,1-dibutoxyethane (Pereira et al., 2012, 2014), and more recently for 1,1 diethoxybutane (Graça and Rodrigues, 2017). The promising results led to the development of a patent (Silva et al., 2010) where the authors suggest the PermSMBR technology for the continuous production of oxygenated compounds as esters, acetals, ethers, and carbonate, focusing on the examples of the ethyl lactate and 1,1-diethylacetal synthesis.

3.1 Modeling a PermSMBR Process

The mathematical model used to describe the PermSMBR is very similar to that described for the SMBR (in Section 2.1), as expected since it results from the integration of the pervaporation membrane reactor and the SMBR. However, pervaporation data, as the flux of the species (J_{ik}) that permeate through the membranes, must be taken into account. The assumptions of the present mathematical model are: axial dispersion flow for the bulk fluid phase, LDF approximation for the interparticle and intraparticle mass transfer rates, membrane concentration polarization due to the global membrane resistance (diffusive transport in the boundary layer combined with the membrane resistance), liquid velocity variations due to reaction, adsorption/desorption and species permeation through the membranes, isothermal operation and constant porosity, and length of the packed bed. Accordingly, for the integrated PermSMBR model Eqs. (1) and (3) should be replaced by Eqs. (33) and (34), considering additionally the following equations:

Bulk fluid mass balance to component i in column k

$$\frac{\partial C_{ik}}{\partial t}+\frac{\partial(u_k C_{ik})}{\partial z}+\frac{(1-\epsilon)}{\epsilon}\frac{3}{r_p}K_{L,ik}\left(C_{ik}-\overline{C}_{p,ik}\right)$$
$$=D_{ax,k}\frac{\partial}{\partial z}\left(C_T\frac{\partial x_{ik}}{\partial z}\right)-\frac{A_m}{\epsilon}J_{ik} \tag{33}$$

The total mass balance, which allows to determine the interstitial fluid velocity variation, is given by:

$$\frac{du_k}{dz}=-\frac{(1-\epsilon)}{\epsilon}\frac{3}{r_p}\sum_{i=1}^{NC}K_{L,ik}V_{mol,i}\left(C_{ik}-\overline{C}_{p,ik}\right)-\frac{A_m}{\epsilon}\sum_{i=1}^{NC}J_{ik}V_{mol,i} \tag{34}$$

where A_m is the membrane-specific area (membrane area per volume) and the permeate flux is given by:

$$J_{ik}=k_{ov,ik}\left(x_{ik}a_{ik}P_i^0-y_{ik}P_{perm}\right) \tag{35}$$

P_{perm} and P_i^0 are the permeate and saturation pressure, respectively, x_{ik} is the molar composition of compound i in the liquid phase, y_{ik} is its molar composition in the vapor phase in the permeate, and a_{ik} is its activity coefficient usually determined by the UNIFAC or UNIQUAC models.

The global membrane mass transfer coefficient, $k_{ov,ik}$, combines the resistance due to the diffusive transport in the boundary layer with the membrane resistance (Wijmans et al., 1996):

$$\frac{1}{k_{ov,ik}} = \frac{1}{Q_{memb,ik}} + \frac{a_{ik} P^{0}_{ik} V_{mol,i}}{k_{bl,ik}} \tag{36}$$

where $Q_{memb,ik}$ is the permeance of the membrane in relation to component i in the column k.

For laminar flow and Graetz number, $d^2_{int} v_s/(D_m L_m)$, much greater that one, the mass transfer coefficient for transport in the boundary layer, $k_{bl,ik}$, is determined by the Lévêque correlation (Lévêque, 1928):

$$Sh = 1.62 Re^{0.33} Sc^{0.33} \left(\frac{d_{int}}{L_m}\right)^{0.33}, \quad (\mathrm{Re} < 2300) \tag{37}$$

where $Sh = k_{bl,ik} d_{int}/D_m$, $Re = \rho d_{int} v_s/\eta$, and $Sc = \eta/(\rho D_m)$ are the Sherwood, Reynolds, and Schmidt numbers, respectively, D_m is the solute diffusivity in the boundary layer, d_{int} is the inside diameter of the tubular membrane, L_m is the membrane length, ρ is the density, and η is the viscosity.

The mole fraction of component i in the vapor phase (permeate side) of the column k, y_{ik}, is defined as:

$$y_{ik} = \frac{J_{ik}}{\sum_{i=1}^{NC} J_{ik}} \tag{38}$$

Nevertheless, the coupled PermSMBR requires other changes, since each fixed-bed column is followed by a selective membranes module. Therefore, the equations of the bulk fluid mass balance and the total mass balance for SMBR (Eqs. 1 and 3) are the same for the coupled PermSMBR, but the following equations should be included in the respective mathematical model:

Mass balance to component i in the retentate side of the membrane m

$$\frac{\partial C_{im}}{\partial t} + \frac{\partial (u_m C_{im})}{\partial z} = D_{ax,m} \frac{\partial}{\partial z}\left(C_T \frac{\partial x_{im}}{\partial z}\right) - A_m J_{im} \tag{39}$$

where J_{im} is obtained by Eq. (35).

Fluid velocity variation in membrane m

$$\frac{du_m}{dz} = -A_m \sum_{i=1}^{NC} J_{im} V_{mol,i} \tag{40}$$

Initial and Danckwerts boundary conditions are given by:

$$t=0: \quad C_{im} = C_{ik} = \overline{C}_{p,ik} = C_{ik,0} \text{ and } q_{ik} = q_{ik,0} \tag{41}$$

$$z=0: \quad u_k C_{ik}\Big|_{z=0} - D_{ax,k} \frac{\partial C_{ik}}{\partial z}\Big|_{z=0} = u_k C_{ik,F} \tag{42}$$

$$u_m C_{im}\big|_{z=0} - D_{ax,m} \frac{\partial C_{im}}{\partial z}\big|_{z=0} = u_m C_{im}\big|_{z=L} \tag{43}$$

$$z=L: \quad \frac{\partial C_{ik}}{\partial z}\Big|_{z=L} = 0 \tag{44}$$

$$\frac{\partial C_{i,m}}{\partial z}\Big|_{z=L_m} = 0 \tag{45}$$

where the subscripts F and 0 refer to the feed and initial states, respectively.

The mass balance to the nodes can be described by Eqs. (13)–(19); however, each section must consider also the following equations:

$$C_{b,i,k|z=L} = C_{i,m|z=0} \tag{46}$$

$$C_{i,m|z=L} = C_{b,i,k+1|z=0} \tag{47}$$

$$Q_{k|z=L} = Q_{m|z=0} \tag{48}$$

$$Q_{m|z=L_m} = Q_{k+1|z=0} \tag{49}$$

The assessment of the PermSMBR feasibility for different operating conditions is performed by analyzing the respective performance parameters which are the same used to evaluate the SMBR process, described by Eqs. (22)–(26), in order to compare both technologies.

3.2 Design and Methodology

The methodology used to design and define the operating parameters for the PermSMBR process is the same described for SMBR (Section 2.2). After defining the limit values for the fluid/solid interstitial velocities ratios in sections 1 and 4 based on Equilibrium Theory, values for the fluid/solid interstitial velocities ratios in sections 2 and 3 are determined for different operating parameters, like configurations and switching times, attaining the respective RSRs. For that, geometrical parameters should be set. For all the SMBR studies before illustrated, the specifications of the Licosep

12-26 SMBR pilot-scale unit with 12 Superformance SP 23.0 × 2.6 (length × ID, cm) columns packed with commercial ion exchange resin Amberlyst-15 were considered, as earlier stated. Thus, for the PermSMBR case studies will be presented in this chapter, 12 columns will also be considered; however, each column comprises 13 commercial hydrophilic tubular membranes (Pervatech BV) (25.45 cm of length and 0.7 cm of diameter) packed with the resin Amberlyst-15 inside the membrane tube, according to the integrated PermSMBR design. The number of membranes was set by imposing the same mass of catalyst and effective cross-sectional area of the SMBR, taking into account the commercial dimensions of the membranes selected.

In the case of the coupled PermSMBR, the setup includes 12 chromatographic reactors (fixed-bed columns), each column is followed by a membrane module comprising 13 parallel hydrophilic tubular membranes (Pervatech BV). The geometry and number of fixed-bed columns as well as of membranes were set in order to have the same amount of catalyst/adsorbent and the same membrane area of the integrated PermSMBR, respectively. In Table 16 is displayed a summary of the characteristics of the SMBR and PermSMBR setups.

Regarding the preliminary studies, in addition to the reaction kinetic, equilibrium, and the multicomponent adsorption parameters required for the SMBR research, pervaporation data from the selective membranes should also be taken into account for the PermSMBR process study.

Table 16 Characteristics of the SMBR and PermSMBR Setups

			Coupled PermSMBR	
Parameters	**SMBR**	**PermSMBR**	**Fixed-Bed Columns**	**Pervaporation Module**
Solid weight (A15), g	47.6	47.6	47.6	—
Length of the bed, cm	23	25.45	23	25.45
Internal diameter, cm	2.6	0.7	2.6	0.7
Bulk density, kg/m^3	390	374	390	—
Bed porosity	0.4	0.424	0.4	—
Number of fixed-bed/membranes per module	12 Fixed-bed columns	13 Membranes per 12 modules	12 Fixed-bed columns	13 Membranes per 12 modules

3.3 Membranes and Pervaporation Studies

Pervaporation membranes have been widely applied for the separation of liquid mixtures. The separation is performed by partial vaporization of the species through the membrane based on its affinity with the membrane material. Hence, the species having higher affinity adsorb and diffuse through the membrane, while the other species with lower affinity are retained by the membrane. The hydrophilic membranes are the most commonly used in the industry mainly for organic solvent dehydration since they are very selective to water, being generally made of polymeric materials such as polyvinyl alcohol, polyamides, polymaleimides, Nafion, and polyacrylonitrile (PAN). Its versatility, by changing the respective active layer with different chemical compositions, allows to extract a significant amount of water with considerable flux and selectivity values (Jyoti et al., 2015). They can be classified as organic, inorganic, and organic–inorganic hybrid membranes. The large number of publications concerning pervaporation studies using inorganic membranes, such as zeolite and silica, evidence membrane's interesting characteristics. According to the literature, these membranes show better stability under acidic and high temperature conditions than polymeric membranes, being the best alternative for the dehydration of the reaction medium (Pereira et al., 2010).

Moreover, membrane pervaporation processes offer additional advantages like high selectivity, low energy consumption, compact and modular design, as well as moderate cost to performance ratio. Thus, this separation technology is considered as one of the most interesting for organic–water and organic–organic separations, taking advantage over distillation columns which present significant thermodynamic limitations (Luis and Van der Bruggen, 2015). The separation by pervaporation involves a transport mechanism based on physical–chemical interactions between the membrane material and the permeating molecules, allowing to operate at lower temperatures than distillation process which depends on the relative volatility. Therefore, pervaporation is strongly recommended for separating of heat-sensitive products and azeotropic mixtures.

Due to the ability to reduce significantly the concentration and the flow rates to be treated and, consequently, the energy and costs requirement (Ahmad and Lone, 2012), the pervaporation-based hybrid processes have been intensively investigated, mainly for the intensification of esterification reactions, such as lactic acid with ethanol, acetic acid with ethanol/methanol/isopropanol/isobutanol/butanol, and propionic acid with isopropanol, among several others (Jyoti et al., 2015). That way, membrane pervaporation can be integrated with chemical reaction, helping to shift the equilibrium toward the product formation by continuous water permeation. Therefore, the multicomponent study by pervaporation in the absence of reaction is fundamental to evaluate the different important parameters involved in the process, such as selectivity, permeate fluxes, driving force, and separation factor, and to understand the influence of variables like the temperatures and feed mole composition in the membrane performance.

A pervaporation study for the compounds involved in the ethyl lactate system (Pereira et al., 2010) will be briefly described as instance to follow in order to get the pervaporation data required for the PermSMBR investigation.

3.3.1 Preliminary Studies

Commercial hydrophilic silica membrane from Pervatech BV (The Netherlands) was evaluated for the dehydration of ethanol, ethyl lactate, and lactic acid from 48°C to 72°C. First, preliminary tests were carried out with the same feed mixture (ethanol/water) and different absolute feed pressures in order to evaluate the quality of the membrane. No significant imperfections were detected since the total flux and selectivity were kept practically constant.

The concentration polarization phenomena are associated to the accumulation of the retained species near the membrane surface and, consequently, their concentration will be higher in the boundary layer (adjacent to the membrane surface) than in the bulk (Mulder, 1995). Generally, concentration polarization results from the depletion of the most permeable component in the vicinity of the feed/membrane interface and it is owed to the slow diffusion of the solute from the bulk of the feed to the boundary layer (Neel, 1995). The membrane flux and separation efficiency may be severely influenced due to polarization (Strathmann, 2000). This way, the influence of mass transfer in the boundary layer was investigated with a binary mixture (ethanol/water, $x_{w,F}=0.182$ molar fraction) by measuring the total flux at different feed flow rates. According to Fig. 24, this effect is negligible for flow rates higher than approximately 19 L/h, since the total permeate flux remains constant from this value. So, all pervaporation experiments were carried out at 20 L/h in this work, ensuring the absence of mass transfer resistance.

The performance of the membrane for the ethanol/lactic acid/ethyl lactate/water system was then evaluated experimentally by measuring

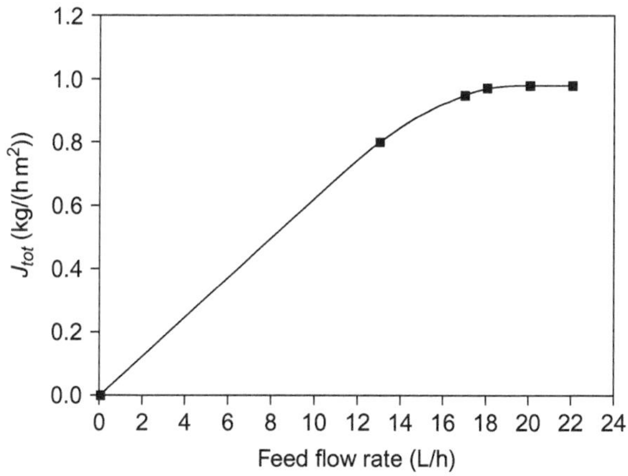

Fig. 24 Evaluation of mass transfer limitations in the boundary layer by feeding a mixture of ethanol/water ($x_{w,F}=0.182$), $P_{perm}=25$ mbar, $T=48°C$. *Adapted from Pereira CSM, Silva VMTM, Pinho SP, Rodrigues AE: Batch and continuous studies for ethyl lactate synthesis in a pervaporation membrane reactor.* J Membr Sci *361(1–2):43–55, 2010.*

pervaporation data for diverse mixtures compositions in the absence of reaction and mass transfer limitations at different temperatures.

After that, the influence of feed composition at different operating temperatures on the total permeation flux and permeate composition was assessed for nonreactive pairs, where the results showed that the total permeation flux increases with feed water composition and temperature.

3.3.2 Pervaporation Transport and Parameters Estimation

Applying the solution-diffusion model (Wijmans and Baker, 1995) to describe the mass transport through the membrane which is given by Eq. (50) (in the absence of mass transfer resistance in the boundary layer):

$$J_i = Q_{memb,i}\left(x_{i,F}a_i p^0_{i,F} - y_{i,perm}P_{perm}\right) \tag{50}$$

The last term $(x_{i,F}a_i p^0_{i,F} - y_{i,perm}P_{perm})$ gives the driving force of the component i.

The separation factor, α, is given by:

$$\alpha_{perv} = \frac{y_{w,perm}\left(1 - x_{w,F}\right)}{x_{w,F}\left(1 - y_{w,perm}\right)} \tag{51}$$

where the subscripts F and w correspond to the feed and water, respectively.

Table 17 Overall Activation Energies and Preexponential Factors for Each Compound (Pereira et al., 2010)

Component	$E_{p,i}$ (kJ/mol)	$Q_{0,i}$ (mol/(s m^2 Pa))
Ethanol	−22.60	2.36×10^{-12}
Ethyl lactate	−10.42	1.99×10^{-9}
Water	−13.93	2.97×10^{-8}

The activation energy required to the phase change of the permeate species was obtained by fitting the experimental results to the Arrhenius equation to describe the permeance dependence on temperature:

$$Q_{memb,i} = Q_{memb,0} \exp\left(\frac{-E_P}{RT}\right) \quad (52)$$

where $Q_{memb,0}$ is the preexponential factor, T is the absolute temperature, R is the ideal gas constant, and E_P is the activation energy of permeation which is a combination of the activation energy of diffusion (E_D) and the heat of adsorption of the permeate in the membrane (ΔH):

$$E_P = E_D + \Delta H \quad (53)$$

Since temperature influences both membrane permeability and the driving force for mass transport, the activation energy of permeation should be evaluated from the slope $\ln(Q_{memb,i})$ vs $1/T$ (Feng and Huang, 1997). This way, an average of all permeance values for each species and temperature was calculated and the permeation activation energies as well as the preexponential factors were determined from the slope and intercept of the corresponding linear regressions. The values of the respective parameters are presented in Table 17. Despite the permeation flux increases with temperature due to increasing driving forces, the membrane permeability decreases as negative values of the activation energy values indicate, which means that the permeation of all species investigated is governed by the adsorption, according to Eq. (53). Moreover, the results reveal that the microporous silica membranes do not permeate lactic acid and present reduced permeation for ethanol and ethyl lactate, while for water it presents high flux and selectivity.

The data attained can be used to simulate a pervaporation membrane reactor being very useful to design and optimize different configurations of pervaporation-based hybrid processes like the PermSMBR.

3.4 PermSMBR for Sorption-Enhanced Reactions

PermSMBR arose aiming to improve the SMBR process performance by using two different separation technologies (SMB and membrane pervaporation) combined with reaction to increase the productivity and reduce the eluent consumption for the adsorbent regeneration. Wherefore, this novel approach is a very promising technology for oxygenated compounds production which involves equilibrium-limited reaction being a clean and economic alternative to conventional processes and it is even competitive comparatively to other reactive separation technologies (Silva et al., 2011).

The first applications of PermSMBR technology were the synthesis of a blending agent for fuels and biofuels that can be added to diesel up to 10% reducing in 34.6% the emissions of particulate mass and 3.3% the NO_x emissions from fuel combustion, the 1,1-diethoxyethane and the synthesis of ethyl lactate, an important green solvent in the chemical industry (Silva et al., 2011). The great results led to extend the investigation for other acetals like 1,1-dibutoxyethane (1,1-DBE) (Pereira et al., 2012) and 1,1-diethoxybutane (1,1-DEB) (Graça and Rodrigues, 2017). Along this section, these studies will be focused, discussed, and compared with the results achieved with the SMBR.

Similar to the SMBR, a common procedure is usually adopted in order to attain the best performance of the PermSMBR and, according with that, the procedure is now summarized in the following scheme (Fig. 25).

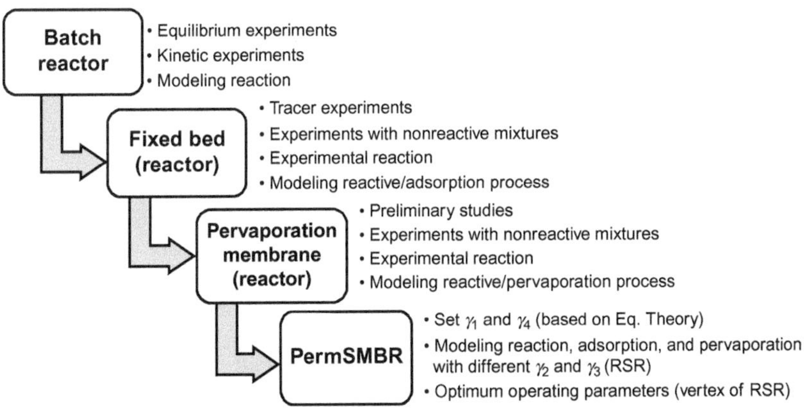

Fig. 25 Summary scheme of the procedure required to study a PermSMBR process.

3.4.1 *Synthesis of Acetals*

3.4.1.1 Case Study (P.1): Synthesis of 1,1 Diethoxyethane

The synthesis of 1,1-DEE involves a liquid-phase thermodynamically limited reaction between ethanol and acetaldehyde according to Fig. 4, which was produced successfully by means of SMBR technology using ethanol as eluent/desorbent and Amberlyst-15 as catalyst/adsorbent.

Basic data required for this study, as kinetic and multicomponent adsorption parameters, were already presented before in Section 2.3.1. Regarding pervaporation data, the permeance acquired for water and ethanol presented in Section 3.3 using a similar selective silica membrane (Pervatech, BV) was taken into account. For the PermSMBR study of the present system (1,1-DEE), acetaldehyde permeance was considered the same of ethanol (almost negligible). However, a preventive measure was adopted to avoid acetaldehyde losses in sections 2 and 3 where the reaction takes place, by deactivating the vacuum in the columns of that sections since the vapor pressure of acetaldehyde (2.6 bar at 50°C) is much higher than ethanol (0.27 bar at 50°C).

Reactive/separation region (Fig. 26) was determined for both SMBR and PermSMBR setting a criteria of 95% for extract and raffinate purities as well as acetaldehyde conversion, ensuring the total regeneration in sections 1 and 4 ($\gamma_1 = 5.302$ and $\gamma_4 = 1.515$). The operating parameters used were the same for both processes with the exception of the switching time which was set to 3.7 min for SMBR and to 4.1 min for PermSMBR in order to keep the same γ_1 and γ_4. In addition, the regeneration regions were also achieved setting the conditions in sections 2 and 3 for the same criteria of purity and reaction conversion (95%) and also is presented in Fig. 26. In this case, the vertex of the region ($\gamma_2 = 2.727$ and $\gamma_3 = 3.712$) indicates the minimum desorbent consumption (minimum desorbent flow rate required) to accomplish the criteria established. According to the RSR and the performance parameters (Table 18), the PermSMBR improves significantly the synthesis of the 1,1-DEE compared with SMBR, as expected, achieving more than 50% of the productivity and reducing around 40% of the eluent consumption.

The regeneration region is larger for PermSMBR process leading to a reduction on the consumption of the desorbent of about 69%, at optimal operating conditions (2.06 L_{Eth}/kg_{DEE} with SMBR and 0.64 L_{Eth}/kg_{DEE} in PermSMBR) (Pereira and Rodrigues, 2013). The lowest value was achieved by reducing the extract flow rate to zero, which means that the

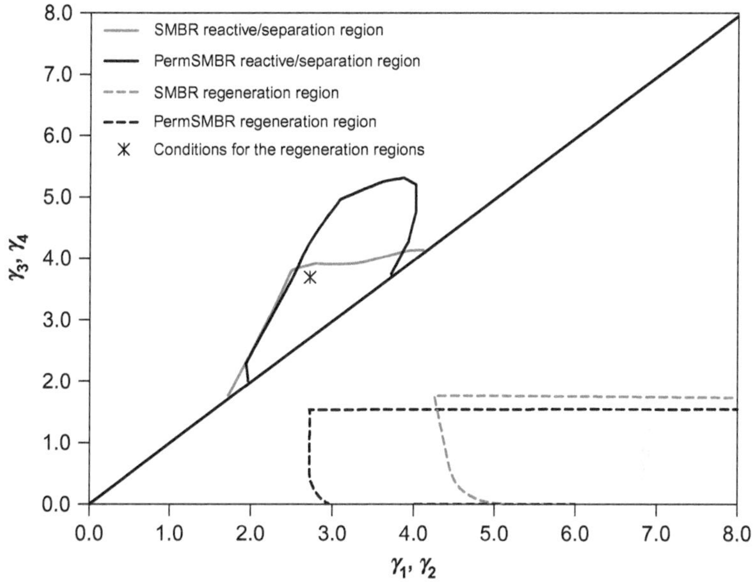

Fig. 26 Reactive/separation and regeneration region for 1,1-DEE synthesis by PermSMBR (t^* =4.1 min) and SMBR (t^* =3.7 min) processes; configuration 3-3-3-3 at T=50°C. *Reprinted from Pereira CSM, Rodrigues AE: Process intensification: new technologies (SMBR and PermSMBR) for the synthesis of acetals.* Catal Today *218–219:148–152, 2013 by permission of Elsevier.*

Table 18 Operating and Performance Parameters for SMBR and PermSMBR Technologies for the Synthesis of 1,1-DEE (T=50°C and Configuration 3-3-3-3)

Parameters	SMBR	PermSMBR
Q_D (mL/min)	50	
Q_{Rec} (mL/min)	20	
Q_F (mL/min)	17.7	24.9
$x_{Acet,F}/x_{Eth,F}$	0.51/0.49	
γ_1; γ_4	5.302; 1.515	
t^* (min)	3.7	4.1
P_{perm} (mbar)	—	5
$Prod$ ($kg_{DEE}/(L_{ads}\,day)$)	27.4	41.95
$DesC$ (L_{Eth}/kg_{DEE})	2.47	1.47

best configuration for 1.1-DEE synthesis is the PermSMBR with three sections (PermSMBR-3s). In fact, water is the most adsorbed component and for which the membranes have more affinity. So, even if the extract stream is removed, water is depleted by pervaporation avoiding the contamination of the raffinate stream.

3.4.1.2 Case Study (P.2): Synthesis of 1,1 Dibutoxyethane

The synthesis of 1,1-DBE was carried out in a batch reactor from a feed mixture composed by butanol and acetaldehyde in liquid phase over ion exchange Amberlyst-15 (Graça et al., 2010). The reaction equilibrium constant, kinetic parameters, as well as the respective operating conditions can be seen in Table 19. The adsorption data acquired at 25°C for Amberlyst-15 by frontal chromatography experiments in a fixed-bed column are described by the multicomponent equilibrium adsorption Langmuir isotherms (Graça et al., 2011a) and are given in Table 20. An approximation was made considering that the adsorption parameters for 50°C and 70°C are the same as those at 25°C, and it was validated by predicting the experimental values of two breakthrough curves at 50°C with reasonably accuracy.

The pervaporation data were measured using two commercial hydrophilic silica membranes (from Pervatech) obtaining a water permeance of $(4.43 \pm 0.66) \times 10^{-6}$ mol/(Pa s m^2) at 50°C and $(3.93 \pm 0.94) \times 10^{-6}$ mol/(Pa s m^2) at 70°C in the range of 0.08–0.38 of water molar fraction, while the butanol permeances are $(9.84 \pm 2.21) \times 10^{-9}$ mol/(Pa s m^2) at 50°C

Table 19 Kinetic Results From Batch Reactor for 1,1-DBE Synthesis (Graça et al., 2010)

Parameter	Experimental Conditions	Result
Equilibrium constant	20–50°C; 6 atm	$K_{eq} = 9.59 \times 10^{-3} \exp\left(\frac{1755.3}{T(K)}\right)$
Reaction rate	10–50°C; 6 atm	$r(\text{mol/g/min}) = k_c \frac{a_A a_B - \frac{a_C a_D}{K_{eq} a_A}}{(1 + K_{s,D} a_D)^2}$
Kinetic constant		$k_c(\text{mol/g/min}) = 2.39 \times 10^9 \exp\left(\frac{-6200.9}{T(\text{K})}\right)$
Adsorption constant		$K_{s,D} = 2.25 \times 10^{-4} \exp\left(\frac{3303.1}{T(\text{K})}\right)$

Table 20 Multicomponent Adsorption Parameters for Langmuir Isotherms of 1,1-DBE System Over Amberlyst-15 at 25°C (Graça et al., 2011a)

Component	Q_{ads} (mol/L_{res})	K (L/mol)
Butanol (A)	8.50	7.50
Acetaldehyde (B)	15.1	0.50
1,1-DBE (C)	5.80	0.40
Water (D)	44.9	12.1

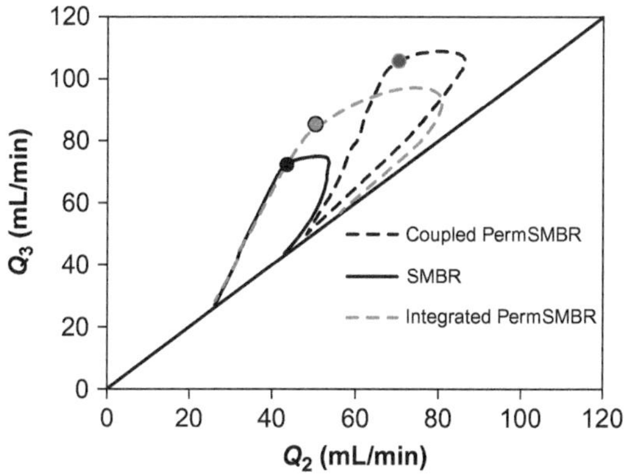

Fig. 27 Reactive separation region for the SMBR (t^*=3.1 min), integrated PermSMBR (t^*=3.4 min), and coupled PermSMBR processes (t^*=3.1 min); configuration 3-3-3-3 at 50°C; Q_D=105 mL/min and Q_{Rec}=21 mL/min. The points represent the respective optimal operating conditions/performance. *Reprinted from Pereira CSM, Silva VMTM, Rodrigues AE: Coupled PermSMBR—process design and development for 1,1-dibutoxyethane production.* Chem Eng Res Des *92(11):2017–2026, 2014 by permission of Elsevier.*

and $(6.48 \pm 2.07) \times 10^{-9}$ mol/(Pa s m^2) at 70°C. The acetaldehyde flux is insignificant with a permeance of $(7.93 \pm 4.24) \times 10^{-10}$ mol/(Pa s m^2) at 70°C and 1,1-DBE does not permeate through the membranes.

The RSR attained for SMBR, integrated PermSMBR, and coupled PermSMBR can be observed in Fig. 27. The comparison of the three reactors for the 1,1-DBE synthesis is performed at the operating conditions shown in Table 21. Butanol was used as desorbent since acetaldehyde oligomerization onto the resin surface can occur in the absence of butanol and because of that acetaldehyde cannot be used as desorbent. A different switching time was imposed for the integrated PermSMBR (3.4 min) in order to have the same liquid and solid velocities in sections 1 and 4.

All regions were built for criteria of 95% for extract and raffinate purities and for acetaldehyde conversion using a configuration of three columns per

Table 21 Operating Conditions for the SMBR and PermSMBR Processes at 50°C

Parameters/Process	**SMBR**	**PermSMBR**
Configuration	3-3-3-3	3-3-3-3 (integrated); 3(3)-3(3)-3(3)-3(3) (coupled)
P_{perm} (mbar)	—	5
t^* (min)	3.1	3.4 (integrated); 3.1 (coupled)
Feed (mole fraction)		$x_{Acet,F}=0.51$; $x_{But,F}=0.49$
Q_D (mL/min)		105 ($\gamma_1=7.997$)
Q_{Rec} (mL/min)		21 ($\gamma_4=1.333$)

section (3-3-3-3). In the coupled PermSMBR case the configuration used was 3(3)-3(3)-3(3)-3(3) that corresponds to three columns plus three membrane module per section, in which one chromatographic column is followed by a membrane module. According to this configuration, the less adsorbed product (1,1-DBE) is collected in the raffinate stream that is located at the outlet of section 2, while the most adsorbed product (water) is removed just by pervaporation. Comparing the RSR of the SMBR with the integrated PermSMBR, the best performance was attained by the PermSMBR with a larger RSR, indicating that it is possible to operate with higher feed flow rates toward higher productivities. The vertex of each region indicates the maximum feed, being 34.9 mL/min for the PermSMBR, while for SMBR the maximum feed possible to operate is 29 mL/min. The respective optimal productivities and desorbent consumptions obtained with the different technologies are 53.15 $kg_{DBE}/(L_{ads}$ day) and 2.69 L_{But}/kg_{DBE} for the SMBR, while for the PermSMBR are 64.15 $kg_{DBE}/(L_{ads}$ day) and 2.15 L_{But}/kg_{DBE}, respectively, meaning that the performance is improved about 20% by using integrated PermSMBR technology for 1,1-DBE synthesis. This way, besides the reduction of the eluent consumption the associated costs are also reduced since the products obtained in the final of the process are less diluted than in the SMBR.

Looking to the RSR of the coupled configuration (Pereira et al., 2014), its performance seems to be even better than the integrated PermSMBR (69.14 $kg_{DBE}/(L_{ads}$ day) and 1.97 L_{But}/kg_{DBE}). Nonetheless, the comparison is not made in the same conditions of the integrated configuration where the switching time was set to 3.4 in order to keep both liquid and solid velocities in sections 1 and 4 of the SMBR, while in the coupled arrangement the value was the same of the SMBR 3.1 min. After each switching time in the coupled reactor, all the inlets and outlets switch one column and one membrane module ahead in the fluid direction to simulate the countercurrent solid movement. Hence, the solid velocity must be determined by other way, taking into account the residence time in the membranes modules to correct the switching time. However, the residence time in the membranes modules changes along the setup depending on the section flow rates. Therefore, an approximation was done considering an average residence time of the membrane module which resulted a value of 2.06 min and consequently a switching time of 5.16 min. Again, the RSR was determined considering the last value (5.16 min) and the maximum performance was achieved for that operating conditions: 48.52 $kg_{DBE}/(L_{ads}\,day)$ and 3.00 L_{But}/kg_{DBE}, being even worse than SMBR process.

As in the previous case studies, a sensitivity analysis was performed to the switching time values for the different PermSMBR configurations in order to perform a fair comparison. Accordingly, the optimal operating conditions for each switching time value were attained leading to the values: 2.5 min for integrated PermSMBR and 3.0 min for the coupled configuration, leading to the respective performance parameters presented in Table 22. The results obtained show very similar performance between the different PermSMBR configurations and both better than the SMBR performance, improving the productivity about 30% and reducing the desorbent consumption in 24%.

The effect of the number of membranes was also investigated in the coupled configuration, but no significant performance improvement was detected.

Considering the other advantage of the PermSMBR which can be reduced from four to three sections (from a configuration 3-3-3-3 to 6-3-3), eliminating the extract stream once that water is the most adsorbed and permeated product, both integrated (Pereira et al., 2012) and coupled (Pereira et al., 2014) configurations were investigated.

Regarding the integrated configuration, the process with three sections was evaluated keeping the operating parameters presented in Table 21;

Table 22 Performance Parameters of the SMBR, Integrated PermSMBR, and Coupled PermSMBR According to the Respective Optimal Switching Time

Process	SMBR	Integrated PermSMBR	Coupled PermSMBR
t^* (min)	3.1	2.5	3.0
Prod ($kg_{DBE}/(L_{ads}\,day)$)	53.2	69.7	69.4
DesC (L_{But}/kg_{DBE})	2.69	2.04	2.07

however, the desorbent flow rate was changed for each feed flow rate in order to achieve the values that accomplish the performance criteria established (95%). Those conditions led to an overload permeation through the membranes and consequently the maximum productivity attained, for that criteria, was 8.04 $kg_{DBE}/(L_{ads}\,day)$. Thereupon, the PermSMBR-3s was evaluated at a higher temperature, 70°C, allowing to increase the water flux through the membranes. Assuming that the adsorption equilibrium is not strongly affected at this temperature, SMBR and PermSMBR with four sections were also evaluated at 70°C in order to compare all processes in a fair way. Thus, the coupled configuration with three sections was also assessed for the same temperature conditions, and all results for the different processes are displayed in Table 23. Comparing the performance parameters, the integrated configuration with three sections seems to be the most advantageous technology presenting similar productivities to the coupled setup but with significant desorbent consumption reduction, about 53%. Moreover, eliminating one section will allow to reduce the associated costs involved in the subsequent water/butanol separation unit, according to Fig. 28, compensating some of the additional investment costs.

Additionally, the influence of the recycle flow rate was also studied taking into account the integrated PermSMBR-3s, which is the best solution for 1,1-DBE synthesis. Considering the same criteria (95%), optimal operating conditions were achieved for each recycle flow rate and the influence of this parameter on the performance parameters was assessed being represented in Fig. 29. The results show that the productivity is not significantly affected until 24 mL/min; by other hand the desorbent consumption decreases considerably (about 33%) from 16 to 24 mL/min. Higher recycle flow rates will not ensure the complete regeneration of the desorbent in section 3, limiting the maximum feed flow rate possible to operate for the required criteria (95%). In conclusion, the optimal recycle flow rate is

Table 23 Performance Parameters of the SMBR, Integrated PermSMBR, Integrated PermSMBR-3s, and Coupled PermSMBR-3s for the 1,1-DBE Synthesis at 70°C

Process	SMBR	Integrated PermSMBR	Integrated PermSMBR-3s	Coupled PermSMBR-3s
Prod ($kg_{DBE}/(L_{ads}$ day))	53.95	72.1	71.0	73.4
DesC (L_{But}/kg_{DBE})	2.64	1.84	0.42	0.89

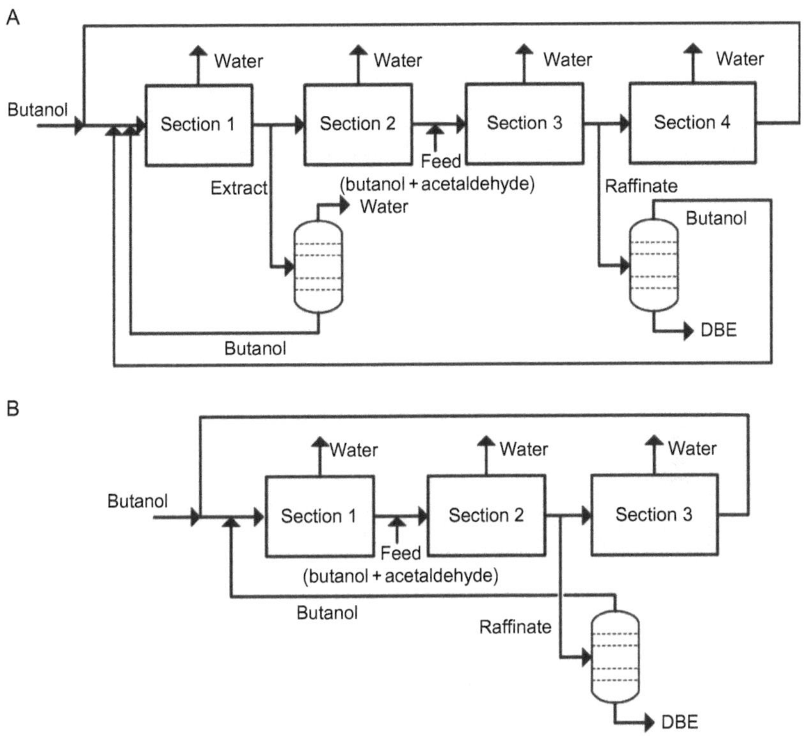

Fig. 28 PermSMBR unit with subsequent separation units for the synthesis of 1,1-DBE: (A) PermSMBR with four section; (B) PermSMBR with three sections (PermSMBR-3s). *Reprinted from Pereira CSM, Silva VMTM, Rodrigues AE: Green fuel production using the PermSMBR technology.* Ind Eng Chem Res *51(26):8928–8938, 2012 by permission of American Chemical Society.*

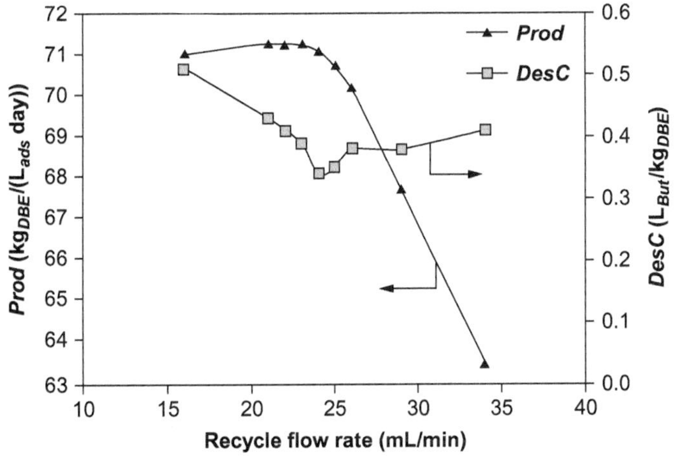

Fig. 29 Influence of recycle flow rate on the integrated PermSMBR-3s performance for the optimal operating points. *Adapted from Pereira CSM, Silva VMTM, Rodrigues AE: Green fuel production using the PermSMBR technology.* Ind Eng Chem Res *51(26):8928–8938, 2012.*

24 mL/min corresponding to a productivity of 70.8 $kg_{DBE}/(L_{ads}\,day)$ and a desorbent consumption of 0.33 L_{But}/kg_{DBE}.

3.4.1.3 Case Study (P.3): Synthesis of 1,1-Diethoxybutane

A summary of the fundamental kinetic and adsorption parameters required for the study of 1,1-DEB synthesis by means of PermSMBR is gathered in Tables 24 and 25, respectively. An approximation was performed by considering that the kinetic parameters for Amberlyst-15 ion exchange resin are similar to that obtained for Amberlyst-47 ion exchange resin.

The pervaporation parameters were measured from 50°C to 70°C using a commercial silica membrane from Pervatech, similar to the membrane used for the ethyl lactate pervaporation study (presented in Section 3.3), and they are also shown in Table 26.

The integrated PermSMBR technology was assessed for 1,1-DEB production by packing Amberlyst-15 inside the membrane tube and considering the geometrical specifications described in Section 3.2 (Table 16). The SMBR RSR was used as basis for the study of the PermSMBR; hence, the operating parameters are the same that those found for SMBR process with the exception of the switching time, which was corrected to 4.4 in order to keep the same interstitial velocities of liquid and solid in sections 1 and 4. In addition, a vacuum source at 5 mbar was introduced in the permeate side. All operating conditions are summarized in Table 27. Both SMBR and PermSMBR RSR can be observed in Fig. 30 from which it is possible to compare both process performance considering a criteria of 95% for extract

Table 24 Kinetic Results From Batch Reactor for 1,1-DEB Synthesis (Rahaman et al., 2015)

Parameter	Experimental Conditions	Result
Equilibrium constant	20–50°C; 6 bar	$K_{eq}=7.73\times10^{-2}\exp\left(\dfrac{1036.80}{T(\mathrm{K})}\right)$
Reaction rate	20–50°C; 6 bar	$r(\mathrm{mol/g/min})=k_c\dfrac{a_Aa_B-\dfrac{a_Ca_D}{K_{eq}a_A}}{(1+K_{S,D}a_D)^2}$
Kinetic constant		$k_c(\mathrm{mol/g/min})=3.21\times10^5\exp\left(\dfrac{-4561.06}{T(\mathrm{K})}\right)$
Adsorption constant		$K_{s,D}=1.454\times10^{-4}\exp\left(\dfrac{2920.68}{T(\mathrm{K})}\right)$

Table 25 Multicomponent Adsorption Parameters for Langmuir Isotherms of 1,1-DEB System Over Amberlyst-15 at 25°C (Regufe et al., 2016)

Component	Q_{ads} (mol/L_{res})	K (L/mol)
Ethanol (A)	16.06	4.07
Butyraldehyde (B)	3.01	0.11
1,1-DEB (C)	6.31	2.47
Water (D)	53.07	4.55

Table 26 Overall Activation Energies and Preexponential Factors for Each Compound of the 1,1-DEB System From the Pervaporation Study (Graça and Rodrigues, 2017)

Component	$Q_{0,i}$ (mol/(s m^2 Pa))	$E_{p,i}$ (kJ/mol)
Ethanol (A)	2.18×10^{-10}	−11.46
Butyraldehyde (B)	6.26×10^{-3}	36.92
1,1-DEB (C)	—	—
Water (D)	2.69×10^{-9}	−17.21

Table 27 Optimal Operating and Performance Parameters for SMBR and PermSMBR (Synthesis of 1,1-DEB)

Parameters/Process	SMBR	PermSMBR
Configuration	3-3-3-3	
T (°C)	50	
P_{perm} (mbar)	—	5
t^* (min)	3.9	4.4
Feed (mole fraction)	$x_{But,F}=0.51$; $x_{Eth,F}=0.49$	
Q_D (mL/min)	44.4 ($\gamma_1=4.278$)	
Q_{Rec} (mL/min)	21 ($\gamma_4=0.696$)	
X (%)	96.3	95.8
PR (%)	95.1	95.1
PX (%)	99.8	98.6
$Prod$ (kg_{DEB}/(L_{ads} day))	32.2	34.3
$DesC$ ($L_{Ethanol}$/kg_{DEB})	1.86	1.73

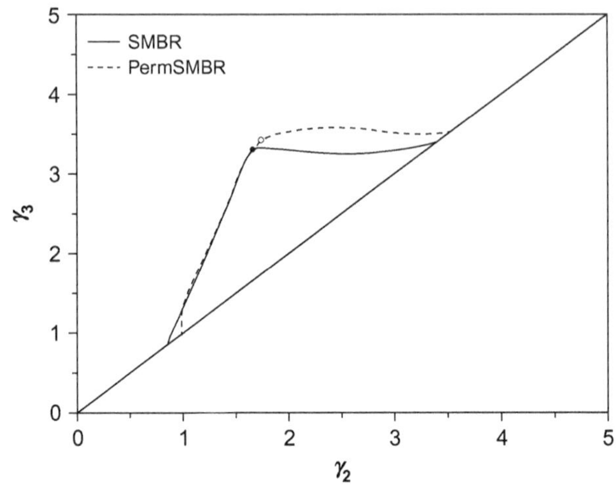

Fig. 30 Reactive/separation regions for the synthesis of 1,1-DEB by means of SMBR and PermSMBR technologies ($T=50°C$; configuration $=3$-3-3-3). *Reprinted from Graça NS, Rodrigues AE: Application of membrane technology for the enhancement of 1,1-diethoxybutane synthesis.* Chem Eng Process *117:45–57, 2017 by permission of Elsevier.*

and raffinate purities as well as for the butyraldehyde conversion. PermSMBR has the ability to fulfill the criteria established (95%) for higher values of liquid and solid velocities ratio in sections 2 and 3 and consequently higher feed flow rates can be processed in PermSMBR achieving higher productivity and lower desorbent consumption.

Analyzing the internal concentration profiles at the cyclic steady state inside the PermSMBR region, shown in Fig. 31, allows a better understanding of the PermSMBR unit behavior in relation to SMBR process in the same operating point. The main difference between both profiles in that region is the accumulation of water along the SMBR unit contaminating the raffinate outlet stream and so low productivity is obtained. Integrating the water pervaporation-selective membranes with SMBR (PermSMBR) leads to a more effective continuous water removal through the membranes, and therefore, it is possible to operate at higher feed flow rate without raffinate contamination.

Comparing the maximum performance parameters of each process attained from the vertex of the respective RSR, which are presented in Table 27, the results reveals an improvement of 6.7% on the productivity and a saving of 7% in the desorbent requirement by using PermSMBR technology for 1,1-DEB synthesis.

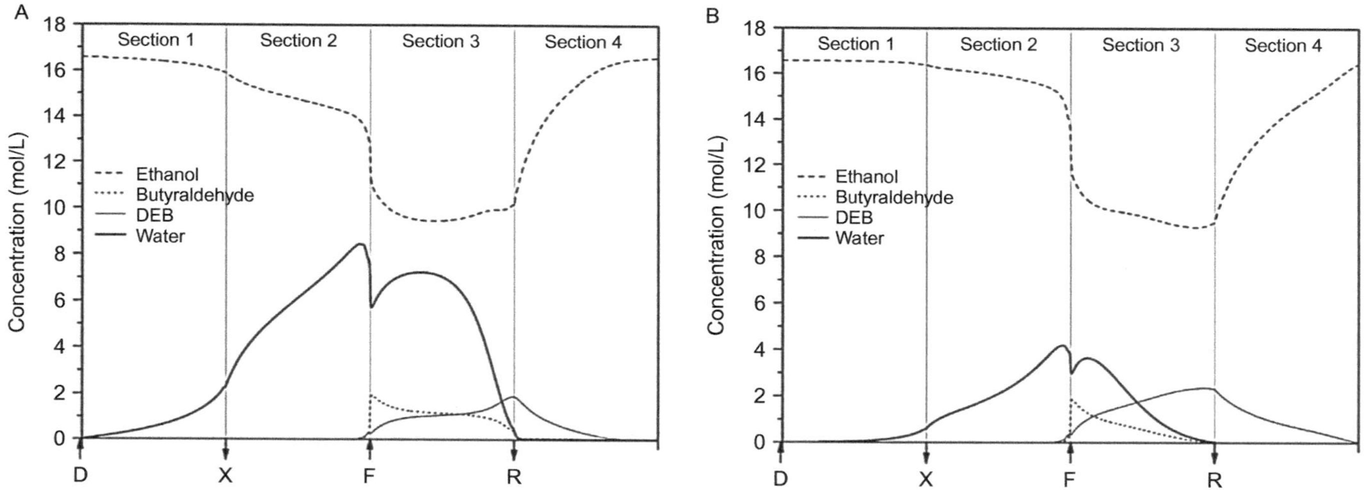

Fig. 31 Internal concentration profiles at steady state of (A) SMBR and (B) PermSMBR; $T = 50°C$ ($\gamma_2 = 2.5$ and $\gamma_3 = 3.47$). *Reprinted from Graça NS, Rodrigues AE: Application of membrane technology for the enhancement of 1,1-diethoxybutane synthesis.* Chem Eng Process *117:45–57, 2017 by permission of Elsevier.*

3.4.2 *Synthesis of Esters*

3.4.2.1 Case Study (P.4): Synthesis of Ethyl Lactate (EL)

The ethyl lactate synthesis by means of an SMBR process at 50°C was described in Section 2.3.3 as well as the respective fundamental data for that study. In addition, a pervaporation study for this system was illustrated in Section 3.3. The main drawbacks associated to the ethyl lactate production are very slow reaction rate, water content present in the feed solution (due to the lactic acid solution usually containing 10%–15% of water), and higher mass transfer limitations due to lactic acid viscosity. Therefore, the ethanol was selected as eluent/desorbent and the optimal operating parameters achieved for the SMBR were used for the study of the PermSMBR: configuration 3-3-4-2, desorbent and recycle flow rates of 58 and 27 mL/min ($\gamma_1 = 3.654$ and $\gamma_4 = 1.161$), respectively, and a feed of lactic acid solution of 85% in water. In order to use the same liquid and solid velocities (in sections 1 and 4) the switching time is corrected from 2.1 to 2.323 min. First, a simulation of the PermSMBR process was performed without vacuum in the permeate side and using a feed and extract flow rates of 7 and 39 mL/min in both cases to observe the differences between the SMBR and PermSMBR performances which led to very similar values as expected, since the PermSMBR is operating in the absence of vacuum and hence in the absence of permeate flux. Afterward the vacuum was set to 10 mbar in the permeate side of the PermSMBR unit and the RSR was determined keeping the values of $\gamma_1 = 3.654$ and $\gamma_4 = 1.161$ and solving successively the cyclic steady-state mathematical model for several γ_2 and γ_3. The same criteria of 95% for extract and raffinate purities as well as for lactic acid conversion were set. A comparison between the SMBR and PermSMBR RSR is made in Fig. 32. Like in the previous cases presented for acetals compounds (Section 3.4.1), the synthesis of ethyl lactate was enhanced by the integration of water pervaporation-selective membranes in the SMBR setup, leading to a productivity improvement of about 34% and desorbent saving of approximately 28% (24.2 $kg_{EL}/(L_{ads}\,day)$ and 3.41 L_{Eth}/kg_{EL}, respectively). Accordingly, as the products are less diluted in ethanol the associated costs will be also reduced. These results are owned to the better resin regeneration into section 1 since water is removed more efficiently, by adsorption and additionally by pervaporation, so the lactic acid conversion increases in sections 2 and 3 allowing to process higher feed flow rates (12.1 mL/min for PermSMBR and 8.8 mL/min for SMBR) toward better performance.

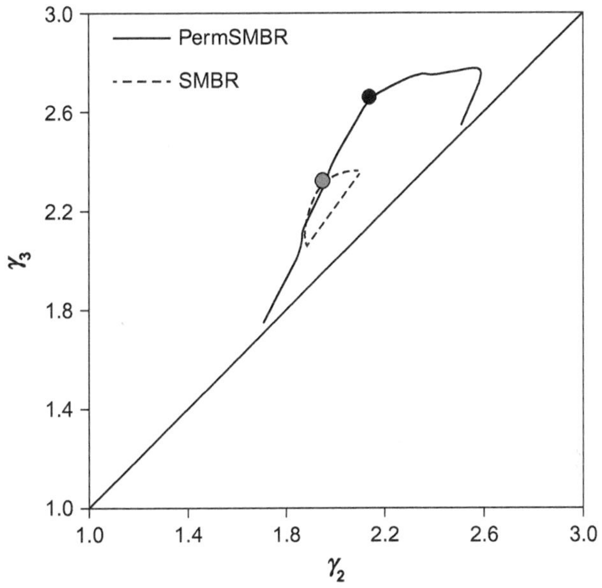

Fig. 32 Reactive/separation regions for SMBR ($t^*=2.1$ min) and PermSMBR ($t^*=2.323$ min) processes (synthesis of ethyl lactate); configuration 3-3-4-2 and $T=50°C$. *Reprinted from Silva VMTM, Pereira CSM, Rodrigues AE: PermSMBR—a new hybrid technology: application on green solvent and biofuel production.* AIChE J *57(7):1840–1851, 2011 by permission of John Wiley & Sons.*

Evaluating the technology with just three sections, the extract stream is eliminated since the most adsorbed compound is the same for which the membranes are selective to the configuration change from 3-3-4-2 to 6-4-2. The same operating conditions described before were considered and the feed and desorbent flow rates were set at 8 and 25 mL/min, respectively, which resulted in a desorbent consumption of 2.00 L_{Eth}/kg_{EL}, lower than that obtained for the four section PermSMBR at optimal conditions (3.41 L_{Eth}/kg_{EL}); nevertheless, the productivity is worst and the criteria of raffinate purity are not fulfilled unless the pressure in the permeate side is reduced from 10 to 6 mbar (16.5 $kg_{EL}/(L_{ads}\,day)$ and 1.96 L_{Eth}/kg_{EL} with ethyl lactate purity of 96.2%). The water is removed more proficiently, enabling a better regeneration of the adsorbent in section 1 and consequently higher raffinate purity in section 3.

Furthermore, a comparison between the different SMBR-based technologies with the reactive distillation process for the ethyl lactate synthesis using Amberlyst-15 can be performed (Table 28). The best performance parameters for RD technology reported by the authors (Asthana et al., 2005) were

Table 28 Performance Parameters for SMBR-Based Technologies and RD for the Ethyl Lactate Synthesis

Parameters/Process	SMBR	PermSMBR	PermSMBR-3s	RD
Prod ($kg_{EL}/(L_{ads}$ day))	18.1	24.2	41.2	40.6
DesC (L_{Eth}/kg_{EL})	4.75	3.41	0.86	2.17

Reprinted from Silva VMTM, Pereira CSM, Rodrigues AE: PermSMBR—a new hybrid technology: application on green solvent and biofuel production. *AIChE J* 57(7):1840–1851, 2011 by permission of John Wiley & Sons.

40.6 $kg_{EL}/(L_{ads}$ day) and 2.17 L_{Eth}/kg_{EL} with the following conditions: feed solution composed by ethanol and 88 wt.% of lactic acid solution in a molar ratio of 3.6 (alcohol/acid), setting a bottom temperature of 128°C achieved 95% of lactic acid conversion and 95% of ethyl lactate purity. Looking into Table 28, where the best performance parameters of the SMBR and PermSMBR at 50°C are presented (for that criteria), RD seems to be the best solution. Nevertheless, the low temperature (50°C) used in that processes is limiting the productivities that can be enhanced by increasing the operating temperature. Thus, the same study was performed for PermSMBR-3s, eliminating the extract stream, at 70°C avoiding the permeation overloading. For that, the following operating conditions were used: feed, desorbent, and recycle flow rates of 20.5, 34.0, and 23.5 mL/min with a permeate pressure of 6 mbar and switching time of 2.323 min. The results are also presented in Table 28 and show that the performance of PermSMBR-32 is improved considerably. The eluent saving increases more than 60% when compared with RD process for a very similar productivity. In summary, according to this research, the PermSMBR-3s is the most effective technology solution for the production of ethyl lactate.

4. FINAL REMARKS

Useful methodologies for studying the SMBR and the PermSMBR were presented by illustrating different oxygenated compounds synthesis, particularly acetals and esters. All information gathered along this chapter aimed at the demystification of the complexity related to the results attained from these kind of processes.

The SMBR is a very interesting and competitive technology which, working in a continuous mode, presents a great ability to increase the conversion of equilibrium-limited reaction, favoring the formation of the target product by the continuous removal of other products by selective adsorption. Hence, higher productivities are achieved for the demanded purity criteria compared with the conventional processes. Moreover, this concept was already experimentally demonstrated more than once. Therefore, this multifunctional reactor can be an attractive solution for numerous processes that involve temperature-sensitive compounds or azeotropic mixtures since it ensures the products separation through its affinity toward the solid phase being able to operate at lower temperatures than other multifunctional rectors, as reactive distillation columns. However, the outlet streams are generally diluted in the desorbent used to regenerate the solid phase, requiring complementary units to purify the respective products.

Regarding the PermSMBR, all examples detailed proved that this is an extremely effective process for the production of organic oxygenated compounds. The excellent performance attained for the different systems revealed the process versatility and, besides that, the process feasibility. This recent approach brings great advantages for the process intensification since the concept comprises two different separation techniques and chemical reaction by combining the SMBR with pervaporation-selective membranes. The capacity of the SMBR to increase the conversion of equilibrium-limited reactions is reinforced by pervaporation process in the PermSMBR leading to higher productivities, for the same purity requirements, due to the continuous removal of water by adsorption and pervaporation simultaneously (in case of water-selective membranes). Other resilient characteristic is the significant reduction of the desorbent requirement since the solid-phase regeneration is not so critical. Moreover, the PermSMBR enables different configurations and can be reduced from four to three sections (PermSMBR-3s) by removing the outlet stream of the product for which the membranes are selective. This arrangement usually requires higher temperature to avoid the membranes overloading; nonetheless, the costs associated with the higher energy demand can be redressed by the products concentration at the outlet streams.

The coupled PermSMBR-3s presented similar productivities to the integrated configuration also with three sections but proved to be even more efficient in terms of desorbent consumption. Consequently, and taking into account its easier maintenance and higher mechanical and chemical stability in relation to the integrated PermSMBR, the coupled PermSMBR can be considered as a first choice for the oxygenated compounds synthesis involving equilibrium-limited reactions.

REFERENCES

Ahmad SA, Lone SR: Hybrid process (pervaporation–distillation): a review, *Int J Sci Eng Res* 3:1–5, 2012.

Asthana N, Kolah A, Vu DT, Lira CT, Miller DJ: A continuous reactive separation process for ethyl lactate formation, *Org Process Res Dev* 9(5):599–607, 2005.

Boennhoff KDC and Obenaus FDCD: Non-petroleum diesel fuels—based on lower alkyl acetal and/or cyclododecatriene. DE 2911411, 1980.

Boennhoff K and Sausen ED: Method for enhancing the ignition performance of dialkoxyalkanes used as diesel fuel, in particular 1,1-diethoxyethane. DE3136030, 1983.

Constantino DSM, Pereira CSM, Faria RPV, Ferreira AFP, Loureiro JM, Rodrigues AE: Synthesis of butyl acrylate in a fixed-bed adsorptive reactor over Amberlyst 15, *AIChE J* 61(4):1263–1274, 2015a.

Constantino DSM, Pereira CSM, Faria RPV, Loureiro JM, Rodrigues AE: Simulated moving bed reactor for butyl acrylate synthesis: from pilot to industrial scale, *Chem Eng Process Process Intensif* 97:153–168, 2015b.

Constantino DSM, Faria RPV, Pereira CSM, Loureiro JM, Rodrigues AE: Enhanced simulated moving bed reactor process for butyl acrylate synthesis: process analysis and optimization, *Ind Eng Chem Res* 55(40):10735–10743, 2016.

Faria RPV, Pereira CSM, Silva VMTM, Loureiro JM, Rodrigues AE: Glycerol valorisation as biofuels: selection of a suitable solvent for an innovative process for the synthesis of GEA, *Chem Eng J* 233:159–167, 2013.

Feng X, Huang RY: Liquid separation by membrane pervaporation: a review, *Ind Eng Chem Res* 36(4):1048–1066, 1997.

Gandi GK, Silva VMTM, Rodrigues AE: Process development for dimethylacetal synthesis: thermodynamics and reaction kinetics, *Ind Eng Chem Res* 44(19):7287–7297, 2005.

Gandi GK, Silva VMTM, Rodrigues AE: Synthesis of 1,1-dimethoxyethane in a fixed bed adsorptive reactor, *Ind Eng Chem Res* 45(6):2032–2039, 2006.

Glueckauf E: Theory of chromatography. Part 10—formulae for diffusion into spheres and their application to chromatography, *Trans Faraday Soc* 51:1540–1551, 1955.

Graça NS, Rodrigues AE: Application of membrane technology for the enhancement of 1,1-diethoxybutane synthesis, *Chem Eng Process* 117:45–57, 2017.

Graça NS, Pais LS, Silva VMTM, Rodrigues AE: Oxygenated biofuels from butanol for diesel blends: synthesis of the acetal 1,1-dibutoxyethane catalyzed by Amberlyst-15 ion-exchange resin, *Ind Eng Chem Res* 49(15):6763–6771, 2010.

Graça NS, Pais LS, Silva VMTM, Rodrigues AE: Dynamic study of the synthesis of 1,1-dibutoxyethane in a fixed-bed adsorptive reactor, *Sep Sci Technol* 46(4):631–640, 2011a.

Graça NS, Pais LS, Silva VMTM, Rodrigues AE: Analysis of the synthesis of 1,1-dibutoxyethane in a simulated moving-bed adsorptive reactor, *Chem Eng Process* 50(11–12):1214–1225, 2011b.

Gyani VC, Reddy B, Bhat R, Mahajani S: Simulated moving bed reactor for the synthesis of 2-ethylhexyl acetate. Part I: experiments and simulations, *Ind Eng Chem Res* 53(41):15811–15823, 2014.

Jyoti G, Keshav A, Anandkumar J: Review on pervaporation: theory, membrane performance, and application to intensification of esterification reaction, *J Eng* 24:1–24, 2015.

Kawase, Suzuki TB, Inoue K, Yoshimoto K, Hashimoto K: Increased esterification conversion by application of the simulated moving-bed reactor, *Chem Eng Sci* 51(11):2971–2976, 1996.

Kawase M, Inoue Y, Araki T, Hashimoto K: The simulated moving-bed reactor for production of bisphenol A, *Catal Today* 48(1–4):199–209, 1999.

Laborde M: *Producción de acetal a partir de bioetanol,* Buenos Aires, 2003, Ediciones CYTED.

Lévêque A: *Les Lois de la transmission de chaleur par convection,* 1928, Dunod.

Lode F, Houmard M, Migliorini C, Mazzotti M, Morbidelli M: Continuous reactive chromatography, *Chem Eng Sci* 56(2):269–291, 2001.
Luis P, Van der Bruggen B: The driving force as key element to evaluate the pervaporation performance of multicomponent mixtures, *Sep Purif Technol* 148:94–102, 2015.
Mazzotti M, Kruglov A, Neri B, Gelosa D, Morbidelli M: A continuous chromatographic reactor: SMBR, *Chem Eng Sci* 51:1827–1836, 1996.
Mazzotti M, Storti G, Morbidelli M: Optimal operation of simulated moving bed units for nonlinear chromatographic separations, *J Chromatogr A* 769(1):3–24, 1997.
Minceva M, Rodrigues AE: Two-level optimization of an existing SMB for p-xylene separation, *Comput Chem Eng* 29(10):2215–2228, 2005.
Minceva M, Gomes PS, Meshko V, Rodrigues AE: Simulated moving bed reactor for isomerization and separation of p-xylene, *Chem Eng J* 140(1–3):305–323, 2008.
Mulder M: Polarization phenomena and membrane fouling, *Membr Sci Technol* 2:45–84, 1995.
Neel J: Pervaporation, *Membr Sci Technol* 2:143–211, 1995.
Nicoud RM, Fuchs G, Adams P, et al: Preparative scale enantioseparation of a chiral epoxide: comparison of liquid chromatography and simulated moving bed adsorption technology, *Chirality* 5(4):267–271, 1993.
Nord KE, Haupt D: Reducing the emission of particles from a diesel engine by adding an oxygenate to the fuel, *Environ Sci Technol* 39(16):6260–6265, 2005.
Ostaniewicz-Cydzik AM, Pereira CSM, Molga E, Rodrigues AE: Reaction kinetics and thermodynamic equilibrium for butyl acrylate synthesis from n-butanol and acrylic acid, *Ind Eng Chem Res* 53(16):6647–6654, 2014.
Pereira CSM, Rodrigues AE: Process intensification: new technologies (SMBR and PermSMBR) for the synthesis of acetals, *Catal Today* 218–219:148–152, 2013.
Pereira CSM, Gomes PS, Gandi GK, Silva VMTM, Rodrigues AE: Multifunctional reactor for the synthesis of dimethylacetal, *Ind Eng Chem Res* 47(10):3515–3524, 2007.
Pereira CSM, Pinho SP, Silva VMTM, Rodrigues AE: Thermodynamic equilibrium and reaction kinetics for the esterification of lactic acid with ethanol catalyzed by acid ion-exchange resin, *Ind Eng Chem Res* 47(5):1453–1463, 2008.
Pereira CSM, Zabka M, Silva VMTM, Rodrigues AE: A novel process for the ethyl lactate synthesis in a simulated moving bed reactor (SMBR), *Chem Eng Sci* 64(14):3301–3310, 2009.
Pereira CSM, Silva VMTM, Pinho SP, Rodrigues AE: Batch and continuous studies for ethyl lactate synthesis in a pervaporation membrane reactor, *J Membr Sci* 361(1–2):43–55, 2010.
Pereira CSM, Silva VMTM, Rodrigues AE: Green fuel production using the PermSMBR technology, *Ind Eng Chem Res* 51(26):8928–8938, 2012.
Pereira CSM, Silva VMTM, Rodrigues AE: Coupled PermSMBR—process design and development for 1,1-dibutoxyethane production, *Chem Eng Res Des* 92(11): 2017–2026, 2014.
Perkins LR, Geankoplis CJ: Molecular diffusion in a ternary liquid system with the diffusing component dilute, *Chem Eng Sci* 24:1035–1042, 1969.
Rahaman M, Graça NS, Pereira CSM, Rodrigues AE: Thermodynamic and kinetic studies for synthesis of the acetal (1,1-diethoxybutane) catalyzed by Amberlyst 47 ion-exchange resin, *Chem Eng J* 264:258–267, 2015.
Ray AK, Carr RW, Aris R: The simulated countercurrent moving bed chromatographic reactor: a novel reactor-separator, *Chem Eng Sci* 49(4):469–480, 1994.
Regufe MJ, Faria RPV, Ribeiro AM, Loureiro JM, Rodrigues AE: Synthesis of the biofuel additive 1,1-diethoxybutane in a fixed-bed column with Amberlyst-15 wet, *Chem Eng Technol* 39(8):1509–1518, 2016.

Rodrigues AE and Silva VMTM: Industrial process for acetals production in a simulated moving bed reactor. US 7488851, 2009.

Rodrigues AE, Pereira CSM, Santos JC: Chromatographic reactors, *Chem Eng Technol* 35(7):1171–1183, 2012.

Rodrigues AE, Pereira CSM, Minceva M, Santos JC: Chapter 1—principles of simulated moving bed. In *Simulated moving bed technology*, Oxford, 2015a, Butterworth-Heinemann, pp 1–30.

Rodrigues A, Pereiria C, Minceva M, et al: Simulated moving bed technology: principles. In *Design and process applications*, Oxford, 2015b, Elsevier Science.

Ruthven DM: *Principles of adsorption and adsorption processes,* New York, 1984, Wiley & Sons.

Scheibel EG: Correspondence. liquid diffusivities viscosity of gases, *Ind Eng Chem* 46:2007–2008, 1954.

Schwarzer S, Hoffmann U: Experimental reaction equilibrium and kinetics of the liquid-phase butyl acrylate synthesis applied to reactive distillation simulations, *Chem Eng Technol* 25:975–980, 2002.

Silva VMTM, Rodrigues AE: Synthesis of diethylacetal: thermodynamic and kinetic studies, *Chem Eng Sci* 56(4):1255–1263, 2001.

Silva VMTM, Rodrigues AE: Dynamics of a fixed-bed adsorptive reactor for synthesis of diethylacetal, *AlChE J* 48:625–634, 2002.

Silva VMTM, Rodrigues AE: Novel process for diethylacetal synthesis, *AlChE J* 51(10):2752–2768, 2005.

Silva VMTM, Pereira CSM, Rodrigues AE: Simulated moving bed membrane reactor, new hybrid separation process and uses thereof. US 20120108841, 2010.

Silva VMTM, Pereira CSM, Rodrigues AE: PermSMBR—a new hybrid technology: application on green solvent and biofuel production, *AlChE J* 57(7):1840–1851, 2011.

Stankiewicz A: Reactive separations for process intensification: an industrial perspective, *Chem Eng Process* 42:137–144, 2003.

Storti G, Mazzotti M, Morbidelli M, Carrà S: Robust design of binary countercurrent adsorption separation processes, *AlChE J* 39(3):471–492, 1993.

Strathmann H: *Membrane separation processes, 4. concentration polarization and membrane fouling,* Weinheim, 2000, Wiley-VCH Verlag GmbH & Co. KGaA.

Tonkovich ALY, Carr RW: A simulated countercurrent moving-bed chromatographic reactor for the oxidative coupling of methane: experimental results, *Chem Eng Sci* 49(24):4647–4656, 1994.

Vignes A: Diffusion in binary solutions. Variation of diffusion coefficient with composition, *Ind Eng Chem Fundam* 5:189–199, 1966.

Wijmans J, Baker R: The solution-diffusion model: a review, *J Membr Sci* 107(1-2):1–21, 1995.

Wijmans JG, Athayde AL, Daniels R, Ly JH, Kamaruddin HD, Pinnau I: The role of boundary layers in the removal of volatile organic compounds from water by pervaporation, *J Membr Sci* 109(1):135–146, 1996.

Zhang Z, Hidajat K, Ray AK: Application of simulated countercurrent moving-bed chromatographic reactor for MTBE synthesis, *Ind Eng Chem Res* 40(23):5305–5316, 2001.

INDEX

Note: Page numbers followed by "*f*" indicate figures, and "*t*" indicate tables.

D

E

F

G

H

I

K

L

M

T

W

Z

FSC
www.fsc.org
MIX
Papier | Fördert gute Waldnutzung
FSC® C083411

Kolibri 360 GmbH
Ferdinand-Jühlke-Straße 7
99095 Erfurt, Deutschland
produktsicherheit@kolibri360.de